AF676432

Advanced Interconnect and Packaging

Advanced Interconnect and Packaging

Editor

Wensheng Zhao

MDPI • Basel • Beijing • Wuhan • Barcelona • Belgrade • Manchester • Tokyo • Cluj • Tianjin

Editor
Wensheng Zhao
School of Electronics and
Information
Hangzhou Dianzi University
Hangzhou
China

Editorial Office
MDPI
St. Alban-Anlage 66
4052 Basel, Switzerland

This is a reprint of articles from the Special Issue published online in the open access journal *Micromachines* (ISSN 2072-666X) (available at: www.mdpi.com/journal/micromachines/special_issues/Advanced_Interconnect_and_Packaging).

For citation purposes, cite each article independently as indicated on the article page online and as indicated below:

LastName, A.A.; LastName, B.B.; LastName, C.C. Article Title. *Journal Name* **Year**, *Volume Number*, Page Range.

ISBN 978-3-0365-6733-4 (Hbk)
ISBN 978-3-0365-6732-7 (PDF)

© 2023 by the authors. Articles in this book are Open Access and distributed under the Creative Commons Attribution (CC BY) license, which allows users to download, copy and build upon published articles, as long as the author and publisher are properly credited, which ensures maximum dissemination and a wider impact of our publications.
The book as a whole is distributed by MDPI under the terms and conditions of the Creative Commons license CC BY-NC-ND.

Contents

About the Editor

Wensheng Zhao

Wensheng Zhao received a B.E. degree in electronic science and technology from the Harbin Institute of Technology, Harbin, China, in 2008, and a Ph.D. degree in electronic science and technology from Zhejiang University, Hangzhou, China, in 2013.

During his Ph.D. program, he visited the National University of Singapore, Singapore, from 2010 to 2013. He was a Visiting Scholar with the Georgia Institute of Technology, Atlanta, GA, USA. He is currently a Full Professor with Hangzhou Dianzi University, Hangzhou. He has published three books, three chapters, and more than 110 journal articles (including more than 50 IEEE articles). His current research interests include IC interconnect and packaging, electromagnetic devices, and electronic design automation. Dr. Zhao is a Senior Member of IEEE and the Chinese Institute of Electronics. He is also an Associate Editor of IEEE Access and Microelectronics Journal, and a Guest Editor of *Micromachines* and *Electronics*.

micromachines MDPI

Editorial

Editorial for the Special Issue on Advanced Interconnect and Packaging

Wen-Sheng Zhao

Zhejiang Provincial Key Lab of Large-Scale Integrated Circuit Design, School of Electronics and Information, Hangzhou Dianzi University, Hangzhou 310018, China; wshzhao@hdu.edu.cn

Unlike transistors, the continuous downscaling of feature size in CMOS technology leads to a dramatic rise in interconnect resistivity and concomitant performance degradation. At nanoscale technology nodes, interconnect delay and reliability become the major bottlenecks faced by modern integrated circuits. To resolve these interconnect problems, various emerging technologies, including airgap, nanocarbon, optical, and through-silicon via (TSV), have been proposed and investigated. For example, by virtue of TSV technology, dies can be stacked to increase the integration density. More importantly, 3D integration and packaging also offer the most promising platform to implement "More-than-Moore" technologies, providing heterogenous materials and technologies on a single chip.

This Special Issue seeks to showcase research papers on new developments in advanced interconnect and packaging, i.e., on the design, modeling, fabrication, and reliability assessment of emerging interconnect and packaging technologies. Additionally, there are two interesting papers on carbon nanotube interconnects and interconnect reliability issues.

In particular, Liu et al. successfully realize the thermocompression bonding of Pt–Pt metal electrodes through process exploration and form a packaging interconnection that meets the requirements [1]. Sun et al. achieve low-temperature assembly by reflowing 13.5Sn–37.5Bi–45In–4Pb quaternary eutectic solder paste and a SAC 305 solder ball together at 140 °C for 5 min [2]. Liu et al. investigate the flow process of nano glass powder melted at a high temperature [3]. To reduce the wettability of the glass paste on the Au electrode, they grow a silicon dioxide isolation layer on the surface of golden lead via chemical vapor deposition. Wang et al. investigate the chip-level hermetic package for a high-temperature graphene pressure sensor and demonstrate that the combination of Cu–Sn and Au–Au is extremely suitable for hermetic packaging [4]. Wu et al. provide theoretical support for the application of thin coatings at high temperatures and in harsh environments [5].

Wang et al. study the electrical performance of graphene-based on-chip spiral inductors by virtue of a physics-based equivalent circuit model [6]. Kim et al. propose a novel interposer channel structure with vertical tabbed vias to achieve high-speed signaling in high-bandwidth memory and demonstrate that the proposed channel structure could reduce dynamic power consumption [7]. Kim designs noise suppression structures that generate an electromagnetic bandgap and studies the mechanism of the proposed structure based on dispersion analysis [8]. Zheng et al. study the average power handling capability of corrugated slow-wave transmission lines [9]. Du et al. develop a mathematical degradation model for evaluating the degradation of vacuum packaged MEMS sensors and perform a temperature-accelerated test of MEMS gyroscopes with different vacuums [10]. Zhao et al. introduce recent studies on the physics-based modeling of the electromigration aging of interconnects [11].

Pan et al. present a novel wideband bandpass filter based on the integration of a substrate-integrated waveguide and a spoof surface plasmon polariton [12]. Wei et al. present a new method for the analytical approximation of spatial current/field profiles of frequency-selective surfaces and demonstrate that the transmission line loss has little

Citation: Zhao, W.-S. Editorial for the Special Issue on Advanced Interconnect and Packaging. *Micromachines* **2023**, *14*, 171. https://doi.org/10.3390/mi14010171

Received: 6 January 2023
Accepted: 8 January 2023
Published: 10 January 2023

Copyright: © 2023 by the author. Licensee MDPI, Basel, Switzerland. This article is an open access article distributed under the terms and conditions of the Creative Commons Attribution (CC BY) license (https://creativecommons.org/licenses/by/4.0/).

influence on the current distribution [13]. Bai et al. propose a broadband frequency-selective rasorber based on spoof surface plasmon polaritons [14]. Tian et al. investigate the deadbeat current controllers for an isolated bidirectional dual-active-bridge dc–dc converter, including in the peak current mode and middle current mode [15]. Kuo et al. develop a flexible blood oxygen sensing system with a 3 × 3 array and use a flip chip package to integrate the sensing chip [16]. Finally, Xu et al. review the advantages, recent developments, and dilemmas of carbon nanotube-based interconnects from the perspective of different interconnect lengths and through-silicon via applications [17].

To conclude, I would like to take this opportunity to thank all the authors for submitting their papers to this Special Issue, as well as the reviewers for the effort and time they expended to improve the quality of the published papers. I also want to recognize Mr. Dikies Zhang from the *Micromachines* publishing office for his endless assistance and help in disseminating this Special Issue.

In view of the success reached in terms of the number and quality of papers published, we plan to open a second Volume where we hope to continue the conversation regarding the latest advances in interconnect and packaging.

Funding: This research was funded by National Natural Science Foundation of China (grant number 62222401) and Zhejiang Provincial Natural Science Foundation (grant number LXR22F040001).

Conflicts of Interest: The authors declare no conflict of interest.

References

1. Liu, J.; Wang, J.; Li, M.; Zhang, H. High quality Pt-Pt metal bonding for high temperature packaging. *Micromachines* **2022**, *13*, 1543. [CrossRef] [PubMed]
2. Sun, L.; Guo, Z.; Zhao, X.; Liu, Y.; Tu, K.; Liu, Y. A new low-temperature solder assembly technique to replace eutectic Sn-Bi solder assembly. *Micromachines* **2022**, *13*, 867. [CrossRef] [PubMed]
3. Liu, Y.; Chen, J.; Zheng, G. Study on the wetting mechanism between hot-melt nano glass powder and different substrates. *Micromachines* **2022**, *13*, 1683. [CrossRef] [PubMed]
4. Wang, J.; Zhang, H.; Chen, X.; Li, M. Hermetic packaging based on Cu-Sn and Au-Au dual bonding for high-temperature graphene pressure sensor. *Micromachines* **2022**, *13*, 1191. [CrossRef] [PubMed]
5. Wu, C.; Wang, J.; Liu, X.; Li, M.; Zhu, Z.; Qi, Y. Au wire ball welding and its reliability test for high-temperature environment. *Micromachines* **2022**, *13*, 1603. [CrossRef] [PubMed]
6. Wang, D.W.; Yuan, M.J.; Dai, J.Y.; Zhao, W.S. Electrical modeling and characterization of graphene-based on-chip spiral inductors. *Micromachines* **2022**, *13*, 1829. [CrossRef] [PubMed]
7. Kim, H.; Lee, S.; Song, K.; Shin, Y.; Park, D.; Park, J.; Cho, J.; Ahn, S. A novel interposer channel structure with vertical tabbed vias to reduce far-end crosstalk for next-generation high-bandwidth memory. *Micromachines* **2022**, *13*, 1070. [CrossRef] [PubMed]
8. Kim, Y. Design of power/ground noise suppression structures based on a dispersion analysis for packages and interposers with low-loss substrates. *Micromachines* **2022**, *13*, 1433. [CrossRef] [PubMed]
9. Zheng, Z.; Tang, M.; Zhang, H.; Mao, J. Average power handling capability of corrugated slow-wave transmission lines. *Micromachines* **2022**, *13*, 961. [CrossRef] [PubMed]
10. Du, G.; Dong, X.; Huang, X.; Su, W.; Zhang, P. Reliability evaluation based on mathematical degradation model for vacuum packaged MEMS sensor. *Micromachines* **2022**, *13*, 1713. [CrossRef] [PubMed]
11. Zhao, W.S.; Zhang, R.; Wang, D.W. Recent progress in physics-based modeling of electromigration in integrated circuit interconnects. *Micromachines* **2022**, *13*, 883. [CrossRef] [PubMed]
12. Pan, D.; You, B.; Wen, X.; Li, X. Wideband substrate integrated waveguide chip filter using spoof surface plasmon polariton. *Micromachines* **2022**, *13*, 1195. [CrossRef] [PubMed]
13. Wei, S.; Wen, W. Antenna current calculation based on equivalent transmission line model. *Micromachines* **2022**, *13*, 714. [CrossRef] [PubMed]
14. Bai, J.; Yang, Q.; Liang, Y.; Gao, X. Broadband frequency selective rasorber based on spoof surface plamson polaritons. *Micromachines* **2022**, *13*, 1969. [CrossRef] [PubMed]
15. Tian, C.; Wei, S.; Xie, J.; Bai, T. Dynamic enhancement for dual active bridge converter with a deadbeat current controller. *Micromachines* **2022**, *13*, 2048. [CrossRef] [PubMed]

16. Kuo, W.C.; Wu, T.C.; Wang, J.S. Design and application of a flexible blood oxygen sensing array for wearable devices. *Micromachines* **2022**, *13*, 1742. [CrossRef] [PubMed]
17. Xu, B.; Chen, R.; Zhou, J.; Liang, J. Recent progress and challenges regarding carbon nanotube on-chip interconnects. *Micromachines* **2022**, *13*, 1148. [CrossRef] [PubMed]

Disclaimer/Publisher's Note: The statements, opinions and data contained in all publications are solely those of the individual author(s) and contributor(s) and not of MDPI and/or the editor(s). MDPI and/or the editor(s) disclaim responsibility for any injury to people or property resulting from any ideas, methods, instructions or products referred to in the content.

micromachines

MDPI

Article

High Quality Pt–Pt Metal Bonding for High Temperature Packaging

Jiazheng Liu [1,2], Junqiang Wang [1,2,*], Mengwei Li [1,2,*] and Haikun Zhang [1,2]

1 Academy for Advanced Interdisciplinary Research, North University of China, Taiyuan 030051, China
2 Notional Key Laboratory of Instrumentation Science & Dynamic Measurement, North University of China, Taiyuan 030051, China
* Correspondence: wangjq210@nuc.edu.cn (J.W.); lmwnuc@163.com (M.L.); Tel.: +86-158-1133-8056 (J.W.); +86-139-3424-8366 (M.L.)

Abstract: Platinum is an ideal material for high-temperature resistant device packaging due to its higher melting point and good electrical properties. In this paper, the thermocompression bonding of Pt–Pt metal electrodes was successfully realized through process exploration, and the package interconnection that meets the requirements was formed. A square bump with a side length of 160 μm and a sealing ring with a width of 80 μm were fabricated by magnetron sputtering. Different pressure parameters were selected for chip-level bonding; the bonding temperature was 350 °C for about 20 min. Analysis of the interface under a scanning electron microscope found that the metal Cr diffused into Pt. It was found that two chips sputtered with 300 nm metal Pt can achieve shear resistance up to 30 MPa by flip-chip bonding at 350 °C and 100 MPa temperature and pressure, respectively. The leakage rate of the sample is less than 2×10^{-3} Pa·cm^3/s, the bonding interface is relatively smooth, and the hot-pressed metal bonding of Pt electrodes with good quality is realized. By comparing the failure rates at different temperatures and pressures, the process parameters for Pt–Pt bonding with higher success rates were obtained. We hope to provide new ideas and methods for the packaging of high-temperature resistant devices.

Keywords: Pt–Pt interconnection; high-temperature resistant packaging; metallic bonding

Citation: Liu, J.; Wang, J.; Li, M.; Zhang, H. High Quality Pt–Pt Metal Bonding for High Temperature Packaging. *Micromachines* **2022**, *13*, 1543. https://doi.org/10.3390/mi13091543

Academic Editor: Wensheng Zhao

Received: 25 August 2022
Accepted: 15 September 2022
Published: 17 September 2022

Publisher's Note: MDPI stays neutral with regard to jurisdictional claims in published maps and institutional affiliations.

Copyright: © 2022 by the authors. Licensee MDPI, Basel, Switzerland. This article is an open access article distributed under the terms and conditions of the Creative Commons Attribution (CC BY) license (https://creativecommons.org/licenses/by/4.0/).

1. Introduction

Sensors working with graphene as a sensitive material have received extensive attention in recent years. The unique thermal properties, electrical properties, and high-temperature resistance [1–3] of graphene show a strong potential for enhancing performance and improving the reliability of devices operating at high temperatures and harsh environments. However, the lack of reliable high-temperature packaging technology impedes the application of graphene in the field of high-temperature MEMS devices. Burla et al. [4] achieved nickel wire bonding for high-temperature packaging, and Ni wire bonds were found to be electrically stable for temperatures up to 550 °C. However, the high-temperature oxidation of nickel limits its practical application. On the other hand, Pt has almost perfect corrosion resistance, making it a better high-temperature encapsulation material than nickel. Brachmann et al. [5] demonstrated a Pt wire bonding method that can withstand an 1100 °C environment. The investigated Pt films were composed of a 50 nm thick e-beam evaporated Cr seed layers and an approximately 1 μm electrodeposited Pt film. This study proved the excellent prospects of Pt in the field of high-temperature packaging. To the best of our knowledge, the current high-temperature packaging mostly uses wire bonding technology, while wire bonding technology can no longer meet the requirements of miniaturization, light weight, high performance, and low power consumption of modern electronic products [6]. Notably, the flip chip is obviously more in line with the future development trend of the electronics industry.

The common electronic packaging method is realized by forming flip chips [7,8] by thermocompression metallic bonding [9] or direct bonding [10]. The essence of metallic bonding is the mutual diffusion of atoms on the surface of two metals [9], which relies on metallic bonds, metal melting, and other factors in order to bond firmly. On the one hand, it combines the excellent properties of the material and, on the other hand, completely utilizes the benefits of the metal film to improve the photoelectric performance of the device. Direct bonding involves cleaning and activating the surface of the bonding sheet [11], directly bonding it at room temperature, and finally combining it with heat treatment to form an interconnected interface. The difference between the two methods is that metallic bonding involves the same kind of metal material, while direct bonding can be carried out with two different materials. Therefore, defects such as the dislocations generated in the process of metal bonding only exist near the interface of metal bonding and do not extend to the entire material and thus hardly affect the performance of the material before bonding. Since a thin oxide film is formed on the surface of the metal in the air, the oxide film blocks the mutual diffusion of atoms on the two metal surfaces, restricting the diffusion of atoms unconditionally. Additionally, the bonding interface of metal bonding cannot add a dielectric layer. Metal bonding is usually achieved by heating and pressing and is different from the eutectic interface interconnection formed between different metals [12]. For example, Au–Au, Cu–Cu [13,14] and Al–Al [15,16] make atomic-level contact under the simultaneous action of heat and pressure. Under the movement of atoms, the two layers of metals undergo diffusion movement, and the diffused atoms connect the two layers of metals together.

In this paper, we have focused on the high-temperature packaging needs of MEMS devices [17–19] and used Pt with excellent performance at high temperatures as the bonding material [20,21] to explore the process required for its packaging and its performance after packaging.

2. Materials and Methods

There are three steps to realize Pt–Pt interconnection: test sample design, fabrication of fine-pitch bumps followed by bumps surface pretreatment, and finally Pt–Pt thermocompression bonding.

2.1. Test Sample Design

The top and low substrates of the bonding are two chips of 6×6 mm^2 and 8×8 mm^2, respectively. The total bonding area is approximate to 5.5 mm^2. The schematic diagram of the mask is shown in Figure 1. The outer side length of the sealing ring of the single repeating unit is 1.31 mm, the inner side length is 1.15 mm, and the side length of the small square of the bonding bump is 160 μm. The side length of the total mask structure is 5×5 mm, and each structure contains 16 repeating units. The figure on the right of Figure 1a shows a single, repeating unit, which is located on a substrate with a size of 8×8 mm. The figure on the right of Figure 1b shows a single, repeating unit, which is located on a substrate with a size of 6×6 mm. The above shapes are made of silicon as the substrate through photolithography, sputtering, and other processes. The lithography was performed using an MA6 model lithography machine produced by SUSS MicroTec in Germany. The top and bottom two substrates are bonded by thermocompression to ensure a firm connection between the sealing ring and the bump, forming a closed space.

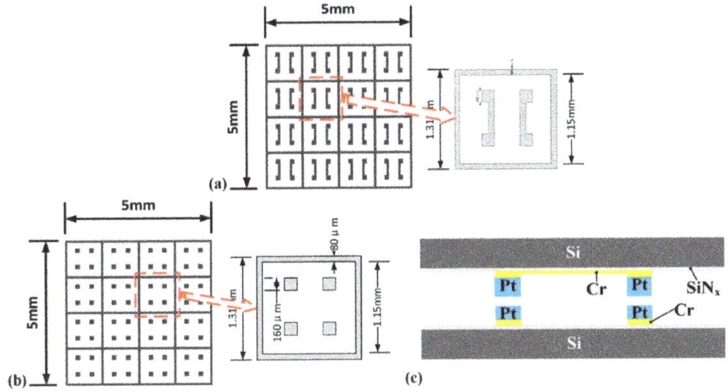

Figure 1. Schematic diagram of mask and bonding structure. (**a**) Mask of the top substrate. (**b**) Mask of the bottom substrate. (**c**) Cross-sectional diagram of bonded structure.

2.2. Substrate Fabrication

The fabrication process of the top substrate is shown in Figure 2. First, the 400 μm thick silicon wafers were routinely cleaned, followed by ultrasonic cleaning with acetone, isopropanol, and water for 5 min each, and finally dried under N2 [Figure 2a] [22]. A 300 nm thick SiNx passivation layer was first deposited on a 400 μm thick Si wafer by plasma-enhanced chemical vapor deposition (PECVD), and a mask structure was created on the SiNx passivation layer using a negative photoresist. Subsequently, the bottom Cr/Pt electrodes with thicknesses of 50 and 300 nm, respectively, were deposited on the SiNx layer by magnetron sputtering [Figure 2b]. The electrode should not be too thick because an excessively thick bonding interface layer may cause the formation of microcracks and lead to poor bonding quality [23,24]. Finally, the negative adhesive peeling was completed in acetone to produce the top substrate. The process for fabrication of the bottom substrate is the same as that of the top substrate.

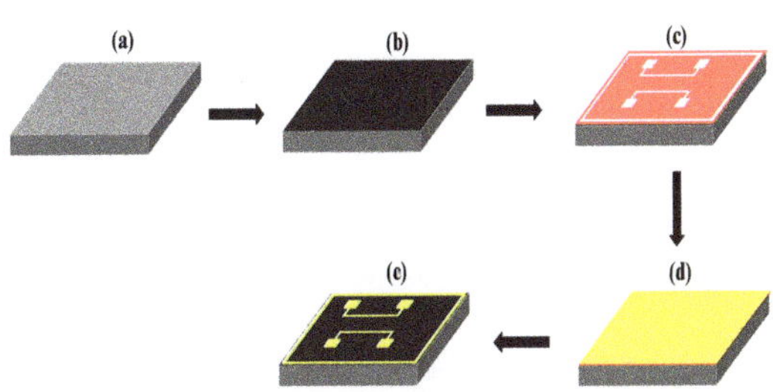

Figure 2. Schematic of the process for fabrication of the top substrate. (**a**) Cleaning. (**b**) SiNx passivation layer deposited by PECVD. (**c**) Negative photoresist masks and lithography. (**d**) Sputtering Cr/Pt electrodes. (**e**) Negative adhesive stripping.

To remove surface oxide film, both the top substrate and bottom substrate were pretreated by Ar (a small amount of H2) plasma with a gas flow rate of 250 sccm under the power of 200 W for 180 s. After the metal surface is activated, the degree of atomic diffusion is increased by heating and pressing to tightly combine the two structures. The bonding adopts the electronic packaging FC150 flip-chip welding machine.

2.3. Pt-Pt Thermocompression Bonding

The relationship between bonding time and temperature is shown in Table 1. The present work attempts to use four sets of bonding parameters to determine the range of parameters that can achieve better Pt–Pt bonding interconnections. There are two purposes for using these four sets of parameters. On the one hand, it can be identified whether the interconnection interface can be formed after bonding, and on the other hand, whether the shear resistance of the sample after bonding is good within this pressure range can be ascertained. Finally, parameters that can form an interconnected interface and have a certain shear resistance are selected. The variables of these four sets of parameters are all pressure, the temperature is 350 °C, and the bonding time is 1200 s. Due to the high melting point of platinum, it is more difficult to bond by thermocompression than other metals such as gold and copper. Therefore, when flip-chip welding is used for thermocompression bonding, the bonding temperature and bonding time are both selected to be close to the upper limit of the instrument.

Table 1. Case of bonding parameter.

Parameter	Force	Temperature	Time
Case1	165 N	350 °C	1200 s
Case2	275 N	350 °C	1200 s
Case3	440 N	350 °C	1200 s
Case4	550 N	350 °C	1200 s

From Figure 1a,b, the bonding area of the bonding pair can be calculated to be about 5.5 mm^2. According to the formula $P = F/S$, the pressure of case 1 is 30 MPa, the pressure of case 2 is 50 MPa, the pressure of case 3 is 80 MPa, and the pressure of case 4 is 100 MPa during thermocompression bonding.

3. Results and Discussion

Four different tests, including interfacial analysis, shear strength analysis, hermeticity detection, and failure analysis were performed to evaluate the bonding performance.

3.1. *Interfacial Analysis*

After polishing at the 10 μm and 1 μm fine levels, the cross-sectional interface of Pt–Pt bonding under SEM can be clearly seen in Figure 3. It can be observed from the backscatter mode of the electron microscope that the bonding interface is relatively flat with no obvious cracks or gaps. This suggests that after the thermocompression bonding described in this article, the Pt–Pt metal electrodes are interconnected.

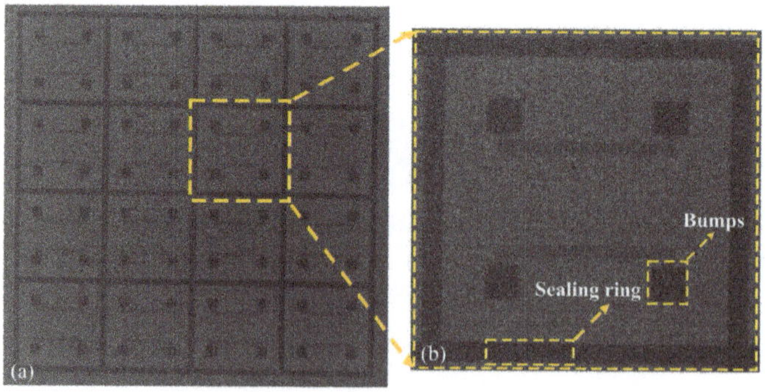

Figure 3. X-ray image. (**a**) Overall SEM image of the chip. (**b**) Enlarged view of a single unit.

Figure 4 is the bonding image of the bump obtained by the SEM mode under the electronic scanning electron microscope. The bright part in the middle is the metal layer. The thickness of the metal layer is measured to be about 700 nm, which is consistent with the thickness of 50 nm Cr and 300 nm Pt plated during sputtering. Further, no obvious cracks were observed in the metal layer, and the whole surface tends to be smooth, forming a good interconnection interface. Interestingly, no obvious delamination phenomenon of Cr and Pt resulting from the diffusion of Cr is observed in the metal region on each side.

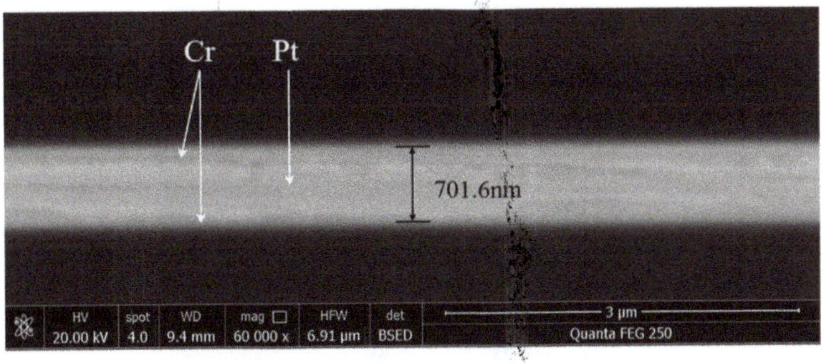

Figure 4. SEM image of the bump.

Figure 5 shows the result of the line scan of the bonding interface. Through the longitudinal line scan of the bonding interface, the content of three elements is analyzed. It can be found that most of the Si elements are distributed on both sides of the metal layer, which is more in line with the actual situation, indicating that the hot-press bonding does not cause the non-metallic surface to diffuse inward. Moreover, the content of Pt element is mainly concentrated in the metal layer area, the middle part is more concentrated, and the two sides are more symmetrical and uniform, which shows that the interconnection interface formed after thermocompression bonding has no displacement, and the Pt–Pt interface is better connected with fewer impurities. Additionally, the widely distributed Cr element is essentially more concentrated between the Si layer and the Pt layer, and is consistent with the SEM image shown in Figure 4. The Cr layer is sandwiched between the metal layer and the dielectric layer as an adhesion layer. After thermocompression bonding, a small amount of diffusion of Cr element occurred between the Pt metal layers.

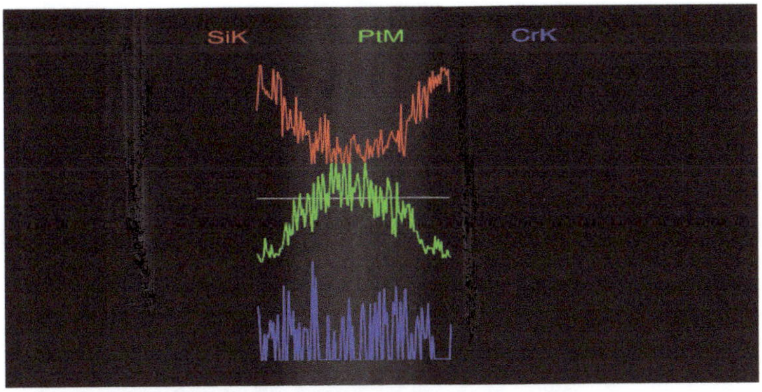

Figure 5. EDS line scan of the bonding interface.

Figure 6 is the distribution of silicon and Pt in the backscattered SEM image of the interconnect interface. The blue dots in Figure 6a represent the distribution of silicon, and the yellow dots in Figure 6b indicate the distributed Pt, and Figure 6c shows the effect of integrating the two SEM images. Through elemental surface scanning analysis, it can be observed that the bonded Pt element is still in the bonding area with no dislocation or drift.

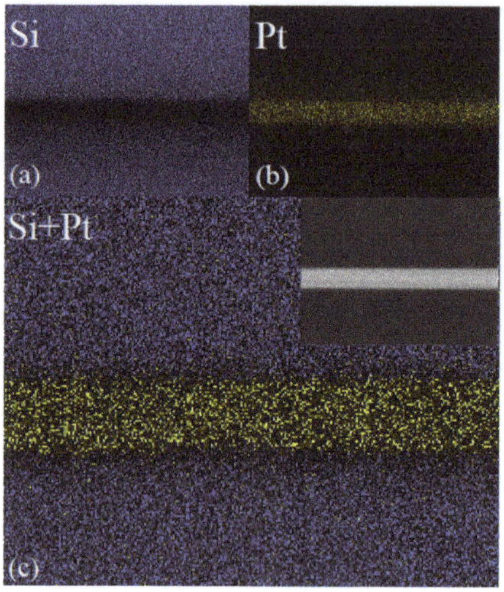

Figure 6. EDS surface scanning element analysis diagram of the bonding interface. (**a**) Distribution diagram of silicon element, (**b**) distribution diagram of platinum element, (**c**) consolidated diagram of element distribution.

In Figure 7a, the green dots in Figure 7b represent the Si element, the red dots in Figure 7c represent the N element, the pink dots in Figure 7d represent the Cr element, and the yellow dots in Figure 7e represent the Pt element; the yellow dots representing the Pt element, Figure 7a, are a visual SEM image of the final integration of each element. The combination of (Si + Cr + Pt = 100) and (Si + N + Cr + Pt = 100) elements were selected for analysis. The delamination of each element in the bonded sample was more obvious and no dislocation diffusion or element drift was observed.

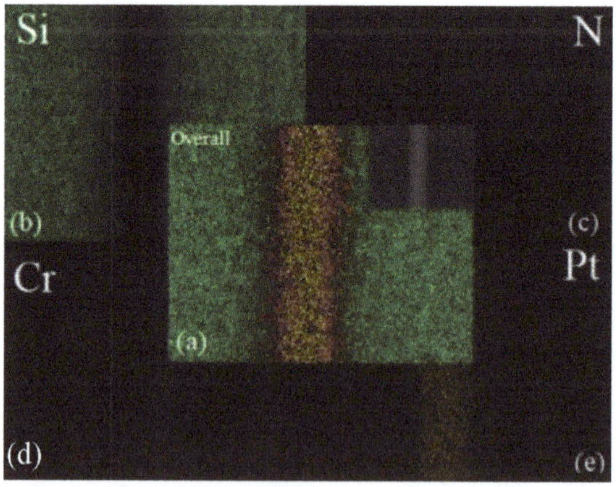

Figure 7. EDS surface scanning element analysis diagram of the bonding interface. (**a**) Consolidated diagram of element distribution, (**b**) distribution diagram of silicon element, (**c**) distribution diagram of nitrogen element, (**d**) distribution diagram of chromium element, (**e**) distribution diagram of platinum element.

3.2. Shear Strength Analysis

Figure 8 is a schematic diagram of the shear force test. The equipment used in the experiment is Dage4000 bond tester which can provide testing of bond shear force and tensile force. The sample in the picture is denoted by the blue part in the middle. After the lower substrate is fixed, the push knife moves in the horizontal direction until the upper and lower bonded substrates are separated. At the moment of separation, the force required to stop the pushing knife is the force of bonding at that moment.

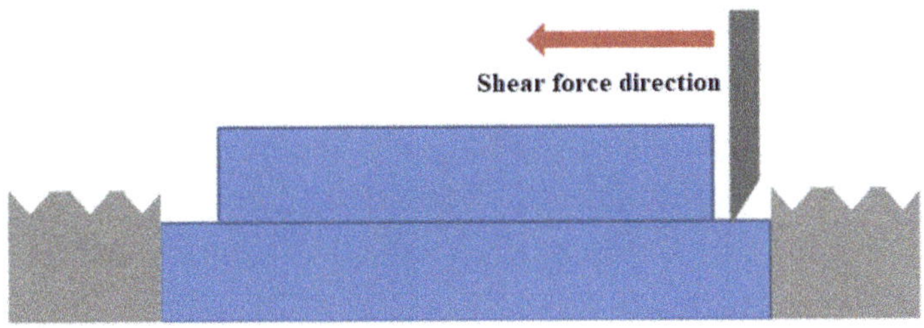

Figure 8. Schematic diagram of shear force test.

Based on theoretical inferences, increasing the pressure can enhance the bonding strength. Importantly, the pressure applied should not exceed the threshold of the substrate's withstanding ability. Excessive pressure may cause an overflow of sputtered metal or cracks in the substrate. The shear resistance test in the effectively bonded sample in this experiment is shown in Figure 9. A total of 12 bonding samples were selected and divided into four groups according to the different bonding pressures: 30 MPa, 50 MPa, 80 MPa, and 100 MPa. Their shear resistance was found to be in the range of 12.2–14.3 MPa, 15.1–17.6 MPa, 16.9–18.9 MPa, and 17.3–31.8 MPa. The bonding pressure is positively correlated within the withstandable range of the substrate, and its shear resistance can reach a maximum of 30 Mpa as the pressure increases. When the bonding pressure is 100 MPa, the measured value is the maximum shear force that the silicon wafer can withstand, and the average shear resistance can reach 25 MPa; thereby, the shear strength of Pt–Pt bonding meets the standard.

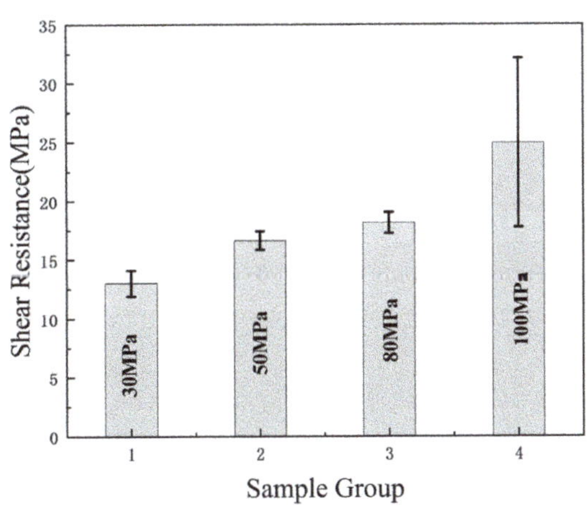

Figure 9. Bonding strength of substrates.

Figure 10 shows the microscope schematic diagrams of the Pt metal sealing ring before thermocompression bonding and the Pt metal sealing ring of the debris after the shear force test. The shear test is generally divided into three fracture modes: IMC mode, solder mode, and mixed IMC/solder mode. Fractures in the IMC mode generally occur in the IMC layer. As can be observed from the Figure 10, the shear test destroyed the bonding electrode and the sealing ring, and it can be seen that the fracture mode in the test is mainly the IMC mode [25,26], indicating that the shear resistance is mainly due to the force exerted on the bonding electrode. This demonstrates that the bonding is effective.

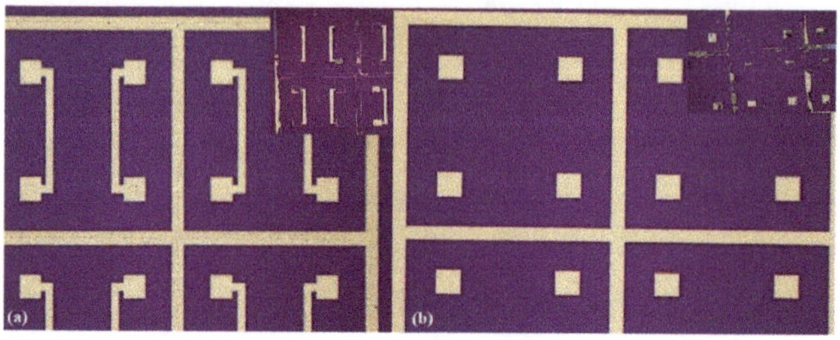

Figure 10. Shear force test comparison image. (**a**) Top substrate. (**b**) Bottom substrate.

Of course, some metals do not perform well in shear tests. This may be caused by uneven sputtering during the coating process due to the influence of experimental factors.

3.3. *Hermeticity Detection*

According to the method defined by the inspection standard (GJB 548B-2005 method 1014.2), the purpose of the test is to determine the hermeticity of microelectronic and semiconductor device packages with internal cavities. The hermeticity test of the four groups of bonded pairs is carried out. First, a detailed inspection was carried out using the ZHP-30D helium mass spectrometer leak detector. The sample is kept under pressure of 4×10^{-5} Pa for 2 h, and the leakage rate is measured with the leak detector after taking it out. The actual leakage rate of the technically required samples is less than 2×10^{-3} Pa·cm^3/s, that is, less than the specified leakage rate value (5×10^{-3} Pa·cm^3/s). As Table 2 shown, the first group of samples through the experiment showed the minimum leakage rate measured to be 30 MPa/165 N, 350 °C, 1200 s: 3.3×10^{-4} Pa·cm^3/s, the maximum leakage rate: 9.8×10^{-4} Pa·cm^3/s, the average leakage rate: 6.55×10^{-4} Pa·cm^3/s, which is less than the leakage rate value required by the specification. Similarly, the second group of samples showed the leakage rate measured under the bonding conditions of 50 MPa/275 N, 350 °C, 1200 s: 5.9×10^{-4} Pa·cm^3/s, 3.3×10^{-5} Pa·cm^3/s, average leakage rate: 3.115×10^{-4} Pa·cm^3/s, which is less than the specified leakage rate value. The third group of samples: 80 MPa/550 N, 350 °C, the maximum leakage rate measured under the bonding condition of 1200 s: 2.77×10^{-5} Pa·cm^3/s, the minimum leakage rate: 1.37×10^{-5} Pa·cm^3/s, the average leakage rate: 1.81×10^{-5} Pa·cm^3/s. The fourth group of samples: 100 MPa/550 N, 350 °C under the bonding conditions of 1200 s, the maximum leakage rate measured: 1.83×10^{-5} Pa·cm^3/s, the minimum leakage rate: 1.29×10^{-5} Pa·cm^3/s, average leakage rate: 1.48×10^{-5} Pa·cm^3/s. In conclusion, the leakage rate values measured by the four groups of different bonding parameters are all within the range of the leakage rate values required by the specification. It can be inferred from Figure 11 that with the increase in the bonding pressure, the average air tightness of the samples will gradually increase, and the average air tightness of the successfully bonded samples can meet the packaging requirements.

Table 2. Case of leakage rate.

Parameter	Condition	Minimum Leakage Rate	Maximum Leakage Rate	Average Leakage Rate
Group 1	30 MPa/165 N	3.3×10^{-4}	9.8×10^{-4}	6.55×10^{-4}
Group 2	50 MPa/275 N	3.3×10^{-5}	5.9×10^{-4}	3.115×10^{-4}
Group 3	80 MPa/550 N	1.37×10^{-5}	2.77×10^{-5}	1.81×10^{-5}
Group 4	100 MPa/550 N	1.29×10^{-5}	1.83×10^{-5}	1.48×10^{-5}

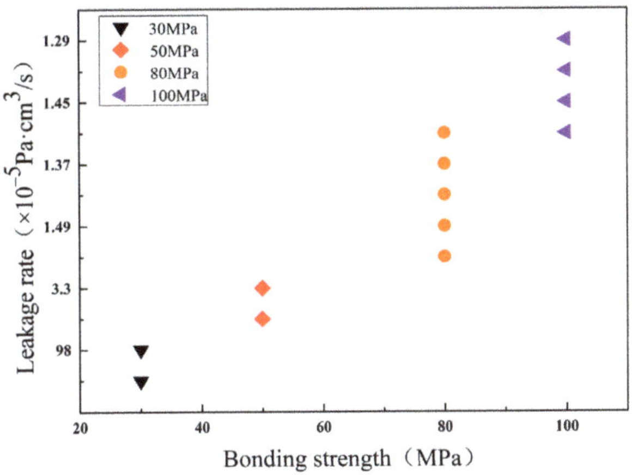

Figure 11. Influence of pressure on bonding leakage rate.

3.4. Failure Analysis

Owing to the high melting point and boiling point of Pt, the diffusion process is greatly restricted by temperature during the flip-chip welding process. Since the maximum welding temperature of flip-chip welding in the bulk silicon process does not exceed 400 °C, this work used a welding temperature of 250 °C–350 °C and a bonding pressure of 30 MPa–100 Mpa to explore the influence of different temperatures and pressures on welding failure rate. Bonding failure may be caused by a variety of reasons, and the failure may be manifested in that the top and bottom substrates do not adhere together or the tests such as shear force and air tightness cannot meet the test requirements. It was found that both the bonding pressure and the bonding temperature are positively correlated to the failure rate as shown in Figure 12. When the bonding temperature is 625 K and the bonding pressure is above 80 MPa, the failure rate is less than 0.4; under the same pressure conditions, when the temperature is less than 625 K, the failure rate is greater than 0.5. It shows that the failure rate is greatly affected by the bonding temperature, which should optimally be above 625 K. When the bonding pressure is less than 50 MPa, the failure rate of the bonding is more than 0.5, suggesting the optimal bonding pressure of Pt–Pt bonding is more than 50 MPa.

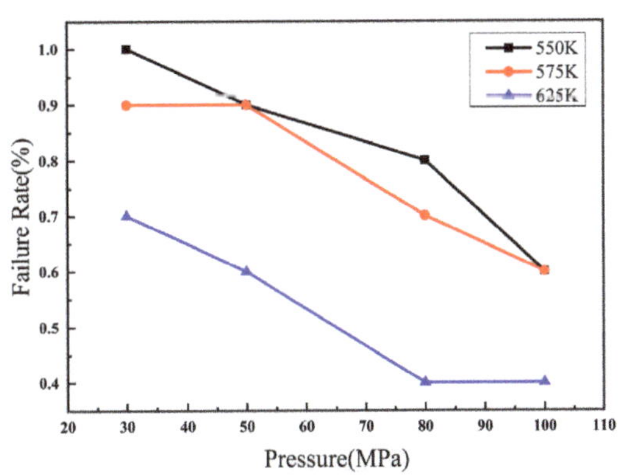

Figure 12. Influence of temperature and pressure on welding failure rate.

4. Conclusions

The Pt–Pt metal interconnection used flip-chip hot-press packaging technology to achieve a relatively stable interconnection at a temperature of 350 °C and pressure above

80 MPa. A series of experiments to evaluate its bonding performance was carried out. The thickness of the metal layer did not change significantly after bonding, and there was no Pt overflow at the bonding interface. In the case of bonding force of 550 N for 20 min, the shear resistance could reach up to 30 MPa. The shear experiment shows that the fractured interface is mostly on the silicon-metal layer, indicating that the interconnection interface after bonding is stable and does not easily fracture, indicating that the Pt–Pt metal bonding has a certain feasibility in packaging. The failure analysis experiment shows that the bonding pressure and temperature have an important influence on the failure rate, which is in line with the basic principle of metal welding. In addition, increasing the metal activity of the Pt metal interface may lower the requirements for bonding temperature and bonding pressure. Thus, these materials are considered worth exploring. We can plasma-treat the metal layer before bonding to improve the metal activity or find equipment that can increase the bonding temperature and bonding pressure to improve the quality of the bonding. In short, Pt–Pt metal bonding is a means of encapsulation that has great potential in the high-temperature environment in the future.

Author Contributions: Data curation and writing—original draft, J.L.; conceptualization, methodology, project administration, writing— review and editing, J.W.; funding acquisition, supervision, M.L.; investigation and validation, H.Z. All authors have read and agreed to the published version of the manuscript.

Funding: This research was funded by the "173" Projects of China (Grant No. 2021JCJQJJ0172, 2020JCJQZD043).

Data Availability Statement: Not applicable.

Acknowledgments: The authors appreciate the facility assistance from Suzhou Institute of Nano-Tech and Nano-Bionics, Chinese Academy of Sciences.

Conflicts of Interest: The authors declare no conflict of interest.

References

1. Ghosh, S.; Nika, D.L.; Pokatilov, E.P.; Balandin, A. Heat conduction in graphene: Experimental study and theoretical interpretation. *New J. Phys.* **2009**, *11*, 095012. [CrossRef]
2. Wang, L.; Meric, I.; Huang, P.Y.; Gao, Q.; Gao, Y.; Tran, H.; Taniguchi, T.; Watanabe, K.; Campos, L.M.; Muller, D.A.; et al. One-dimensional electrical contact to a two-dimensional material. *Science* **2013**, *342*, 614–617. [CrossRef]
3. Yin, Y.; Cheng, Z.; Wang, L.; Jin, K.; Wang, W. Graphene, a material for high temperature devices–intrinsic carrier density, carrier drift velocity and lattice energy. *Sci. Rep.* **2014**, *4*, 1–6. [CrossRef] [PubMed]
4. Burla, R.K.; Chen, L.; Zorman, C.A.; Mehregany, M. Development of nickel wire bonding for high-temperature packaging of SiC devices. *IEEE Trans. Adv. Packag.* **2009**, *32*, 564–574. [CrossRef]
5. Brachmann, E.; Seifert, M.; Ernst, D.; Menzel, S.B.; Gemming, T. Pt-wire bonding optimization for electroplated Pt films on γ-Al_2O_3 for high temperature and harsh environment applications. *Sens. Actuators A Phys.* **2018**, *284*, 129–134. [CrossRef]
6. Kang, S.K.; Gruber, P.; Shih, D.Y. An overview of Pb-free, flip-chip wafer-bumping technologies. *JOM* **2008**, *60*, 66–70. [CrossRef]
7. Wun, J.; Lai, C.; Chen, N.; Bowers, J.E.; Shi, J.-W. Flip-Chip Bonding Packaged THz Photodiode with Broadband High-Power Performance. *Photonics Technol. Lett. IEEE* **2014**, *26*, 2462–2464. [CrossRef]
8. Wang, P.J.; Lee, C.C. Silver flip-chip technology by solid-state bonding. *J. Electron. Packag.* **2010**, *132*, 035001. [CrossRef]
9. Shimatsu, T.; Uomoto, M. Atomic diffusion bonding of wafers with thin nanocrystalline metal films. *J. Vac. Sci. Technol. B Nanotechnol. Microelectron. Mater. Process. Meas. Phenom.* **2010**, *28*, 706–714. [CrossRef]
10. Kim, T.H.; Howlader, M.; Itoh, T.; Suga, T. Room temperature Cu–Cu direct bonding using surface activated bonding method. *J. Vac. Sci. Technol. A Vac. Surf. Film.* **2003**, *21*, 449–453. [CrossRef]
11. Shimbo, M.; Furukawa, K.; Fukuda, K.; Tanzawa, K. Silicon-to-silicon direct bonding method. *J. Appl. Phys.* **1986**, *60*, 2987–2989. [CrossRef]
12. Maszara, W.P.; Goetz, G.; Caviglia, A.; McKitterick, J.B. Bonding of silicon wafers for silicon-on-insulator. *J. Appl. Phys.* **1988**, *64*, 4943–4950. [CrossRef]
13. Ye, Z. Sensor/ROIC integration using oxide bonding. In Proceedings of the 2008 International Linear Collider Workshop (LCWS08 and ILC08), Chicago, IL, USA, 16–20 November 2008.
14. Enquist, P.; Fountain, G.; Petteway, C.; Hollingsworth, A.; Grady, H. Low cost of ownership scalable copper direct bond interconnect 3D IC technology for three dimensional integrated circuit applications. In Proceedings of the 2009 IEEE International Conference on 3D System Integration, San Francisco, CA, USA, 28–30 September 2009.

15. Lin, H.; Stevenson, J.T.M.; Gundlach, A.M.; Dunare, C.C.; Walton, A.J. Direct Al–Al contact using low temperature wafer bonding for integrating MEMS and CMOS devices. *Microelectron. Eng.* **2008**, *85*, 1059–1061. [CrossRef]
16. Hao, F.; Wang, J.; Li, M. Low-Temperature Anodic Bonding for Wafer-Level Al–Al Interconnection in MEMS Grating Gyroscope. *IEEE Trans. Compon. Packag. Manuf. Technol.* **2021**, *11*, 19–24. [CrossRef]
17. Mersinligil, M.; Desset, J.; Brouckaert, J.F. High-temperature high-frequency turbine exit flow field measurements in a military engine with a cooled unsteady total pressure probe. *Proc. Inst. Mech. Eng. Part A J. Power Energy* **2011**, *225*, 954–963. [CrossRef]
18. Yamawaki, S. Verifying heat transfer analysis of high pressure cooled turbine blades and disk. *Ann. N. Y. Acad. Sci.* **2001**, *934*, 505–512. [CrossRef]
19. Yang, Y.; Yang, J.; Fang, D. Research progress on thermal protection materials and structures of hypersonic vehicles. *Appl. Math. Mech.* **2008**, *29*, 51–60. [CrossRef]
20. Seifert, M.; Brachmann, E.; Rane, G.K.; Menzel, S.B.; Oswald, S.; Gemming, T. Pt-RuAl bilayers as a model system for Pt wire bonding of high-temperature RuAl electrodes. *J. Alloys. Compd.* **2019**, *29*, 393–398. [CrossRef]
21. Trumić, B.; Gomidželović, L.; Marjanović, S.; Ivanović, A.; Krstić, V. Pt-based alloys: Investigation of the effect of impurities content on creep rate, rupture time and relative elongation at high temperatures. *Mater. Res.* **2016**, *20*, 191–199. [CrossRef]
22. Wang, J.; Wang, Q.; Liu, Z.; Wu, Z.; Cai, J.; Wang, D. Activation of electroplated-Cu surface via plasma pretreatment for low temperature Cu-Sn bonding in 3D interconnection. *Appl. Surf. Sci.* **2016**, *384*, 200–206. [CrossRef]
23. Hu, X.; Xu, T.; Keer, L.M.; Li, Y.; Jiang, X. Microstructure evolution and shear fracture behavior of aged Sn3Ag0. 5Cu/Cu solder joints. *Mater. Sci. Eng. A* **2016**, *673*, 167–177. [CrossRef]
24. Hu, X.; Xu, T.; Keer, L.M.; Li, Y.; Jiang, X. Shear strength and fracture behavior of reflowed Sn3.0Ag0.5Cu/Cu solder joints under various strain rates. *J. Alloys. Compd.* **2017**, *690*, 720–729. [CrossRef]
25. Bi, X.; Hu, X.; Li, Q. Effect of Co addition into Ni film on shear strength of solder/Ni/Cu system: Experimental and theoretical investigations. *Mater. Sci. Eng. A* **2020**, *788*, 139589. [CrossRef]
26. Wang, H.; Hu, X.; Jiang, X. Effects of Ni modified MWCNTs on the microstructural evolution and shear strength of Sn-3.0Ag-0.5Cu composite solder joints. *Mater. Charact.* **2020**, *163*, 110287. [CrossRef]

micromachines

Article

A New Low-Temperature Solder Assembly Technique to Replace Eutectic Sn-Bi Solder Assembly

Lingyao Sun [1], Zhenhua Guo [1], Xiuchen Zhao [1], Ying Liu [1], Kingning Tu [2,3] and Yingxia Liu [4,*]

1 School of Materials Science and Engineering, Beijing Institute of Technology, Beijing 100081, China; sunlingyao_513@163.com (L.S.); guozhenhua1318@163.com (Z.G.); zhaoxiuchen@bit.edu.cn (X.Z.); yingliu@bit.edu.cn (Y.L.)
2 Department of Materials Science and Engineering, City University of Hong Kong, Hong Kong, China; kntu@cityu.edu.hk
3 Department of Electrical Engineering, City University of Hong Kong, Hong Kong, China
4 Department of Advanced Design and System Engineering, City University of Hong Kong, Hong Kong, China
* Correspondence: yingxliu@cityu.edu.hk

Abstract: We successfully achieved low-temperature assembly by reflowing the 13.5Sn-37.5Bi-45In-4Pb quaternary eutectic solder paste and the SAC 305 solder ball together at 140 °C for 5 min. The wetting angle of the mixed solder joint is 17.55°. The overall atomic percent of Pb in the mixed solder joint is less than 1%, which can be further reduced or eliminated. Moreover, after aging at 80 °C for 25 days, we observed no obvious decrease in shear strength of the fully mixed solder joint, which is the most advantage of this assembly technique over Sn58Bi solder assembly. The Bi phase segregation at the interface is slowed down compared with Sn-Bi solder joint. This low-temperature assembly is promising to be applied in advanced packaging technology to replace the eutectic Sn-Bi solder.

Keywords: low-temperature soldering; 3D IC; Bi aggregation; Sn-Bi solder

Citation: Sun, L.; Guo, Z.; Zhao, X.; Liu, Y.; Tu, K.; Liu, Y. A New Low-Temperature Solder Assembly Technique to Replace Eutectic Sn-Bi Solder Assembly. *Micromachines* **2022**, *13*, 867. https://doi.org/10.3390/mi13060867

Academic Editor: Wensheng Zhao

Received: 8 May 2022
Accepted: 25 May 2022
Published: 31 May 2022

Publisher's Note: MDPI stays neutral with regard to jurisdictional claims in published maps and institutional affiliations.

Copyright: © 2022 by the authors. Licensee MDPI, Basel, Switzerland. This article is an open access article distributed under the terms and conditions of the Creative Commons Attribution (CC BY) license (https://creativecommons.org/licenses/by/4.0/).

1. Introduction

As the downscaling trend of silicon chips is approaching its physical limits, advanced packaging technologies that integrate multiple chips, either vertically or horizontally, provide an alternative approach for developing post-Moore-era electronics [1–3]. The vertical or horizontal stacking of chips in advanced packaging technology requires a low-melting-point solder. Specifically, vertical stacking such as 3D integrated circuit (3D IC) is achieved via multiple reflows, and the use of a low-melting-point solder can prevent the re-melting of solder joints connected in a previous reflow [4]. In addition, packaging sizes generated by horizontal stacking are becoming increasingly large, leading to severe warpage problems [5]. This warpage can be relieved by using a low-melting-point solder during assembly. Thus, low-temperature assembly enables vertical and horizontal advanced packaging to be achieved, supporting continuing advances in microelectronic devices.

Currently, the packaging industry is trying to find an appropriate low-melting-point solder alloy. Tin–bismuth (Sn–Bi) eutectic solder has a melting point of 139 °C, but its brittleness limits its application in mobile electronic devices, especially after aging [6–12]. On the one hand, this is because aging causes Bi atom to segregate at the interface between the solder joint and the substrate [11,12]. As Bi is inherently brittle, this segregation increases the likelihood of interfacial fracture during drop testing. On the other hand, after aging, the IMC at the interface between the solder joint and the substrate will grow and thicken with the aging time. Due to the brittleness of the IMC, excessive thickness will reduce the mechanical properties and reliability of the joint [13–17]. In addition, the wettability of solder is also an important evaluation in electronic packaging industry. Since Bi will reduce the reaction speed of Sn and Cu, the wetting time will increase and the wettability will decrease [18]. Sn-Bi solder has very limited utility in mobile devices due to its brittleness and poor wettability. Tin–indium (Sn-In) eutectic solder has a suitable melting point (118 °C), and bonds well with Cu, Ni, and Au substrates [19]. However, Sn-In solder is too soft, in addition, the widespread use of low-temperature solder with high In content is economically unfavorable because In is very expensive [20]. Most studies on low-melting-point solders have explored whether the addition of minor amounts of a third or fourth element into eutectic Sn-Bi alloys can produce alloys with better mechanical properties. These studies have not been successful, as the brittle nature of eutectic Sn–Bi has proven to be largely unmodifiable [21]. Additionally, there have been attempts to obtain a composite solder with better performance by mixing the two solders [22–27]; however, the reflow temperature is still high. Thus, there is an urgent need for a low-temperature assembly technology to replace eutectic Sn-Bi for use in industrial applications of mobile devices. In this paper, we report a low-temperature assembly technology to do so. Moreover, the assembled solder joint has a relatively stable shear test performance even after a long time of aging.

2. Experimental Section

The prepared Cu solder pads were ultrasonically cleaned with acetone, dilute hydrochloric acid, deionized water, and alcohol to remove surface oil and oxides. The Cu pads we used are 1 mm in diameter. And we developed a solder paste of 13.5Sn-37.5Bi-45In-4Pb quaternary eutectic alloy, which has a melting point of 60 °C. We use a low-temperature flux in the preparation of the 13.5Sn-37.5Bi-45In-4Pb solder paste, which is developed by the Beijing Institute of Nonferrous Metals and Rare Earth Applications. It works well in the range of 50~150 °C. The method of preparing the solder paste will be presented in our following paper. We printed the solder paste on the Cu pads with a designed stencil. A Sn96.5Ag3.0Cu0.5 (SAC305) solder ball was placed on each pad. The printed quaternary eutectic ultra-low-temperature solder paste is a cylinder of 1 mm in diameter and 100 μm in height. The SAC305 solder ball is 600 μm in diameter, as shown in Figure 1. Then, the substrate was reflowed at 100 °C, 120 °C, 140 °C, and 160 °C for 5 min. The reflowing temperature is higher than the melting point of 13.5Sn-37.5Bi-45In-4Pb solder paste and lower than the melting point of SAC305 solder balls. During reflow, the quaternary eutectic solder paste melts into a liquid state, while the SAC305 solder balls remain solid, which then gradually dissolve into the liquid solder. Figure 2 is a schematic diagram showing our low-temperature assembly strategy. After the reflow, we mounted and polished the samples to obtain cross-sectional images, which were observed by scanning electron microscope (SEM, Regulus 8230). The phase composition of the mixed solder before and after aging was determined by the energy dispersive spectrometer (EDS) and X-ray Diffractometer (XRD). The density of 13.5Sn-37.5Bi-45In-4Pb quaternary eutectic solder paste was measured by the Archimedes drainage method.

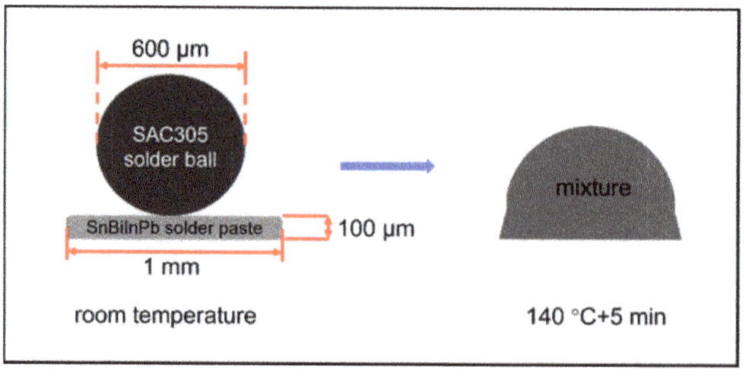

Figure 1. Schematic diagram of the assembly process using SAC305 solder balls and 13.5Sn-37.5Bi-45In-4Pb solder paste.

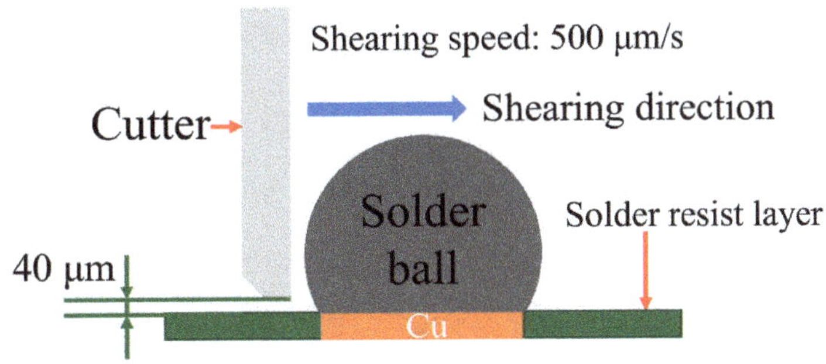

Figure 2. Schematic diagram of solder joint shear strength test.

The shear test was carried out on a PTR-1100 shear testing machine. The shear speed is 500 μm/s with a height of 40 μm, which is shown schematically in Figure 2. Additionally, 10 samples under each reflow condition were tested. The mixed solder bumps, obtained after reflowing at 140 °C for 5 min, were aged at 80 °C for 5 days, 15 days, and 25 days, respectively. Then, the microstructure of the aged solder joint was observed by SEM, and its phase composition was analyzed by EDS. We also tested the maximum shear strength of the fully mixed solder joint after aging by PTR-1100 shear testing machine.

3. Results and Discussion

3.1. *Reflow Results*

We choose a reflow temperature that is between the melting point of SAC305 (at 230 °C) and the melting point of 13.5Sn-37.5Bi-45In-4Pb solder paste (at 60 °C). The DSC curve of 13.5Sn-37.5Bi-45In-4Pb solder paste is shown in Figure 3. Therefore, solid–liquid inter-diffusion happens during the assembly process. During the solid–liquid inter-diffusion, Sn atoms from SAC305 solder ball will dissolve into the molten 13.5Sn-37.5Bi-45In-4Pb solder paste, and the two parts are mixed together. The fully mixed solder joint obtained after reflowing at 100 °C, 120 °C, 140 °C, and 160 °C for 5 min are shown in Figure 4. According to Figure 4a,b, there is still a clear boundary between the two parts after reflowing at 100 °C and 120 °C for 5 min, but a completely mixed and uniform solder bump is achieved after reflowing at 140 °C and 160 °C for 5 min.

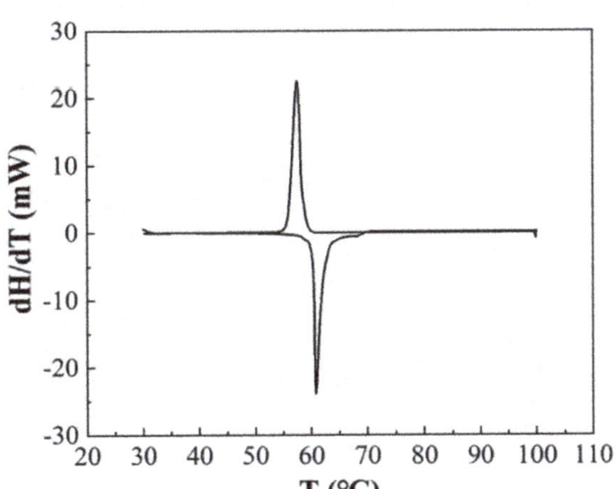

Figure 3. The DSC curve of 13.5Sn-37.5Bi-45In-4Pb solder paste.

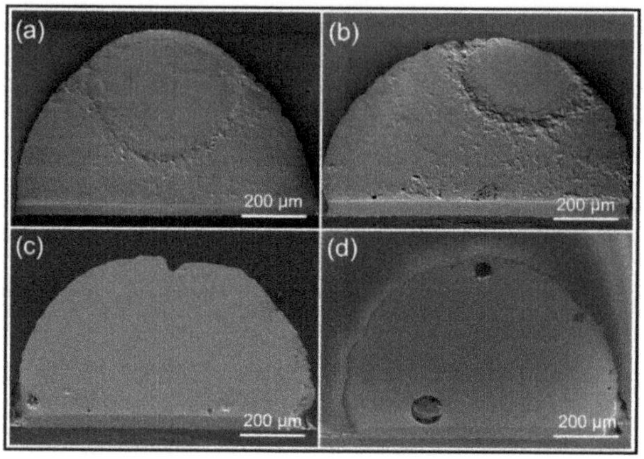

Figure 4. Overall SEM images of 13.5Sn-37.5Bi-45In-4Pb solder paste and SAC305 solder ball after reflowing at (**a**) 100 °C, (**b**) 120 °C, (**c**) 140 °C and (**d**) 160 °C for 5 min, respectively.

Higher magnification SEM images in Figure 5 show the uniformity of the microstructure after reflowing at 140 °C for 5 min. Figure 5a–f are the 500× magnification SEM images of the lower left, left, upper-left edge, upper right-edge, right, and lower right parts of the solder joint, respectively, and (g–i) are the 2000× enlarged view of the connection between the solder joint and the Cu plate, the middle and top of the solder joint. According to Figure 5, the composition of the mixed solder obtained after reflowing at 140 °C for 5 min is completely homogeneous. Therefore, 140 °C is considered the optimum temperature to obtain a fully mixed solder of 13.5Sn-37.5Bi-45In-4Pb and SAC305.

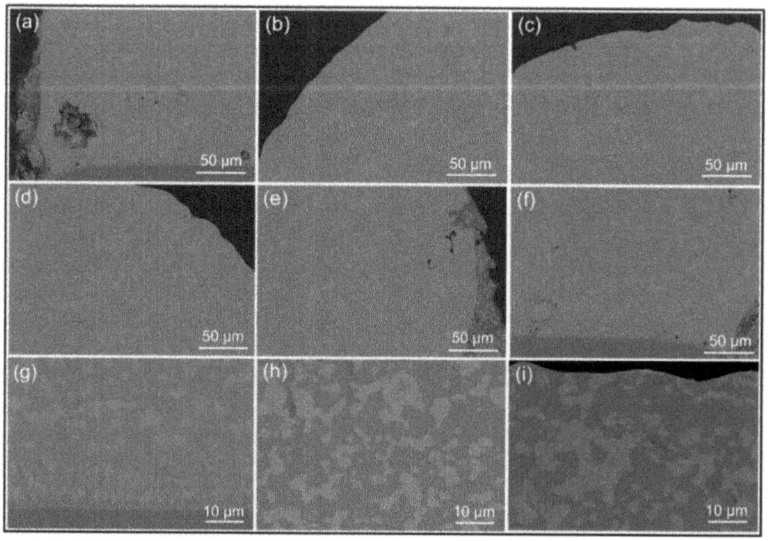

Figure 5. The fully mixed solder joint obtained after reflowing of 13.5Sn-37.5Bi-45In-4Pb solder and SAC305 at 140 °C for 5 min; (**a**–**f**): the 500× magnification of the lower left, left, upper-left edge, upper-right edge, right and lower right parts of the solder joint; (**g**–**i**): the 2000× enlarged view of the connection between the bottom of the solder joint and the Cu plate, the middle and top of the solder joint.

3.2. Composition of the Fully Mixed Solder Joint

We calculated the composition of the fully mixed solder joint with the value of volume and density. The volume of the solder paste and solder ball is calculated from the dimensions marked in Figure 1. The density of SAC305 solder balls is 7.37 g/cm^3 provided by the manufacturer, and the density of 13.5Sn-37.5Bi-45In-4Pb quaternary eutectic solder paste is 8.01 g/cm^3 tested by Archimedes drainage method. Then, the atomic percentage of each element is calculated and listed in Table 1.

Table 1. Calculation of the atomic percentage of each element in the fully mixed solder.

Elements	at. %
Ag	2.02
Cu	0.56
Sn	65.12
Bi	9.81
In	21.43
Pb	1.06

We performed the EDS mapping of the fully mixed solder joint after reflowing at 140 °C for 5 min, and the results are shown in Figure 6. Comparing the atomic percentage table in Figure 6 with Table 1, the difference in the content of each element is not significant and within the measurement error.

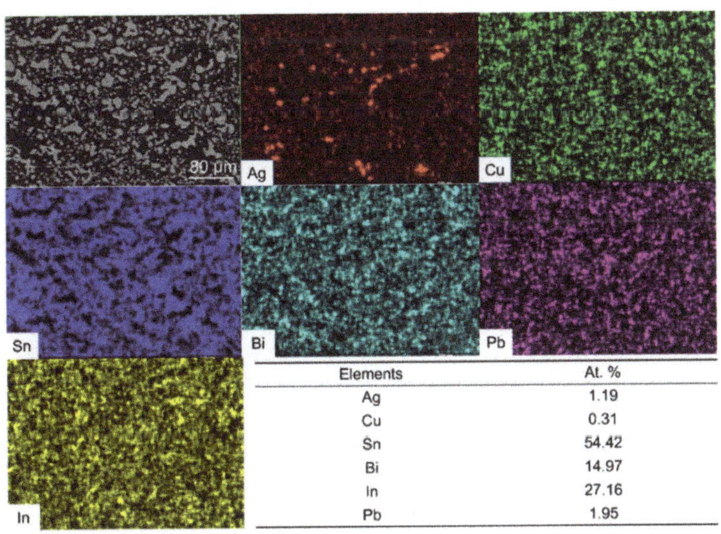

Elements	At. %
Ag	1.19
Cu	0.31
Sn	54.42
Bi	14.97
In	27.16
Pb	1.95

Figure 6. Elemental distribution and atomic percentage results of fully mixed composite solder joint.

From Figures 5 and 6, we can conclude that a well-mixed solder joint can be achieved by the 140 °C reflow of SAC305 solder ball and 13.5Sn-37.5Bi-45In-4Pb solder paste for 5 min. One thing that needs to be noted is the overall Pb atomic concentration in the mixed solder is about 1%. With further optimization, we can reduce the Pb atomic concentration to less than 0.5%. We can reduce the relative Pb content in the fully mixed solder by further reducing the amount of 13.5Sn-37.5Bi-45In-4Pb eutectic solder paste or increasing the amount of SAC305 solder. Specifically, this can be achieved by using thinner stencils (such as 50 μm, 30 μm) or by using larger diameter SAC305 solder ball. We will also try to reduce or even eliminate the Pb used in the low-melting-point solder paste. This means that the technology is promising to be applied to consumer products.

We further investigated the microstructure of the mixed solder joint after reflowing at 140 °C for 5 min with EDS and XRD, as shown in Figures 7 and 8 and Table 2. According to Figure 7 and Table 2, there are four phases in the mixed solder joint, including the Cu_6Sn_5, γ-phase, Bi_3In_5, and Ag_3Sn phases. The Cu_6Sn_5 phase is generated by the interfacial reaction between the solder joint and the Cu substrate, which is illustrated in the darkest area marked as region 1 in Figure 7. The γ-phase is a Sn-rich phase similar to the one marked in the Sn-In phase diagram shown in Figure 9a [28]. Bi_3In_5 is an intermetallic compound phase formed in the reaction between Bi and In [29,30], and so does Ag_3Sn, formed in the reaction between Ag and Sn [31]. Bi-In and Bi-Sn-In phase diagrams are shown in Figure 9b,c. All four phases are not pure and have substitutions from other elements. The XRD results in Figure 8 show that the two main phases in the mixed solder joint, Bi_3In_5 and γ-phase, are consistent with the EDS results [30]. The amount of Cu_6Sn_5 and Ag_3Sn phases is too little to be observed in the XRD results.

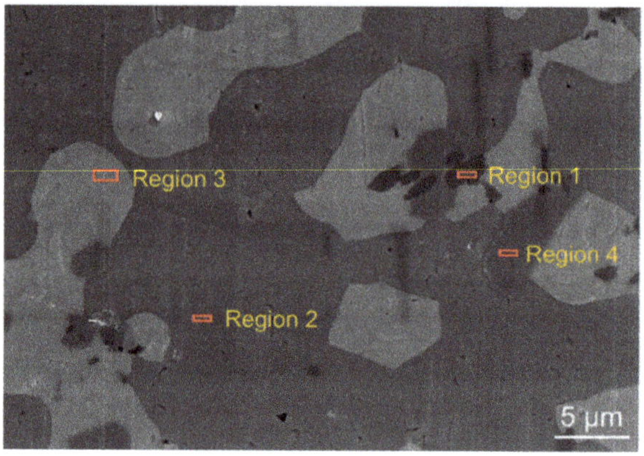

Figure 7. Local magnification SEM image of the fully mixed solder joint.

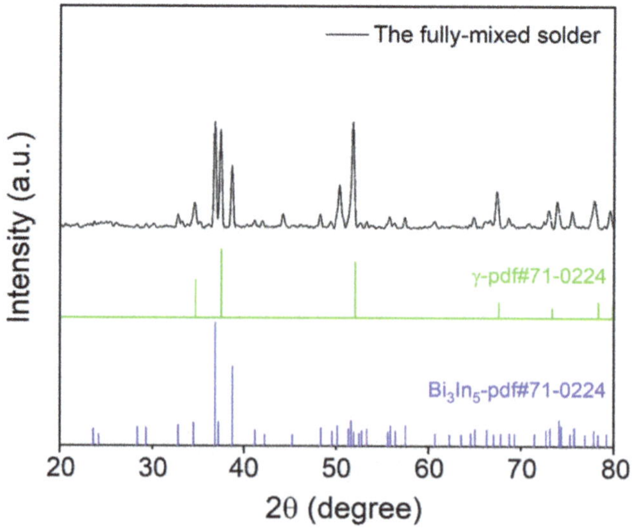

Figure 8. XRD analysis results of the mixed solder joint formed by SAC305 and 13.5Sn-37.5Bi-45In-4Pb after reflowing at 140 °C for 5 minutes.

Table 2. Element content of the regions marked in Figure 7.

Element / Atom (%) / Region	Sn	In	Bi	Pb	Ag	Cu	Phase
1	43.23	1.13	1.57	0.24	0.19	53.64	Cu_6Sn_5
2	78.02	17.23	3.82	0.58	0.35	-	γ
3	3.07	60.12	36.20	0.07	0.31	0.23	Bi_3In_5
4	23.84	9.63	0.39	0.28	65.86	-	Ag_3Sn

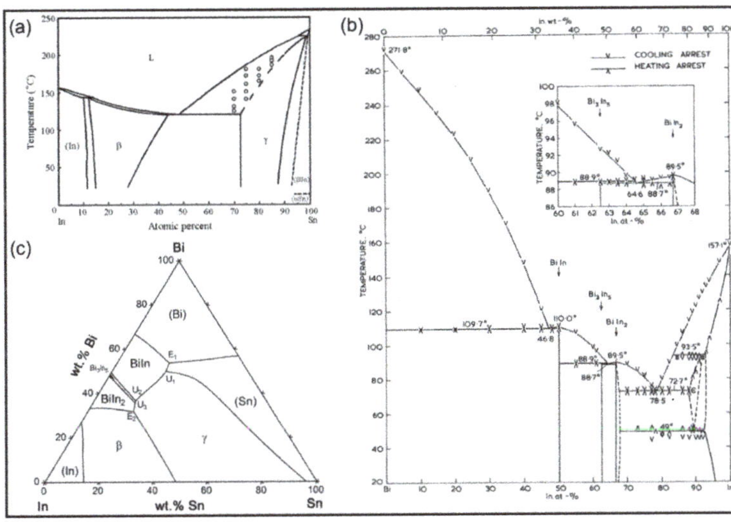

Figure 9. (**a**) The Sn-In phase diagram [28]; (**b**) the Bi-In phase diagram [29]; (**c**) the Bi-In-Sn phase diagram [30].

3.3. *Wettability*

To evaluate the wettability of composite solder joint where SAC305 solder ball is completely mixed with 13.5Sn-37.5Bi-45In-4Pb quaternary eutectic solder paste, we reflow the mixed solder directly on the cleaned Cu substrate. The wetting angle was measured, and that of SAC305 solder ball, Sn58Bi solder paste, and 13.5Sn-37.5Bi-45In-4Pb quaternary eutectic solder paste also were tested, respectively. The results are shown in Figure 10, where the four were compared. The wetting angle of the mixed solder is only 17.05°, while the wetting angles of the 13.5Sn-37.5Bi-45In-4Pb quaternary eutectic solder paste, SAC305 solder ball, and Sn58Bi solder paste are 26.82°, 24.24°, and 29.34°, respectively. Therefore, the wettability of our mixed solder is the most excellent among the four.

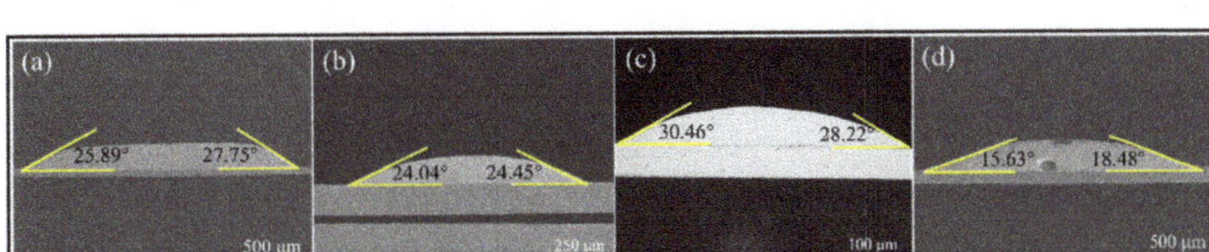

Figure 10. Wetting angle: (**a**) 13.5Sn-37.5Bi-45In-4Pb quaternary eutectic solder paste; (**b**) SAC305 solder ball; (**c**) Sn58Bi solder paste; (**d**) the fully mixed solder of 13.5Sn-37.5Bi-45In-4Pb quaternary eutectic solder paste and SAC305 solder ball.

3.4. *Aging Performance of the Mixed Solder Joint*

The solder bumps obtained after reflowing at 140 °C for 5 min were thermally aged at 80 °C for 5 days, 15 days, and 25 days [32]; the results are shown in Figure 11a–d. The thicknesses of intermetallic compounds (IMC) generated by the reaction between the mixed solder and the interface before aging and after aging for 5, 15, and 25 days were measured as 1.06 μm, 2.39 μm, 3.66 μm, and 4.08 μm, respectively, as shown in Table 3. The composition analysis of IMC before aging and after aging for 25 days was carried out using EDS, as shown in Figure 12. It was found that the IMC was $Cu_6(Sn, In)_5$, with In substitute some Sn atoms in Cu_6Sn_5 IMC.

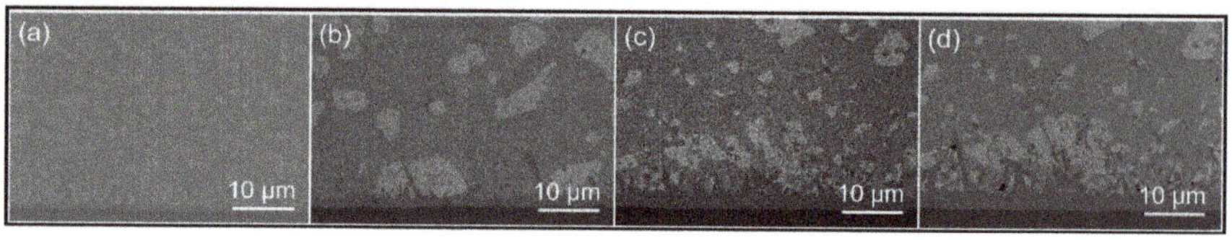

Figure 11. SEM images of the interface between the fully mixed solder joint and the Cu substrate: (**a**) before aging, (**b**) 5 days, (**c**) 15 days, and (**d**) 25 days of aging.

Table 3. The thickness of IMC of the fully mixed solder joints before aging and after aging for 5 days, 15 days, and 25 days.

Aging Time	The Average Thickness of IMC Layer (μm)
0 day	1.06
5 days	2.39
15 days	3.66
25 days	4.08

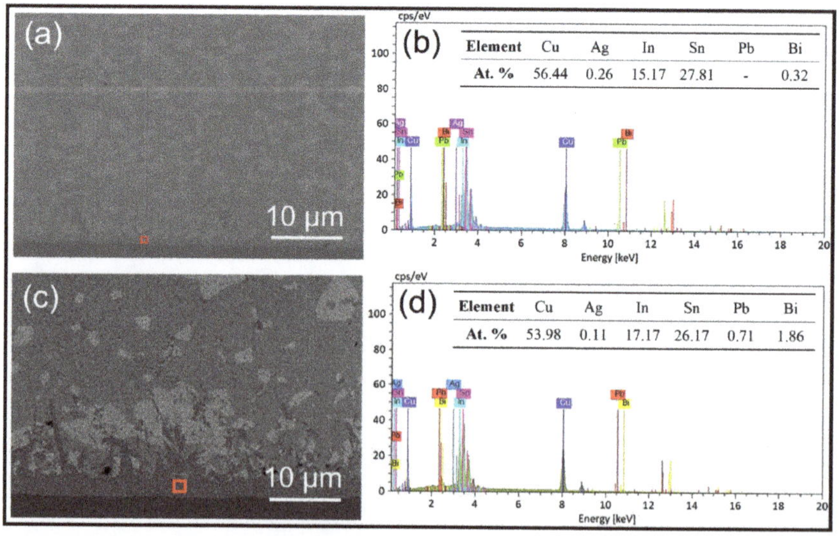

Figure 12. Composition analysis of IMC: (**a**) before aging and (**c**) after 25 days of aging: (**b**) EDS results of the region marked in (**a**); (**d**) EDS results of the region marked in (**c**).

After aging for 15 and 25 days, we observe the coarsening effect of Bi_3In_5 phase. We also observe slight aggregation of Bi_3In_5 phase at the interface. The aggregated phase of Bi_3In_5 is not continuous, which may be due to the spinous shape of intermetallic compound at the solder and Cu interface.

Figure 13 shows the EDS Mapping of the mixed solder joint after 25 days of aging. In Figure 13, the element distribution is uniform throughout the solder joint. Further comparison of the tables in Figures 6 and 13 shows that there are also no significant changes in the atomic percentage of the elements.

We put the results of XRD analysis before and after aging together for comparison, as shown in Figure 14. The main components of the phases before aging and after aging remain the same, which are the γ-phase and Bi_3In_5 phase. Figures 13 and 14 demonstrate that the microstructure of the mixed solder joint is stable after aging.

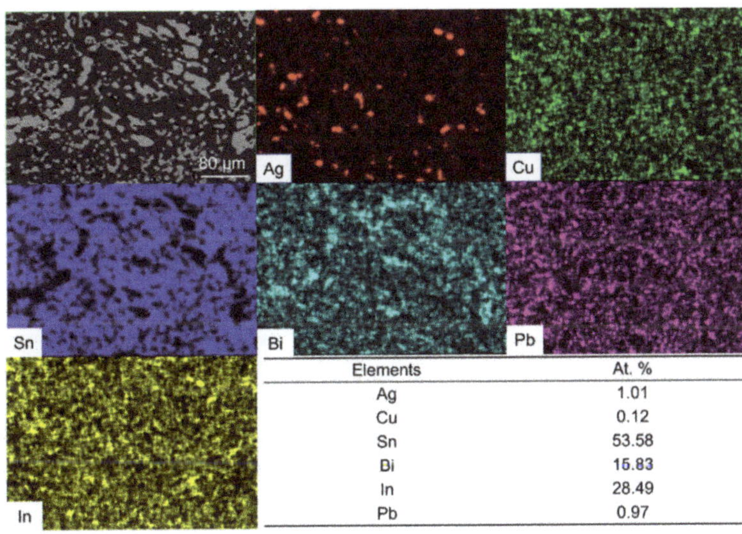

Elements	At. %
Ag	1.01
Cu	0.12
Sn	53.58
Bi	15.83
In	28.49
Pb	0.97

Figure 13. Elemental distribution and atomic percentage results of fully mixed solder joint after 25 days of aging.

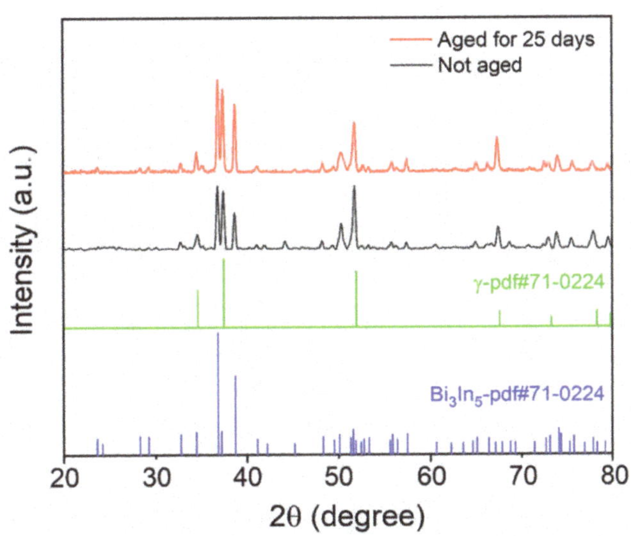

Figure 14. XRD analysis results of the mixed solder joint before and after aging for 25 days.

3.5. Shear Test of the Fully Mixed Solder Joint

The shear strength of the mixed solder joint was tested and compared with that of the original 13.5Sn-37.5Bi-45In-4Pb quaternary solder formed after reflowing at 140 °C for 5 min, as shown in Figure 15a. The shear strength of the original 13.5Sn-37.5Bi-45In-4Pb quaternary solder joint is only 17.62 (±1.57) MPa, while the mixed solder joint is 25.53 (±1.63) MPa, which was much higher than that of the original quaternary solder joint. In addition, the shear strength of SAC305 is 23 MPa [33], so the shear strength of the mixed solder is similar to SAC305.

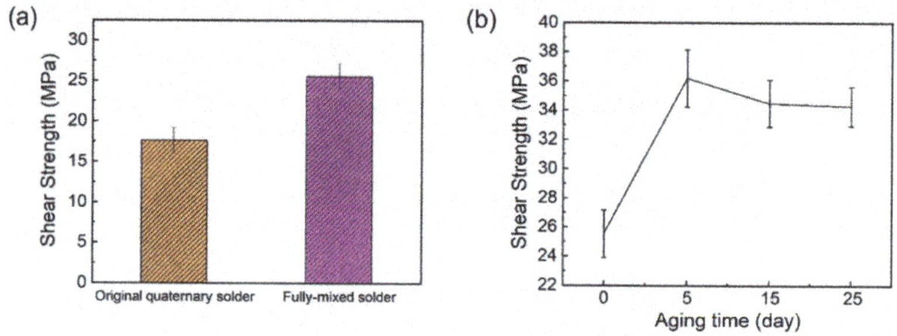

Figure 15. (**a**) Shear strength of the original quaternary solder joint and the fully mixed solder joint; (**b**) Variation of shear strength of fully mixed solder joint with aging time.

The shear strength of the samples after aging for 5, 15, and 25 days after reflowing at 140 °C for 5 min is shown in Figure 15b. The shear strength of the samples before aging was 25.53 (±1.63) MPa, and the shear strength of the samples after aging at 80 °C for 5, 15, and 25 days were 36.18 (±1.96) MPa, 34.47 (±1.61) MPa, and 34.27 (±1.34) MPa, respectively. After 5 days of aging, the shear strength is increased and there is almost no decrease in shear strength with increasing aging time. Because the thickness of the IMC has an important effect on the performance of the solder joint, the IMC needs to be thick enough to ensure a strong and reliable bonding. However, due to the brittleness of the IMC, if it is too thick it will also reduce the joint reliability [14,15,17]. After 5 days of aging, the shear strength increased due to the growth of IMC. Table 3 shows that although the IMC thickens after aging, it is not over-thickened, resulting in good shear strength. It has been shown that the thickness of IMC at the Sn58Bi/Cu interface increases linearly with the square root of the aging time [34–37], that is, the IMC growth at the interface in solder joint follows the empirical diffusion formula:

$$X = (Dt)^{\frac{1}{2}} + X_0 \tag{1}$$

where X is the total IMC thickness, X_0 is the initial IMC thickness, t is the aging time, and D is the diffusivity of the IMC layer [37]. The growth of IMC thickness $(X - X_0)$ is plotted with the square root of aging days $\left(t^{\frac{1}{2}}\right)$, and the results are shown in Figure 16. The growth of IMC with aging time for Sn58Bi was compared with published results [37]. The IMC growth rate of Sn58Bi aged at 135 °C was reported to be 3.0 μm/day$^{1/2}$, while the IMC growth rate of the fully mixed solder was only 0.623 μm/day$^{1/2}$, which is much less than that of Sn58Bi.

The Bi containing low melting temperature solder, which can keep a stable shear strength after aging, is an important finding in our work. Many works have reported the shear strength decrease in Bi containing solder after aging [32,38,39], which is due to the large aggregation of Bi at the interface between substrate and solder [12]. A continuous decrease in the shear strength was reported with an increase in aging time of 7 days, 14 days, and 21 days at an aging temperature of 100 °C. The total decrease of shear strength can be 30~40% after aging for 21 days [8]. In our work, we show a much better shear test performance after aging. We tend to believe it is because the Bi atoms are stabilized by Bi_3In_5 phase in our solder joint. Bi_3In_5 phase is an intermetallic compound phase, and the diffusion of this phase is significantly slower than the diffusion of Bi atoms. Therefore, Bi atoms aggregation slows down, thus a better shear strength can remain and the brittleness of the solder joint can be mitigated after aging. The 140 °C assembly technique brought up in this letter can provide a similar reflow condition to eutectic Sn-Bi solder, while the mechanical performance after aging is much better than Sn-Bi solder. Therefore, the technique is promising to replace the eutectic Sn-Bi solder in advanced electronic packaging.

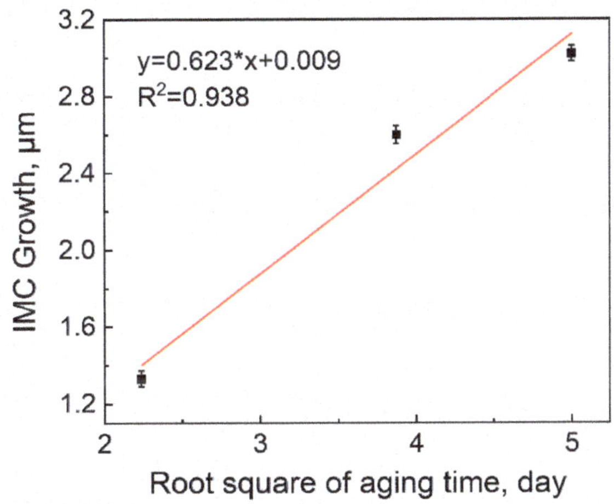

Figure 16. Linear relationship between the thickness of total IMC and the square root of aging days in the fully mixed solder joint.

4. Conclusions

Low-temperature assembly is achieved by reflowing the SAC305 solder ball and 13.5Sn-37.5Bi-45In-4Pb solder paste together at 140 °C for 5 min. The fully mixed solder joint has a uniform microstructure, with two main phases of Bi_3In_5 and γ-phase, and it has very excellent wettability. After 5, 15, and 25 days of aging, the fully mixed solder joint shows a stable microstructure, and there is no continuous aggregation of Bi atoms at the interface between the solder joint and the Cu substrate. More importantly, there is almost no decrease in shear strength after aging for 25 days. The reason is explained as the stabilization effect of Bi atoms by Bi_3In_5 phase, and the IMC did not grow too thick after aging. This technology is promising to replace eutectic Sn-Bi solder in the electronic packaging industry, with a further study on the method of eliminating Pb.

Author Contributions: Conceptualization, Y.L. (Yingxia Liu), Z.G. and L.S.; methodology, Y.L. (Yingxia Liu), X.Z. and L.S.; formal analysis, Y.L. (Yingxia Liu) and L.S.; investigation, L.S.; resources, Y.L. (Yingxia Liu) and Y.L. (Ying Liu); data curation, L.S.; writing—original draft preparation, L.S.; writing—review and editing, Y.L. (Yingxia Liu); visualization, L.S.; supervision, K.T. and Y.L. (Ying Liu); project administration, Y.L. (Yingxia Liu), K.T. and Y.L. (Ying Liu); funding acquisition, Y.L. (Yingxia Liu). All authors have read and agreed to the published version of the manuscript.

Funding: This work was supported by Young Scientists Fund of National Natural Science Foundation of China (grant number 51901022), and the grant from Huawei Technologies (grant number 9239080).

Conflicts of Interest: The authors declare no conflict of interest.

References

1. Gu, S. Material innovation opportunities for 3D integrated circuits from a wireless application point of view. *MRS Bull.* **2015**, *40*, 233–241. [CrossRef]
2. Tu, K.; Liu, Y. Recent advances on kinetic analysis of solder joint reactions in 3D IC packaging technology. *Mater. Sci. Eng. R Rep.* **2018**, *136*, 1–12. [CrossRef]
3. Tu, K. Reliability challenges in 3D IC packaging technology. *Microelectron. Reliab.* **2010**, *51*, 517–523. [CrossRef]
4. Liu, Y.; Li, M.; Kim, D.W.; Gu, S.; Tu, K.N. Synergistic effect of electromigration and Joule heating on system level weak-link failure in 2.5D integrated circuits. *J. Appl. Phys.* **2015**, *118*, 135304. [CrossRef]
5. Mahajan, R.; Sankman, R.; Patel, N.; Kim, D.W.; Aygun, K.; Qian, Z.; Mekonnen, Y.; Salama, I.; Sharan, S.; Iyengar, D.; et al. Embedded Multi-Die Interconnect Bridge (EMIB)—A High Density, High Bandwidth Packaging Interconnect. In Proceedings of the 66th IEEE Electronic Components and Technology Conference (ECTC), Las Vegas, NV, USA, 31 May–3 June 2016; pp. 557–565.
6. Zou, H.; Zhang, Q.; Zhang, Z. Interfacial microstructure and mechanical properties of SnBi/Cu joints by alloying Cu substrate. *Mater. Sci. Eng. A* **2012**, *532*, 167–177. [CrossRef]
7. Song, Q.; Yang, W.; Li, Y.; Mao, J.; Qin, W.; Zhan, Y. Interfacial reaction and mechanical properties of Sn58Bi-XCr solder joints under isothermal aging conditions. *Vacuum* **2021**, *194*, 110559. [CrossRef]
8. Liu, Y.; Ren, B.; Xue, Y.; Zhou, M.; Cao, R.; Zeng, X. Improvement on the mechanical properties of eutectic Sn58Bi alloy with porous Cu addition during isothermal aging. *Mater. Res. Express* **2021**, *8*, 076302. [CrossRef]
9. Mei, Z.; Morris, J.W. Characterization of eutectic Sn-Bi solder joints. *J. Electron. Mater.* **1992**, *21*, 599–607. [CrossRef]
10. Chan, Y.; Lai, J.K.L.; Wu, L.; Poon, N. Residual shear strength of Sn-Ag and Sn-Bi lead-free SMT joints after thermal shock. *IEEE Trans. Adv. Packag.* **2000**, *23*, 708–714. [CrossRef]
11. Liu, P.L.; Shang, J.K. Interfacial embrittlement by bismuth segregation in copper/tin–bismuth Pb-free solder interconnect. *J. Mater. Res.* **2001**, *16*, 1651–1659. [CrossRef]
12. Chen, C.; Ho, C.E.; Lin, A.H.; Luo, G.L.; Kao, C.R. Long-term aging study on the solid-state reaction between 58Bi42Sn solder and Ni substrate. *J. Electron. Mater.* **2000**, *29*, 1200–1206. [CrossRef]
13. Hu, X.; Li, Y.; Min, Z. Interfacial reaction and IMC growth between Bi-containing Sn0.7Cu solders and Cu substrate during soldering and aging. *J. Alloys Compd.* **2014**, *582*, 341–347. [CrossRef]
14. Bi, X.; Hu, X.; Li, Q. Effect of Co addition into Ni film on shear strength of solder/Ni/Cu system: Experimental and theoretical investigations. *Mater. Sci. Eng. A* **2020**, *788*, 139589. [CrossRef]
15. Wang, H.; Hu, X.; Jiang, X. Effects of Ni modified MWCNTs on the microstructural evolution and shear strength of Sn-3.0Ag-0.5Cu composite solder joints. *Mater. Charact.* **2020**, *163*, 110287. [CrossRef]
16. Hu, X.; Xu, T.; Keer, L.M.; Li, Y.; Jiang, X. Microstructure evolution and shear fracture behavior of aged Sn3Ag0.5Cu/Cu solder joints. *Mater. Sci. Eng. A* **2016**, *673*, 167–177. [CrossRef]
17. Zhang, Z.; Hu, X.; Jiang, X.; Li, Y. Influences of mono-Ni (P) and dual-Cu/Ni (P) plating on the interfacial microstructure evolution of solder joints. *Metall. Mater. Trans. A* **2019**, *50*, 480–492. [CrossRef]
18. Zhang, L.; Sun, L.; Guo, Y.H. Microstructures and properties of Sn58Bi, Sn35Bi0.3Ag, Sn35Bi1.0Ag solder and solder joints. *J. Mater. Sci.-Mater. Electron.* **2015**, *26*, 7629–7634. [CrossRef]
19. Morris, J.W.; Goldstein, J.L.F.; Mei, Z. Microstructure and mechanical properties of Sn-In and Sn-Bi solders. *JOM* **1993**, *45*, 25–27. [CrossRef]
20. Han, D.L.; Shen, Y.A.; Huo, F.; Nishikawa, H. Nishikawa, Microstructure Evolution and Shear Strength of Tin-Indium-xCu/Cu Joints. *Metals* **2022**, *12*, 33. [CrossRef]
21. Liu, Y.; Tu, K.N. Low melting point solders based on Sn, Bi, and In elements. *Mater. Today Adv.* **2020**, *8*, 100115. [CrossRef]
22. Xu, R.; Liu, Y.; Sun, F. Effect of isothermal aging on the microstructure, shear behavior and hardness of the Sn58Bi/Sn3.0Ag0.5Cu/Cu solder joints. *Results Phys.* **2019**, *15*, 102701. [CrossRef]
23. Huang, J.Q.; Zhou, M.B.; Zhao, X.F.; Zhang, X.P. Microstructures and shear properties of mixed assembly BGA structure SnAgCu/SnBi(Ag)/Cu joints in board-level packaging. In Proceedings of the 2017 18th International Conference on Electronic Packaging Technology (ICEPT). IEEE, Harbin, China, 16–19 August 2017; pp. 1649–1654.
24. Hadian, F.; Genanu, M.; Owen, R.; Cotts, E.J. The dependence of the microstructure of SnAgCu/SnBiAg mixed assemblies on reflow temperature. In Proceedings of the 2020 SMTA International Conference, Online, 28 September–23 October 2020.
25. Ren, J.; Huang, M.L. Board-level drop reliability and fracture behavior of low temperature soldering Sn–Ag–Cu/Sn–Bi–X hybrid BGA solder joints for consumer electronics. *J. Mater. Sci. Mater. Electron.* **2021**, *32*, 15453–15465. [CrossRef]
26. Wang, F.; Huang, Y.; Du, C. Mechanical properties of SnBi-SnAgCu composition mixed solder joints using bending test. *Mater. Sci. Eng. A* **2016**, *668*, 224–233. [CrossRef]
27. Shen, Y.A.; Zhou, S.; Li, J.; Yang, C.H.; Huang, S.; Lin, S.K.; Nishikawa, H. Sn-3.0Ag-0.5Cu/Sn-58Bi composite solder joint assembled using a low-temperature reflow process for PoP technology. *Mater. Des.* **2019**, *183*, 108144. [CrossRef]
28. Moser, Z.; Dutkiewicz, J.; Gasior, W. *Binary Alloy Phase Diagram*; ASM International: Materials Park, OH, USA, 1990.
29. Kabassis, H.; Rutter, J.W.; Winegard, W.C. Bi-In system: Phase diagram and some eutectic structures. *Met. Sci.* **1984**, *18*, 326–328. [CrossRef]
30. Witusiewicz, V.T.; Hecht, U.; Böttger, B.; Rex, S. Thermodynamic re-optimisation of the Bi-In-Sn system based on new experimental data. *J. Alloys Compd.* **2007**, *428*, 115–124. [CrossRef]

31. Witusiewicz, V.T.; Hecht, U.; Böttger, B.; Rex, S. On the nature of the interface between Ag3Sn intermetallics and sn in Sn-3.5Ag solder alloys. *J. Electron. Mater.* **2007**, *36*, 1615–1620.
32. Lee, C.J.; Min, K.D.; Park, H.J.; Jung, S.B. Mechanical properties of Sn-58 wt% Bi solder containing Ag-decorated MWCNT with thermal aging tests. *J. Alloys Compd.* **2020**, *820*, 153077. [CrossRef]
33. Nai, S.; Wei, J.; Gupta, M. Interfacial intermetallic growth and shear strength of lead-free composite solder joints. *J. Alloys Compd.* **2009**, *473*, 100–106. [CrossRef]
34. Yoon, J.W.; Kim, S.W.; Jung, S.B. IMC growth and shear strength of Sn-Ag-Bi-In/Au/Ni/Cu BGA joints during aging. *Mater. Trans.* **2004**, *45*, 727–733. [CrossRef]
35. Wang, D.; Hao, J.; Zhou, J.; Xue, F.; Tian, S.; Li, S. The influence of thermal aging on reliability of Sn-58Bi interconnects. In Proceedings of the 2017 18th International Conference on Electronic Packaging Technology (ICEPT). IEEE, Harbin, China, 16–19 August 2017; pp. 744–749.
36. Ping, W.U. Effects of Zn addition on mechanical properties of eutectic Sn–58Bi solder during liquid-state aging. *Trans. Nonferrous Met. Soc. China* **2015**, *25*, 1225–1233.
37. Wang, F.; Huang, Y.; Zhang, Z.; Yan, C. Interfacial reaction and mechanical properties of Sn-Bi solder joints. *Materials* **2017**, *10*, 920. [CrossRef] [PubMed]
38. Ahmed, S.; Basit, M.; Suhling, J.C.; Lall, P. IEEE, Effects of Aging on SAC-Bi Solder Materials. In Proceedings of the 15th IEEE Intersociety Conference on Thermal and Thermomechanical Phenomena in Electronic Systems (ITherm), Las Vegas, NV, USA, 31 May–3 June 2016; pp. 746–754.
39. Raeder, C.H.; Felton, L.E.; Tanzi, V.A.; Knorr, D.B. The effect of aging on microstructure, room temperature deformation, and fracture of Sn-Bi/Cu solder joints. *J. Electron. Mater.* **1994**, *23*, 611–617. [CrossRef]

micromachines

Article

Study on the Wetting Mechanism between Hot-Melt Nano Glass Powder and Different Substrates

Yifang Liu [1,2,*], Junyu Chen [1,2] and Gaofeng Zheng [1,2]

1 Department of Instrumental and Electrical Engineering, Xiamen University, Xiamen 361102, China
2 Shenzhen Research Institute of Xiamen University, Shenzhen 518000, China
* Correspondence: yfliu@xmu.edu.cn; Tel.: +86-592-2194957

Abstract: The wettability of molten glass powder plays an essential role in the encapsulation of microelectromechanical system (MEMS) devices with glass paste as an intermediate layer. In this study, we first investigated the flow process of nano glass powder melted at a high temperature by simulation in COMSOL. Both the influence of the different viscosity of hot-melt glass on its wettability on SiO_2 and the comparison of the wettability of hot-melt glass on Au metal lead and SiO_2 were investigated by simulation. Then, in the experiment, the hot-melt glass flew and spread along the length of the Au electrode because of a good wettability, resulting in little coverage of the hot-melt glass on the Au electrode, with a height of only 500 nm. In order to reduce the wettability of the glass paste on the Au electrode, a SiO_2 isolation layer was grown on the surface of golden lead by chemical vapor deposition. It successfully reduced the wettability, so the thickness of the hot-melt glass was increased to 1.95 μm. This proved once again that the wettability of hot-melt glass on Au was better.

Keywords: wettability; hot-melt glass; flow time; coverage thickness; SiO_2 and Au substrate

Citation: Liu, Y.; Chen, J.; Zheng, G. Study on the Wetting Mechanism between Hot-Melt Nano Glass Powder and Different Substrates. *Micromachines* **2022**, *13*, 1683. https://doi.org/10.3390/mi13101683

Academic Editor: Wensheng Zhao

Received: 19 September 2022
Accepted: 4 October 2022
Published: 6 October 2022

Publisher's Note: MDPI stays neutral with regard to jurisdictional claims in published maps and institutional affiliations.

Copyright: © 2022 by the authors. Licensee MDPI, Basel, Switzerland. This article is an open access article distributed under the terms and conditions of the Creative Commons Attribution (CC BY) license (https://creativecommons.org/licenses/by/4.0/).

1. Introduction

Good vacuum packaging [1], even special packaging in a bad environment [2], is an important means to ensure the reliability of MEMS devices. In the middle layer packaging process with nano glass powder, the MEMS sensor can be electrically interconnected with the outside world through the external lead wire of the metal electrode, and the cap, substrate and lead wire can be tightly sealed together by using nano glass powder through hot press bonding. Nano glass powder or the glass frit inter-layer packaging has the advantages of a high tolerance to the surface roughness of the bonding interface, suitable for various materials in the MEMS, electrical insulation characteristics to simplify the electrode lead extraction process and patterning without an additional lithography process by using screen printing [3–5]. It has been widely used in the packaging of the MEMS pressure switch [6,7], MEMS gyroscope [8] and accelerometer [9]. Many scholars only describe the packaging principle, packaging process and packaging results of nano glass powder, but there is no report on both the mechanism of infiltration and flow process of hot-melt glass on the substrate.

After nano glass powder is made on the glass substrate with metal lead through screen printing, during the process of high temperature melting, the wettability of molten nano glass powder on metal lead and the SiO_2 substrate are different due to a different contact angle, surface tension and adhesion work. After cooling and solidification, the adhesion thickness of the glass powder on the metal lead is different from that on the SiO_2 substrate. If the height of the glass powder inter-layer on the Au metal lead is less than 10 μm [10], the package will fail.

In order to improve the results of the direct packaging of nano glass powder in the MEMS structure with metal leads, the wettability of nano glass powder in a hot-melt state was investigated. Firstly, the whole flow process of hot-melt nano glass liquid on silver substrate from the starting point to the material interface wall was simulated by COMSOL.

Then, the wettability of the hot-melt nano glass powder with a different viscosity on the SiO_2 substrate was analyzed and compared by simulation. The wetting effect of hot-melt glass with the same viscosity on SiO_2 and Au substrates were also investigated. Finally, it was verified by experiments that the wettability of hot-melt nano glass on Au metal leads was better than that on SiO_2, which leads to a too small adhesion thickness. By depositing a SiO_2 isolation layer on the metal leads, the wettability of hot-melt nano glass on a Au metal lead was successfully reduced, so as to improve its adhesion thickness on the Au.

2. Simulation Analysis of Wettability

Wettability is the degree of difficulty for a liquid to adhere to a solid when it contacts with a solid. It is usually determined by the contact angle between the solid–liquid interface and the liquid–gas interface θ. When the contact angle is less than 90°, the liquid can wet the solid. When the contact angle is greater than 90°, the liquid is difficult to wet the solid. Zhu Dingyi et al. [11,12] studied the corresponding relationship between liquid surface tension, solid surface tension and the contact angle. Guan C.H. [13] researched the impact of surface roughness on solid–liquid wettability. Li Wei [14] obtained the contact angle between the hot-melt glass and different substrates through experiments, and the better wettability was attained by polishing the surface of the material. In reference [15], the adhesion work was calculated by measuring the contact angle. The viscosity μ of liquid affected the velocity difference of each layer in the flow, which was one of the key factors affecting the fluidity of the liquid. Reference [16] verified that viscosity μ directly affected the fluidity of the hot-melt alloy liquid, and $1/\mu$ was used to characterize the relationship between the wettability and the temperature of the hot-melt alloy. However, the simulations of the wettability of liquids with a different viscosity on the same substrate and liquids with the same viscosity on different substrates have not been reported.

2.1. Simulation Model

The hot-melt glass powder was filled into a silicon pit sputtered with a layer of different substrate materials and heated to reflow to fill the whole pit. Assuming that the bottom radius of the hot-melt glass column was 2 mm and the height was 5 mm, the radius of the sphere equal to its volume was 2.47 mm. Taking the bottom radius of the cylindrical container made of the base material as 3 mm, we got the simulation model as shown in Figure 1.

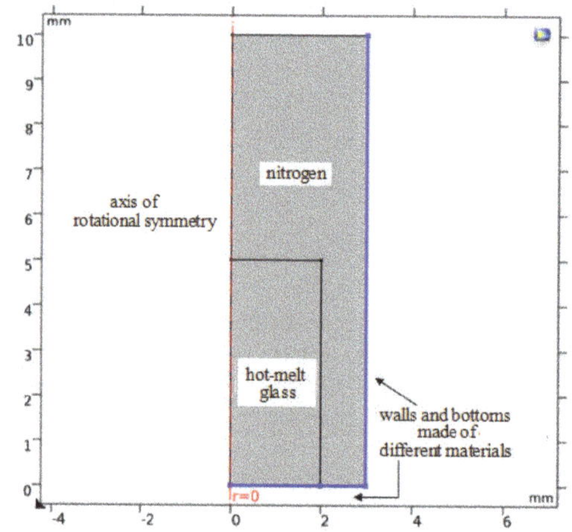

Figure 1. Final simulation model.

The material properties of nano glass powder at room temperature were indicated in Table 1: density, 2.221g/cm^3; viscosity, 1000 Pa·s; and surface tension, 2003.4 mN/m.

Table 1. Material properties of nano glass powder at 950 °C.

Density (g/cm^3)	Viscosity (Pa·s)	Surface Tension (mN/m)
2.221	1000	2003.4

According to the relationship between the surface tension and the temperature in Reference [17], the data in Table 2 were preliminarily sorted out and calculated.

Table 2. Surface tension of silica and gold substrates.

Serial Number	Substrate Material	Density (g/cm^3)	Melting Point (°C)	Surface Tension at 950 °C (mN/m)
1	SiO_2	2.2	1723	457.8
2	Au	19.3	1064	1168

As shown in Figure 2, the relationship between the contact angle θ and the interfacial tension between solid, liquid and gas can be expressed by "Young's formula".

$$\gamma_{sg} = \gamma_{sl} + \gamma_{lg}\cos\theta, \tag{1}$$

γ_{sg}, γ_{sl} and γ_{lg} represent solid–gas interfacial tension, solid–liquid interfacial tension and liquid–gas interfacial tension, respectively.

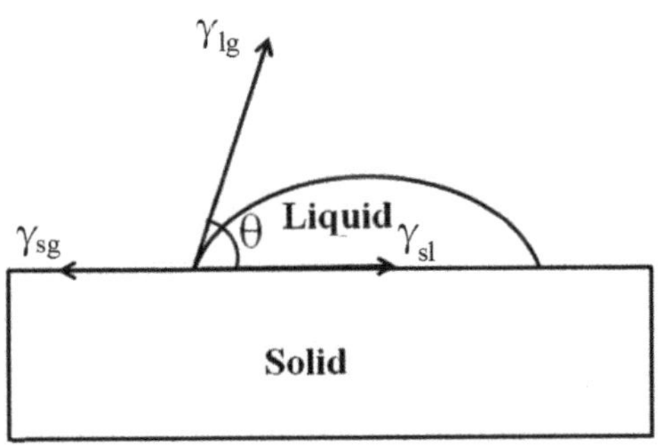

Figure 2. Schematic diagram of surface tension at the junction of contact angle and three phase.

The corresponding relationship between the liquid surface tension, solid surface tension and contact angle [18,19] was expressed by Equation (2):

$$\gamma_{sg} = \frac{\gamma_{lg}}{2} \times \left(\sqrt{1+\sin^2\theta} + \cos\theta\right), \tag{2}$$

According to the data of hot-melted glass in Table 1 and the surface tension data of the substrate material in Table 2, the contact angle formed when the substrate material and the hot-melted glass were infiltrated and could be calculated by Formula (2).

The liquid–gas surface tension γ_{lg} = 2003.4 mN/m. At the same time, Equation (2) was transformed as follows:

$$\left(\sqrt{1+\sin^2\theta} + \cos\theta\right)^2 = 4 \times \left(\frac{\gamma_{sg}}{\gamma_{lg}}\right)^2, \tag{3}$$

$$1 + \cos\theta \cdot \sqrt{1 + (\sin\theta)^2} = 2 \times \left(\frac{\gamma_{sg}}{\gamma_{lg}}\right)^2, \tag{4}$$

According to Equation (2), when the contact angle is 90°, the solid surface tension is 1416.6 mN/m. Thus, to consider the positive and negative values of $\cos\theta$ and convert further:

$$\sin^4\theta = 1 - \left[1 - 2 \times \left(\frac{\gamma_{sg}}{\gamma_{lg}}\right)^2\right]^2 \left(\gamma_{sg} < 1416.6 \cap \theta > 90°\right), \tag{5}$$

$$\sin^4\theta = 1 - \left[2 \times \left(\frac{\gamma_{sg}}{\gamma_{lg}}\right)^2 - 1\right]^2 \left(\gamma_{sg} > 1416.6 \cap \theta < 90°\right), \tag{6}$$

According to Equations (4) and (5), the contact angles between each substrate and hot-melted glass could be obtained from the data in Tables 1 and 2.

Adhesion work is the energy released in the process of adhesion. In the process of adhesion, the surface energy of the solid and liquid is lost, and the surface energy of the solid–liquid interface is generated. The calculation formula of the adhesion work was as follows:

$$W_a = \gamma_{sg} + \gamma_{lg} - \gamma_{sl}, \tag{7}$$

Combined with "Young's formula" (1), we could obtain:

$$W_a = \gamma_{sg} + \gamma_{lg} - \left(\gamma_{sg} - \gamma_{lg}\cos\theta\right) = \gamma_{lg}(1 + \cos\theta), \tag{8}$$

According to Formula (8), the adhesion work between hot-melted glass and different substrates could be obtained. The contact angle and the adhesion work which were calculated are shown in Table 3.

Table 3. Contact angle and adhesion work between hot-melt glass and each substrate [14].

Serial Number	Substrate	Surface Tension at 950 °C (mN/m)	Contact Angle (°)	Adhesion Work (mJ/m^2)
1	SiO_2	457.8	138.2	509.9
3	Au	1168	103.3	1542.5

2.2. Simulation of Wettability of Hot-Melt Glass with Different Viscosity on SiO_2 Substrate

By changing the viscosity of hot-melt glass from 500 Pa·s to 1000 Pa·s, the influence of the viscosity of the hot-melt glass on the flow velocity and wettability of the hot-melt glass was studied with the SiO_2 as a substrate. Taking the yellow light band as the reference point, the relationship between the viscosity of hot-melt glass and the time needed to flow to the junction of the material bottom and the material wall was explored in this paper. On the SiO_2 substrate, the steady state of hot-melt glass with a different viscosity flowing to the junction is shown in Figure 3, which corresponds to a different flow time.

Therefore, when the viscosity of the hot-melt glass was 1000, 900, 800, 700, 600 and 500 Pa·s, respectively, the time of the hot-melt glass flowing to the specified distance on the SiO_2 substrate could also be obtained, as shown in Figure 4. It could be seen that the lower the viscosity of the hot-melt glass, the shorter the flow time to the specified distance, the higher the flow speed and the better the wettability. For the same SiO_2 substrate, the solid surface energy of hot-melt glass with a different viscosity was the same, but the lower the viscosity was, the higher the wettability was.

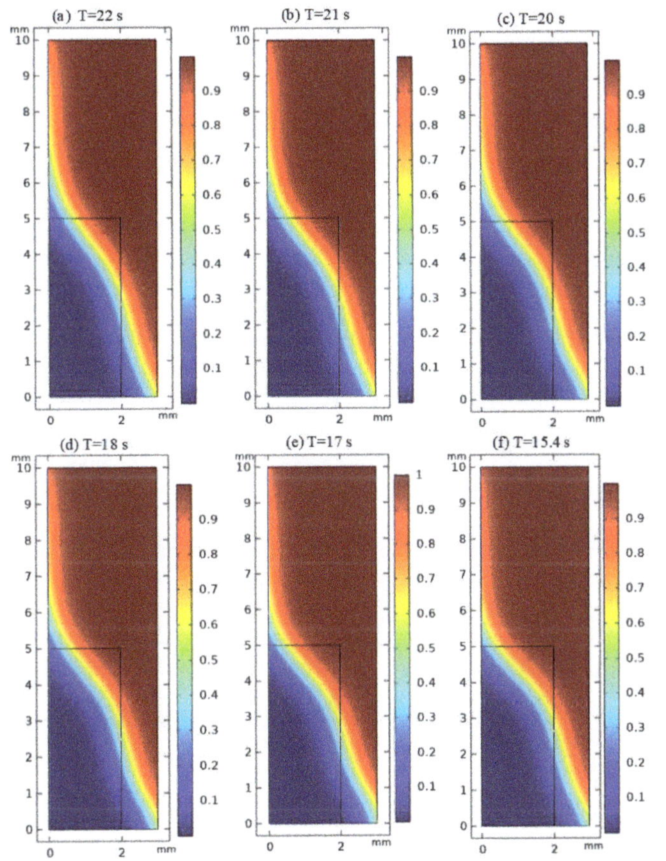

Figure 3. Flow time of hot-melt glass with different viscosity on SiO_2 substrate: (**a**) the viscosity is 1000 Pa·s; (**b**) the viscosity is 900 Pa·s; (**c**) the viscosity is 800 Pa·s; (**d**) the viscosity is 700 Pa·s; (**e**) the viscosity is 600 Pa·s; (**f**) the viscosity is 500 Pa·s.

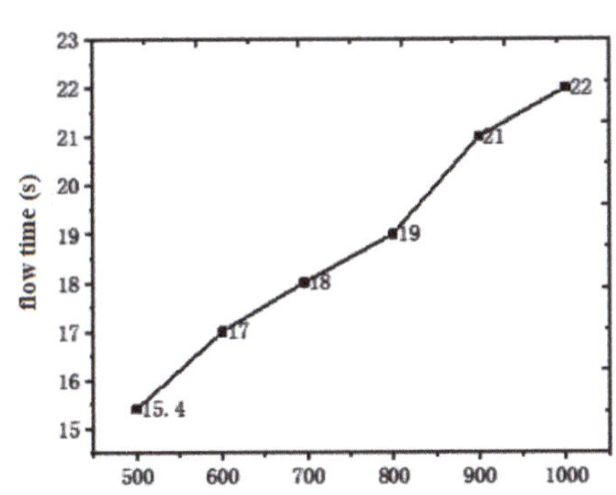

Figure 4. Relationship between flow time and viscosity of hot-melt glass on SiO_2 substrate.

2.3. Comparison of Wettability of Hot-Melt Glass Solution between SiO_2 and Au Substrates

Then, the viscosity of the hot-melt glass was kept at 1000 Pa·s, the simulation was carried out on the SiO_2 and Au substrates and the simulation results, as shown in Figure 5, were obtained. It could be clearly seen from Figure 5 that it took 22 s for the hot-melt glass to flow to the junction on the SiO_2 substrate and 16.5 s on the Au substrate. With the same viscosity, the surface free energy of the liquid was the same. Yet, combined with the parameters in Table 2, the contact angle between the hot-melt glass and the Au substrate was smaller than that of SiO_2, and the adhesion work and surface tension on the Au substrate were larger, so the wettability was higher and the flow velocity was higher.

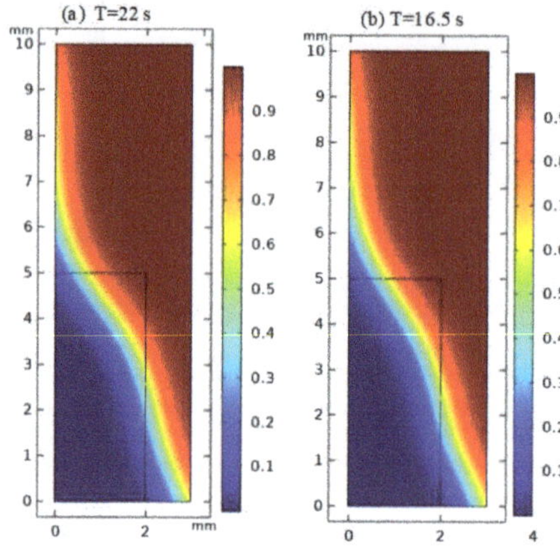

Figure 5. Simulation results of adhesion of hot-melt glass on SiO_2 and Au substrates: (**a**) SiO_2; (**b**) Au.

3. Experimental

The micro pressure switch was packaged with nano glass powder. The hot-melt glass was transparent and the surface morphology was compact and smooth, as shown in Figure 6. However, because the wettability between the hot-melt glass and the Au electrode were stronger than that between the hot-melt glass and the SiO_2 substrate, the hot-melt glass flowed rapidly along the length direction of the Au electrode lead and spread out rapidly, resulting in little coverage of this part of the hot-melt glass. After measurement, the thickness of the hot-melt glass on the Au electrode lead was only 500 nm, as shown in Figure 6b. This thickness was not enough to form a sealed package during bonding.

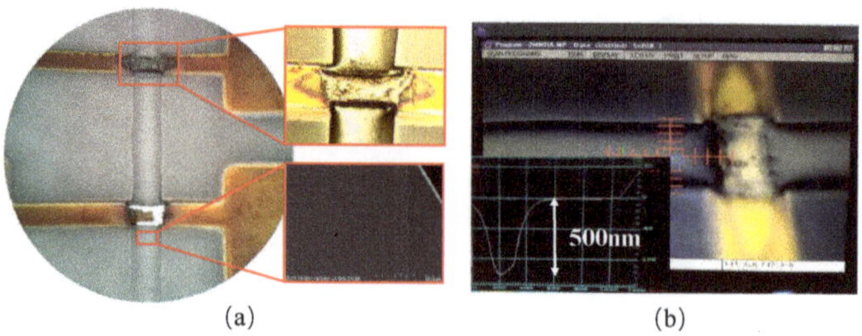

Figure 6. Sintering effect of hot-melt glass: (**a**) slurry morphology; (**b**) measuring diagram of step meter.

The wettability of hot-melt glass to different materials varies greatly [20,21]. From the above simulation and experimental results, it could be seen that the wettability of hot-melt glass on the Au metal lead was good, so the volume of hot-melt glass passing through the Au metal lead decreased sharply. A silicon wafer sputtered when a large area of Au lines was selected and a thin layer of nano glass powder was manually coated on the whole surface and melted at a high temperature. Figure 7b showed that the amount of hot-melt glass on the Au metal leads was very small, and a small part shrank to the metal free area on the silicon wafer. It was proved that the wettability of glass paste on the Au wire was very strong and the adhesion thickness of the glass paste was not as good as that of the silicon or glass. Based on the verification results, it was proposed that a SiO_2 isolation layer should be formed on the surface of the metal lead by chemical vapor deposition to reduce the wettability of the glass slurry in this area.

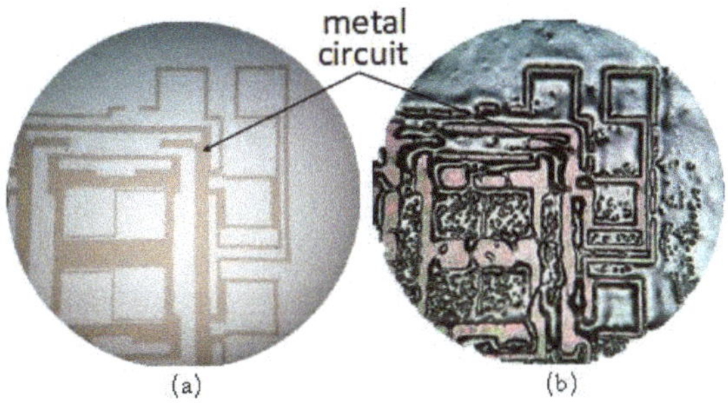

Figure 7. Pre-sintering effect of glass slurry on large-area metal circuit: (**a**) metal circuit; (**b**) morphology of hot-melt glass.

The experimental process and results are shown in Figure 8. The SiO_2 isolation layer successfully reduced the wettability of the hot-melt glass on the Au metal lead, and this part of the hot-melt glass was consistent with that on the glass sheet. The thickness of the hot-melt glass increased from 500 nm to 1.95 μm. It could be seen that there was a significant difference between the thickness of the glass powder on the metal lead covered with a thin layer of SiO_2 and that on the metal lead not covered with SiO_2. This proved once again that the wettability of hot-melt glass on a Au substrate was better.

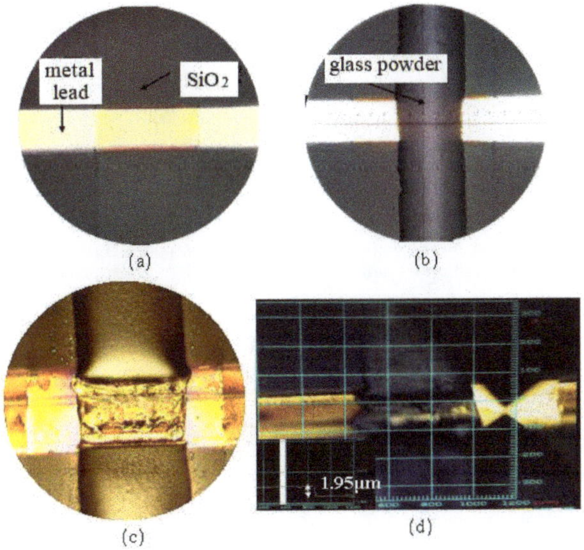

Figure 8. Isolation layer pre-sintering: (**a**) deposition of SiO_2; (**b**) printing glass paste; (**c**) high temperature hot-melt glass; (**d**) characterization of glass powder thickness.

4. Conclusions

The wettability of the molten glass powder was studied by simulation and experimentally. The conclusions obtained in this research are summarized as follows:

1. The smaller the viscosity of the hot-melt glass, the smaller the surface energy of the liquid, the greater the wettability and the higher the flow velocity on SiO_2. When the viscosity of the molten glass slurry decreased from 1000 Pa·s to 500 Pa·s, the time for the hot-melt glass to flow to the specified interface on the silica substrate decreased from 22 s to 15.4 s.
2. The surface tension of Au metal lead was higher than that of SiO_2, the contact angle between the Au metal lead and the hot-melt nano glass was smaller and the wettability of the Au metal lead was stronger. When the molten glass slurry with the same viscosity of 1000 Pa·s flowed on the silica and gold substrates, the time to flow to the designated interface was 22 s and 16.5 s, respectively.
3. Compared with SiO_2, Au had a higher adhesion work, a faster spreading speed and a smaller adhesion thickness in a limited time.
4. By depositing a thin layer of SiO_2 on the Au metal lead, the flattening speed of hot-melt glass could be effectively reduced and the adhesion height of the nano glass powder could be increased from 500 nm to 1.95 μm.

Author Contributions: Conceptualization, Y.L. and G.Z.; methodology, Y.L.; validation, Y.L. and J.C.; formal analysis, J.C.; investigation, J.C.; resources, Y.L.; data curation, J.C.; writing—original draft preparation, Y.L.; writing—review and editing, Y.L. and G.Z.; supervision, Y.L.; project administration, Y.L.; funding acquisition, Y.L. All authors have read and agreed to the published version of the manuscript.

Funding: This research was financially supported by Guangdong Basic and Applied Basic Research Foundation (2022A1515010949; 2022A1515010923).

Conflicts of Interest: The authors declare no conflict of interest.

References

1. Yang, F.; Han, G.W.; Yang, J.; Zhang, M.; Ning, J.; Yang, F.H.; Si, C.W. Research on wafer-level mems packaging with through-glass vias. *Micromachines* **2019**, *10*, 15. [CrossRef]
2. Russu, A.; de Castro, A.J.; Cortes, F.; Lopez-Ongil, C.; Portela, M.; Garcia, E.; Miranda, J.A.; Canabal, M.F.; Arruego, I.; Martinez-Oter, J.; et al. A light compact and rugged ir sensor for space applications. In Proceedings of the Conference on Infrared Sensors, Devices, and Applications IX as Part of SPIE Optics + Photonics Conference, San Diego, CA, USA, 14–15 August 2019.
3. Roshanghias, A.; Bardong, J.; Binder, A. Glass frit jetting for advanced wafer-level hermetic packaging. *Materials* **2022**, *15*, 2786. [CrossRef] [PubMed]
4. Lorenz, N.; Millar, S.; Desmulliez, M.; Hand, D.P. Hermetic glass frit packaging in air and vacuum with localized laser joining. *J. Micromech. Microeng.* **2011**, *21*, 045039. [CrossRef]
5. Knechtel, R. Glass frit bonding: An universal technology for wafer level encapsulation and packaging. *Microsyst. Technol.* **2005**, *12*, 63–68. [CrossRef]
6. Wang, L.Y.; Chen, D.E.; Xiong, H.; Liu, Y.F. Design and optimization of glass frit package structure for micro pressure switch. In Proceedings of the 16th IEEE International Conference on Nano/Micro Engineered and Molecular Systems (IEEE-NEMS), Xiamen, China, 25–29 September 2021.
7. Moriyama, M.; Suzuki, Y.; Totsu, K.; Hirano, H.; Tanaka, S. Metal-bonding-based hermetic wafer-level mems packaging technology using in-plane feedthrough: Hermeticity and high frequency characteristics of thick gold film feedthrough. *Electr. Eng. Jpn.* **2019**, *206*, 44–53. [CrossRef]
8. Chiu, S.R.; Teng, L.T.; Chao, J.W.; Sue, C.Y.; Lin, C.H.; Chen, H.R.; Su, Y.K. An integrated thermal compensation system for mems inertial sensors. *Sensors* **2014**, *14*, 4290–4311. [CrossRef] [PubMed]
9. Burke, C.; Punch, J. An investigation of capped glass frit sealed mems devices in contemporary accelerometers. In Proceedings of the ASME International Mechanical Engineering Congress and Exposition, Lake Buena Vista, FL, USA, 13–19 November 2009.
10. Liu, Y.F.; Chen, D.; Lin, L.W.; Zheng, G.F.; Zheng, J.Y.; Wang, L.Y.; Sun, D.H. Glass frit bonding with controlled width and height using a two-step wet silicon etching procedure. *J. Micromech. Microeng.* **2016**, *26*, 035018.
11. Zhu, D.Y.; Liao, X.M.; Dai, P.Q. Theoretical analysis of reactive solid-liquid interfacial energies. *Chin. Sci. Bull.* **2012**, *57*, 4517–4524. [CrossRef]

12. Zhang, D.; Zhu, D.Y.; Zhang, T.; Wang, Q.F. Kinetics of reactive wetting of graphite by liquid al and Cu–Si alloys. *Trans. Nonferrous Met. Soc. China* **2015**, *25*, 2473–2480. [CrossRef]
13. Guan, C.H.; Lv, X.J.; Han, Z.X.; Chen, C. The wetting characteristics of aluminum droplets on rough surfaces with molecular dynamics simulations. *Phys. Chem. Chem. Phys.* **2020**, *22*, 2361–2371. [CrossRef] [PubMed]
14. Li, W. Study on Preparation and Reflow Process of Nano Glass Powder for Memms Encapsulations. Master's Thesis, Xiamen University, Xiamen, China, 2017.
15. Wang, L.; Wang, L.Q.; Jia, Y.J.; Xing, Y.M. Adhesion performance of asphalt-aggregate based on the grey relational analysis. *Acta Mater. Compos. Sin.* **2017**, *34*, 2070–2078.
16. Ma, Y.Z.; Zhao, Y.X.; Liu, W.S.; Zhang, J.J. Research on viscosity and rheological properties of tungsten-based alloy PEM feed. *Rare Metal Mat. Eng.* **2010**, *39*, 1979–1983.
17. Wang, Y.X.; Zu, C.K.; Chen, J. Effect of components on viscosities and coefficients of thermal expansion of borosilicate glasses. *J. Funct. Mater.* **2012**, *43*, 2985–2988.
18. Lu, Y. Fabrication of Superamphiphobic Surfaces on Titanium Substrates and Research on Wettability Control. Master's Thesis, Dalian University of Technology, Dalian, China, 2013.
19. Zhu, D.Y.; Dai, P.Q.; Luo, X.B.; Zhang, Y.C. Novel characterization of wetting properties and the calculation of liquid-solid interface tension(I). *Sci. Technol. Eng.* **2007**, *7*, 3057–3062.
20. Zhang, X.F.; Lu, A.X.; Zhang, H.X. Research on soakage between ceramic glass bond and diamond at high temperature. *Diam. Abras. Eng.* **2007**, *157*, 44–46+59.
21. Mishinov, S.V.; Stepanov, B.S.; Velmuzhov, A.P.; Shiryaev, V.S.; Lashmanov, E.N.; Potapov, A.M.; Evdokimov, I.I. Wettability of stainless steel with a $Ge_{28}Sb_1Se_{60}$ glass melt and its contact adhesion strength. *J. Non-Cryst. Solids.* **2022**, *578*, 121351. [CrossRef]

micromachines

Article

Hermetic Packaging Based on Cu–Sn and Au–Au Dual Bonding for High-Temperature Graphene Pressure Sensor

Junqiang Wang [1,2,*], Haikun Zhang [1,2], Xuwen Chen [1,2] and Mengwei Li [1,2,*]

1 Academy for Advanced Interdisciplinary Research, North University of China, Taiyuan 030051, China; zhanghaikun0312@163.com (H.Z.); nuc_chenxuwen@163.com (X.C.)
2 Notional Key Laboratory of Instrumentation Science & Dynamic Measurement, North University of China, Taiyuan 030051, China
* Correspondence: wangjq210@nuc.edu.cn (J.W.); lmwnuc@163.com (M.L.); Tel.: +86-158-1133-8056 (J.W.); +86-139-3424-8366 (M.L.)

Abstract: A chip-level hermetic package for a high-temperature graphene pressure sensor was investigated. The silicon cap, chip and substrate were stacked by Cu–Sn and Au–Au bonding to enable wide-range measurements while guaranteeing a high hermetic package. Prior to bonding, the sample was treated with Ar (5% H_2) plasma. The Cu–Sn bonding was firstly performed at 260 °C for 15 min with a pressure of 9.9 MPa, and the corresponding process conditions for Au–Au bonding has increased to 300 °C, 20 min and 19.8 MPa respectively. The average shearing strength was 14.3 MPa, and an excellent leak rate of 1.72×10^{-4} Pa·cm^3/s was also achieved. After high-temperature storage (HTS) at 350 °C for 10 h, the resistance of graphene decreased slightly because the dual bonding provided oxygen-free environment for graphene. The leakage rate of the device slightly increased to 2.1×10^{-4} Pa·cm^3/s, and the average shear strength just decreased to 13.5 MPa. Finally, under the pressure range of 0–100 MPa, the graphene pressure sensor exhibited a high average sensitivity of 3.11 Ω/MPa. In conclusion, the dual bonding that combined Cu–Sn and Au–Au is extremely suitable for hermetic packaging in high-temperature graphene pressure sensors.

Keywords: Cu–Sn bonding; Au–Au bonding; graphene; high-temperature pressure sensor

Citation: Wang, J.; Zhang, H.; Chen, X.; Li, M. Hermetic Packaging Based on Cu–Sn and Au–Au Dual Bonding for High-Temperature Graphene Pressure Sensor. *Micromachines* **2022**, *13*, 1191. https://doi.org/10.3390/mi13081191

Academic Editor: Wensheng Zhao

Received: 11 July 2022
Accepted: 26 July 2022
Published: 28 July 2022

Publisher's Note: MDPI stays neutral with regard to jurisdictional claims in published maps and institutional affiliations.

Copyright: © 2022 by the authors. Licensee MDPI, Basel, Switzerland. This article is an open access article distributed under the terms and conditions of the Creative Commons Attribution (CC BY) license (https://creativecommons.org/licenses/by/4.0/).

1. Introduction

Graphene is a hexagonal, honeycomb-shaped, two-dimensional material with thickness of approximately 0.335 nm. It exhibits excellent properties and has been widely used in pressure tests, energy storage, photoelectric detection, biomedicine, heat dissipation and other fields [1–6]. It also shows great potential in flame retardant materials and medication testing [7,8]. Compared with other two-dimensional materials, the preparation process of a carbon nanotube pressure sensor is complicated, which will cause a poor compatibility with micro-nano manufacturing [9], MoS_2 with low modulus (≈300 GPa) has better flexibility than graphene, but it is difficult to grow large-area and high-quality nanofilm [10]. The overall performance of graphene is superior; high carrier mobility (up to 200,000 cm^2 V·s), Young's modulus (about 1 TPa) and high temperature resistance (up to 2000 °C) [11–13] make it have greater potential in the field of high-temperature pressure sensors. In the past few years, the use of graphene in pressure sensors is gaining momentum. Using single and multilayer graphene sheets, Bunch et al. manufactured the first prototype of a suspended graphene pressure sensor [14]. Sorkin and Zhang studied the mechanical failure of pressure sensors based on graphene nanoflake using atomistic approach. The sensors consist of graphene films suspended over a SiC substrate [15]. Smith et al. demonstrated the piezoresistive effect in graphene while proposing a novel pressure sensor based on a suspended graphene membrane [16,17]. Wang et al. developed a suspended graphene pressure sensor whose sensing unit consisted of porous graphene film arrays [18].

Although graphene pressure sensors are developing rapidly, the suspended or SiN_x-film structure limited pressure range. Previous graphene pressure sensors had no packaging, limiting its high-temperature applications. In our previous research, we fabricated a thin-film-structured graphene pressure sensor using Cu–Sn flip-chip bonding for hermeticity packaging that exhibits high temperature stability [19]. However, this thin-film-structure is relatively suitable for a pressure range lower than 1 MPa, and in the face of the requirements of a wide pressure range, it is necessary to develop a new structure.

Packaging pressure sensors are primarily used to realize sealing and electrical interconnection. This can be realized through different bonding methods, such as anode bonding, solder bonding, metal-to-metal direct bonding, eutectic bonding, and so on. Anode bonding is usually realized under conditions of high pressure and voltage (1000 V) [20]. These conditions can result in the fracture of graphene devices or the breakdown of graphene. Solder bonding is a low-cost, simple process, but it is required to print a wide solder to achieve excellent sealing. This bonding method is not conducive to reducing packaging size and realizes the miniaturization of pressure sensors [21–23]. Common eutectic bonding includes Cu–Sn and Au–Sn bonding, but the melting point of AuSn20, a eutectic alloy of Au and Sn, is only 278 °C, which obviously cannot meet the needs of high-temperature packaging. Direct metal bonding includes Cu–Cu and Au–Au bonding, etc. The oxidation problem of Cu–Cu bonding limits its development, while Au is a chemically inert substance with stable properties and high melting point, so Au–Au bonding is more suitable for high temperature packages. Moreover, metal-to-metal direct bonding and eutectic bonding is characterized by significantly high hermeticity; the stack structure can effectively realize the requirements of miniaturization of devices. This is almost suitable for realizing pressure cavity construction in all pressure sensors [24–28].

In this paper, a dual-bonding of Cu–Sn and Au–Au is proposed for wide-range graphene pressure sensor. The high hermetic package provided effective protection for graphene, and the three-layer stack structure ensured a wide pressure range. The thickness, structure, and layout of the bonding material were designed to obtain a stable interface. The excellent performance of dual bonding is proved by many tests. This bonding method helped realize high-hermeticity sealing connection in a graphene pressure sensor. Excellent high-temperature reliability was achieved, which helped in the integration of multilayer devices.

2. Materials and Methods

There are three steps to realize dual bonding (Cu–Sn and Au–Au) in the chip-level high-temperature graphene pressure sensors: test vehicle design, vehicle fabrication, and Cu–Sn and Au–Au Bonding.

2.1. Test Vehicle Design

The schematic of test structures is shown in Figure 1. The area of silicon cap is 5.2×5.2 mm^2, which consists of two sealing rings, round bump, and pressure cavity. The core chip and substrate were the same area of 6.5×6.5 mm^2. There are only sealing rings in the substrate, and the chip is composed of a cruciform frame structure, wiring, Au electrodes, sealing rings, round bump, and sensitive unit. The silicon cap was firstly bonded to the core chip through Cu–Sn bonding, and the chip was then bonded to substrate via Au–Au bonding. Sealing ring is designed to protect the graphene, while round bump is transmitted displacement. As the pressure acts on the pressure cavity to deform the silicon cap, the root of crossbeam generates strain, resulting in the deformation of the graphene at the corresponding position. Finally, the measurement resistance of graphene changes.

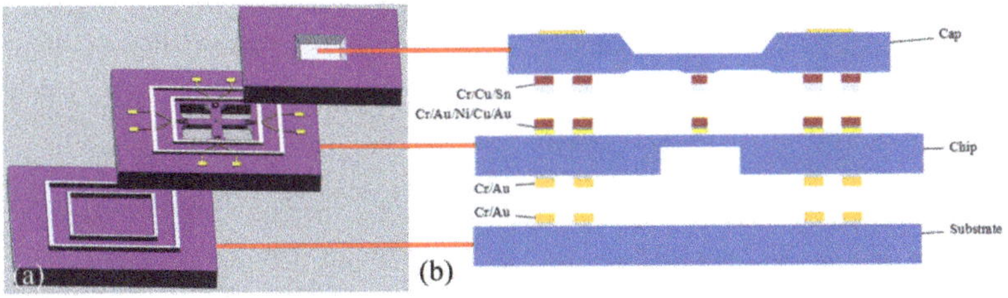

Figure 1. Schematic of test structures. (**a**) Graphene sensor structure (core crossbeam dimensions: 900 μm × 200 μm × 90 μm). (**b**) Bonding structure.

2.2. Vehicle Fabrication

The cap fabrication process is shown in Figure 2. A double-sided 110 nm-thick SiN_x film was first deposited on a new 250 μm-thick silicon wafer following the low-pressure chemical deposition (LPCVD) method (Figure 2a). It protected the area from etch during the process of KOH wet etching. The inductively reactive ion etching (RIE) process was carried out at an etching rate of 50 nm/min, the cavity was characterized by a plane size of 1314 μm × 1314 μm with a center retaining 200 μm × 200 μm convex plate. It was developed on the front face of silicon. Following this, KOH solution (48%) was used to corrode the cavity with a depth of 12 μm at 85 °C, and etching rate was maintained at 1 μm/min (Figure 2b). The LPCVD method was followed to deposit 110 nm SiN_x films to protect the frontal pressure cavity. The Cr–Au layers with thicknesses of 50 and 300 nm were prepared following the magnetron sputtering (MS150, FHR, Inc.) for connection package shell (Figure 2c). Similar to the frontal pressure cavity, RIE and KOH solution were used to etch back pressure cavity, the size of the cavity was 970 μm × 970 μm with the depth of 220 μm (Figure 2d). Finally, The Cr–Cu–Sn layers with thicknesses of 50, 2000, and 2000 nm, respectively, were deposited to form the final bonding structures of the bumps and sealing ring. The equipment was subjected to the process of thermal evaporation (Figure 2e). A scanning electron microscopy (SEM) image of the front side of the final silicon cap is shown in Figure 3b, and the backside is shown in Figure 3a.

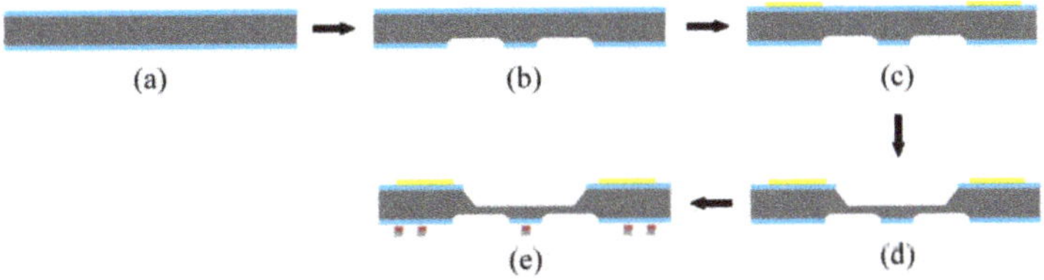

Figure 2. Schematic of the process for fabrication of the silicon cap. (**a**) Depositing the double-side SiNx. (**b**) Etching the front pressure cavity. (**c**) Sputtering the back metal ring. (**d**) Etching the back pressure cavity. (**e**) Evaporating the bump and sealing rings.

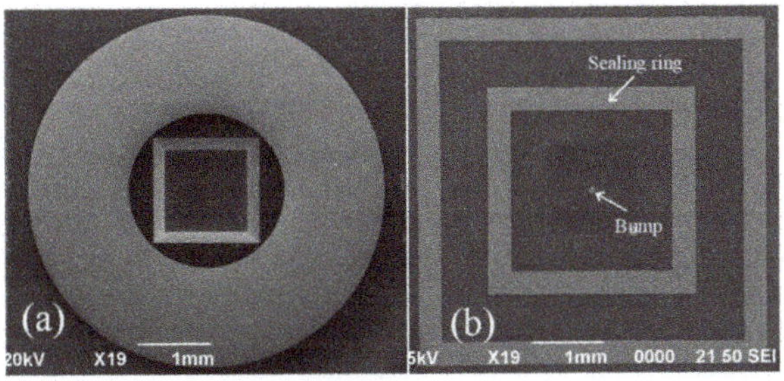

Figure 3. SEM images of the silicon cap. (**a**) Backside. (**b**) Frontside.

The core chip fabrication process is shown in Figure 4. Firstly, a 200 nm-thick SiN_x layer was deposited on one side as an insulation layer following plasma-enhanced chemical vapor deposition (PECVD) (Figure 4a). The Cr–Au layers with thicknesses of 15, 25 nm were deposited to form the bottom electrode (Figure 4b). Following this, the substrate and core chip sealing ring were simultaneously fabricated because the same pattern, and the substrate has been prepared in advance of the insulation layer. The Cr–Au layers with thicknesses of 50, 300 nm were deposited to form the back sealing rings by magnetron sputtering (Figure 4c). A 310-μm-deep pressure cavity with planar dimensions of 847 μm × 847 μm was then created at the back of silicon following the inductively coupled plasma (ICP) technique. The etching rate was maintained at 9.6 μm/min (Figure 4d). The next step involves the preparation of the key force-sensitive unit. This is graphically realized by the O_2 plasma etching process. The process was conducted using wet transferred single layer graphene as the force-sensitive material (Figure 4e). Subsequently, SiN_x with a thickness of 200 nm was deposited formed an insulation layer to protect the surface wiring by PECVD. It is preventing the electrode from conducting with the sealing ring. On the one hand, SiN_x can effectively avoid graphene doping with water or air. On the other hand, SiN_x produced n-type doping of graphene and improves the stability of sensor [29] (Figure 4f). Meanwhile, the RIE method was used to etch the insulation layer on the electrode to ensure that there are no problems with the electrical connection (Figure 4g). The Cr-Au layers with thicknesses of 25 and 100 nm were deposited to form the top electrode by magnetron sputtering, graphene was located between the bottom electrode and the top electrode, which can ensure a low metal–graphene contact resistance (Figure 4h). The Cr–Au–Ni–Cu–Au layers with thicknesses of 50, 100, 100, 1500 and 3 nm, respectively, were deposited to form the final bonding structures of the bumps and sealing ring (Figure 4i). Finally, the ICP method was carried out to release the front cruciform frame structure. An SEM image of the front side of the core chip is shown in Figure 5a, and the backside is shown in Figure 5b.

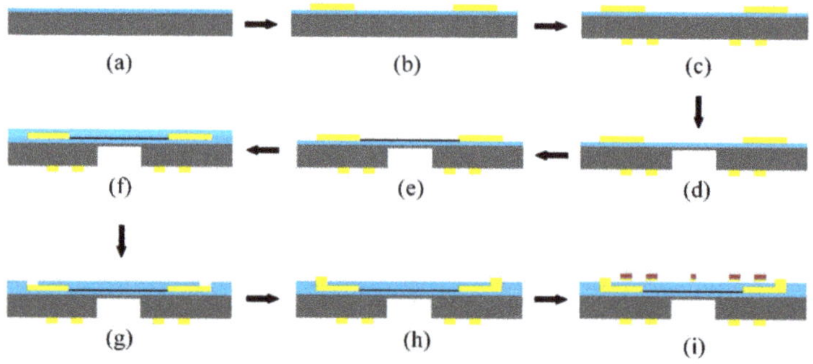

Figure 4. Schematic of the process for fabrication of the core chip. (**a**) Depositing the front SiN_x. (**b**) Sputtering the bottom electrodes. (**c**) Sputtering the back sealing rings. (**d**) Etching the back pressure cavity. (**e**) Transferring and patterning the graphene. (**f**) Depositing the insulation layer. (**g**) Etching the insulation layer on electrode. (**h**) Sputtering the top electrodes. (i) Evaporating the front bump and sealing rings.

2.3. *Cu–Sn and Au–Au Bonding*

Silicon cap and core chip surface needs to be pretreated for cleaning before bonding. The optimal pretreatment time set to be 120 s respectively as plasma power of 200 W and Ar (5% H_2) gas flow rate of 200 sccm were fixed. After pretreatment, Cu–Sn bonding was performed with pressure of 9.9 MPa at 260 °C for 15 min in flip-chip bonder (FC150, SET, Inc.) [19]. The bonding conditions are shown in Figure 6a. The sample in an atmosphere of formic acid (cyan line) for 3 min, it was then subjected to an atmosphere of N_2 (red line) until the end of the bonding process. To ensure the cleanliness of the subsequent Au–Au bonding surface, flip-chip bonder was cleaned before Cu–Sn bonding. Following this, Au–Au bonding was realized. Compared with Cu–Sn bonding, the pretreatment time is reduced to 60 s, other factors remain unchanged, and the corresponding process conditions for Au–Au bonding has increased to 300 °C, 20 min and 19.8 MPa [30,31]. The bonding conditions are shown in Figure 6b.

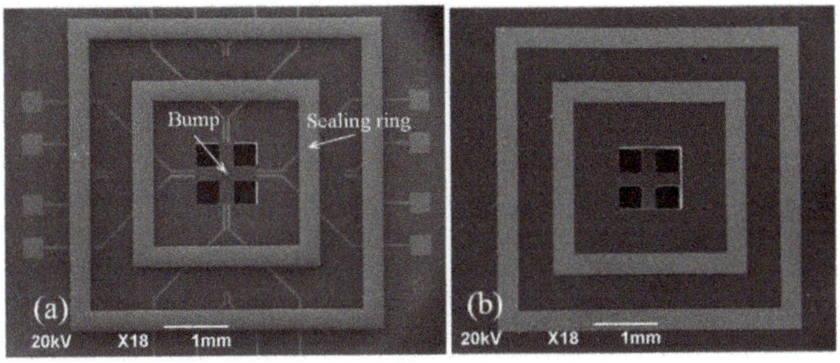

Figure 5. SEM images of the core chip. (**a**) Frontside. (**b**) Backside.

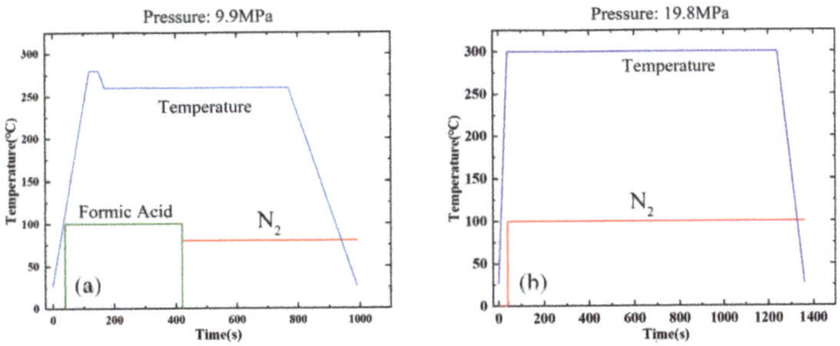

Figure 6. Bonding conditions. (**a**) Cu–Sn bonding. (**b**) Au–Au bonding.

3. Results and Discussion

Four different tests were conducted to evaluate the bonding performance of the bonded device, including interface analysis, shear strength and hermeticity detection, high-temperature reliability, and static test.

3.1. Au Pretreatment Optimization

The pretreatment method followed before the formation of the Au–Au bonding can stimulate the Au surface activity, the Ar + H_2 plasma treatment method could be used to increase the bonding strength. The strength achieved was higher than that achieved following the conventional Ar plasma treatment method [32]. To determine whether to conduct pretreatment, 10 samples were prepared, and divided into two groups; one group was pretreatment, and the other group was not. The result is shown in Figure 7. It can be seen that the shear strength decreases with increasing standing time (Time interval between plasma pretreatment and start of bonding). Meanwhile, the bonding strength of pretreatment sample is far greater than without pretreatment. Au surfaces roughness was characterized by atomic force tests (Figures 8a and 8b, respectively). It was observed that the pretreatment surface was rougher than without pretreatment surface. The square roughness (Ra) increased from 2.6 nm to 4.7 nm.

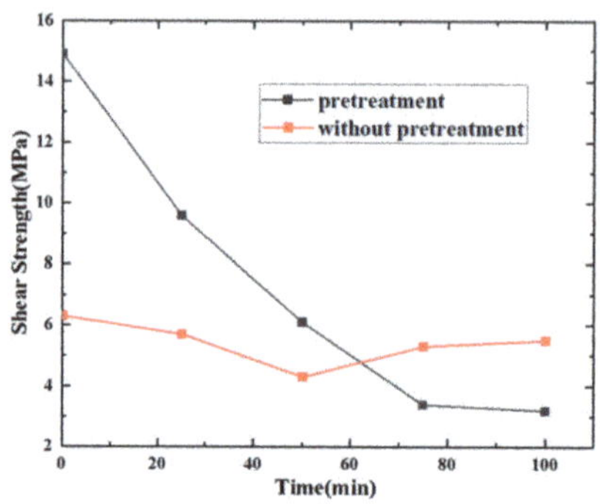

Figure 7. Bonding strength of samples with standing time.

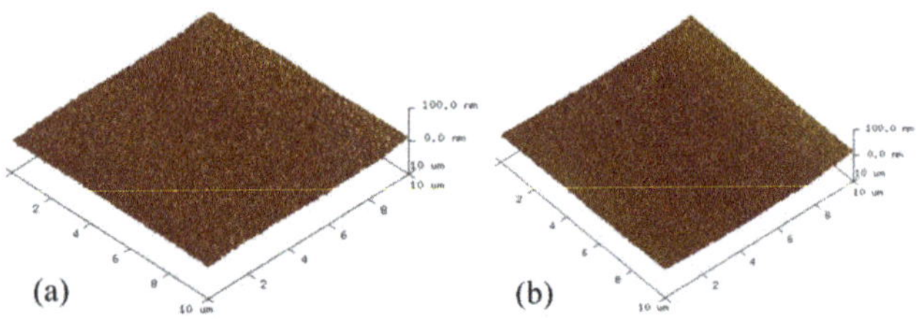

Figure 8. AFM images of Au surface. (**a**) Before pretreatment. (**b**) After pretreatment.

3.2. Interfacial Analysis

An X-ray image recorded for the dual-bonded microstructure is shown in Figure 9a. The images of the corresponding round bumps and sealing ring are shown in Figure 9b,c, respectively. Overflow of Sn was not observed for Cu–Sn bonding. Au did not melt for Au–Au bonding, and therefore no obvious spillover effect was observed. A cross-sectional SEM image of the microstructure of a distinct three-layers bonding interface is shown in Figure 10. The upper sealing ring and round bump could be seen, while the lower Au–Au bonding was not apparent. This can be attributed to the fact that Au units were thinner. A cross-sectional SEM image of Cu–Sn bonding interfacial microstructure is shown in Figure 11a. The results obtained using the energy-dispersive spectroscopy (EDS) (Genesis, EDAX, Inc.) technique are presented in Figure 11b. It shows the bonding interface can be divided into three layers. The upper and lower layers contain Cu (thicknesses of 0.83 and 0.79 μm, respectively). The middle layer is intermetallic compound (IMC) layer (with no obvious cracks and gaps) grown during the process of solid–liquid diffusion with thickness of 2.56 μm. Results from EDS analysis revealed that the ratio of copper atoms to Sn atoms was approximately 3:1. This intermediate layer was determined to be Cu_3Sn. The densities of Cu_3Sn and Cu were similar. When Cu_3Sn grows on the Cu surface, the extent of volume change is small. This results in the generation of dense structures, reasonable metal structures, and few Cu_6Sn_5. Kirkendall effect can be avoided [33]. Au–Au bonding interface is shown in Figure 12. Obvious cracks and gaps in the middle metal layer were not observed. The interface was relatively flat, and width of metal layer did not change significantly. Thus, an excellent interconnecting interface was formed.

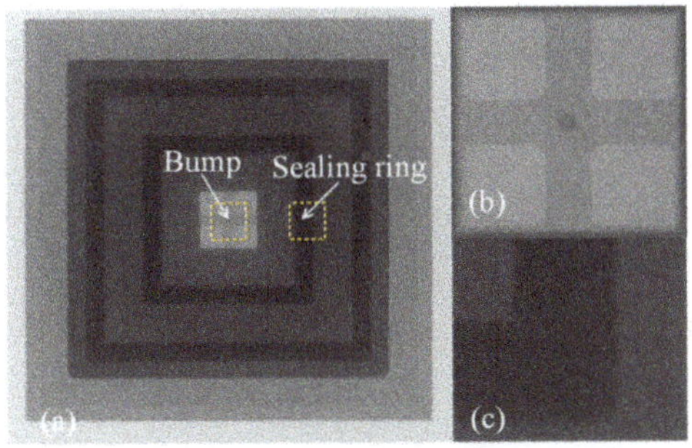

Figure 9. X-ray images. (**a**) Overall situation. (**b**) Round bump. (**c**) Sealing ring.

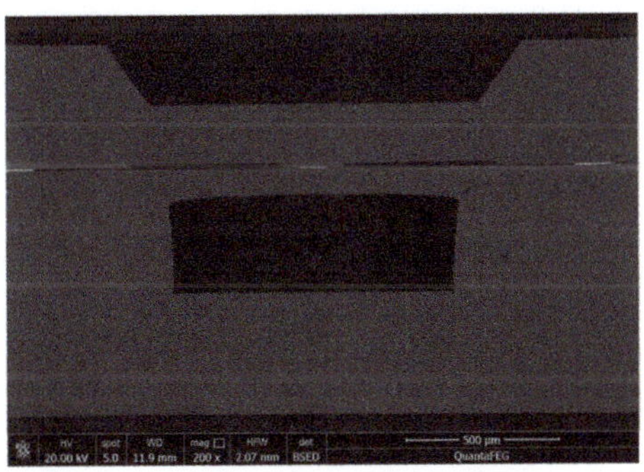

Figure 10. Dual bonding microscopic structure.

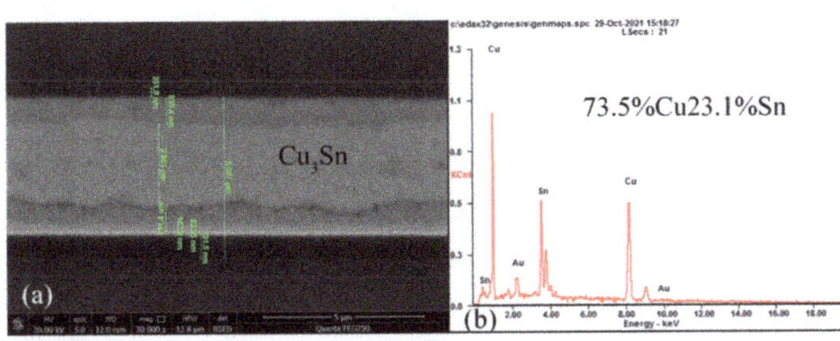

Figure 11. Cu–Sn bonding interfacial structure and composition. (**a**) Cross-sectional SEM image. (**b**) EDS spectrum.

3.3. *Shear Strength and Hermeticity Detection*

The shear strength was qualitatively assessed by conducting shear experiments using shear force tester (DAGE4000, Nordson DAGE, Inc.). The overall shear strengths of Cu–Sn, Au–Au, and dual-bonding devices were tested. The results are shown in Figure 13. A total of 15 bonded samples were tested and divided into three groups. The shear strength of Cu–Sn, Au–Au, and dual bonded were found to be in the range of 19.4–24.7 MPa, 10.8–17.3 MPa, and 11.2–16.5 MPa, the average shear strength was 22.5 MPa, 14.8 MPa, and 14.3 MPa. Shear test showed that shear strength of Cu–Sn bonding was higher than that of Au–Au bonding. Thus, the dual-bonding fracture finally occurred at Au–Au bonding. For Cu–Sn bonding, the fracture usually occurs in the Cu_3Sn layer. This indicates that Cu_3Sn is the weakest layer formed during Cu–Sn bonding. However, Au–Au bonding fracture partially occurs at the original interface present between Au and Au, and the other part occurs at the interface between Cr and Si or Cr and Au. SEM images of the fracture surfaces of the bonded samples are shown in Figure 14.

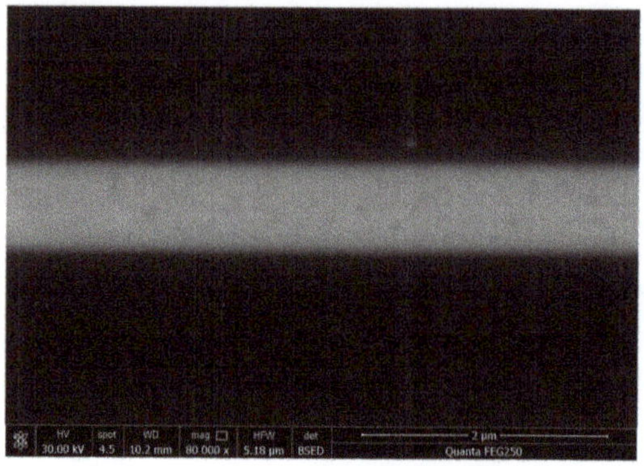

Figure 12. Au–Au bonding cross-sectional SEM image.

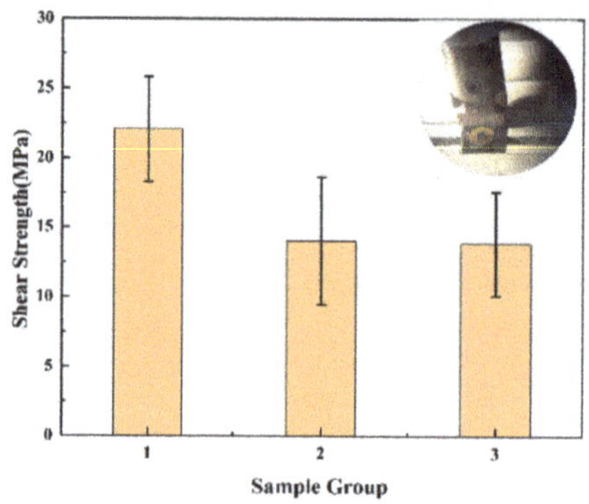

Figure 13. Bonding strength of the Cu–Sn bonding, Au–Au bonding, and dual bonding.

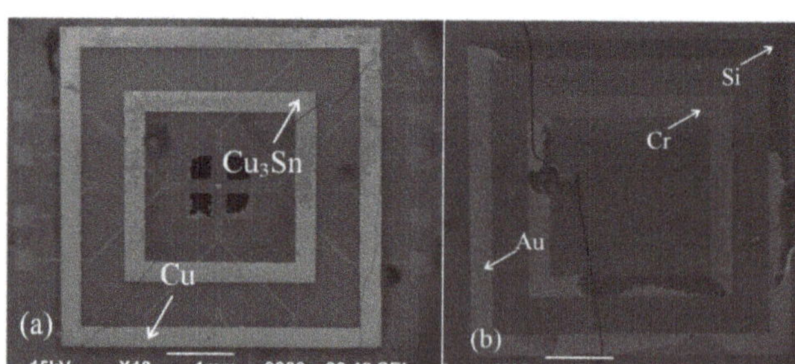

Figure 14. SEM images of the fracture surfaces for bonded device. (**a**) Cu–Sn bonding. (**b**) Au–Au bonding.

High-hermeticity packaging can provide both a pressure reference value and sensitive membrane protection for graphene pressure sensors. According to the MIL-STD-883K method 1014.15, with a volume of 0.25 mm^3, the leak rate limit corresponding to the hermetic cavity was 5×10^{-3} Pa·cm^3/s. Five samples were first placed under an atmosphere of helium at a pressure of 0.4 MPa over a period of 3 h. The helium that became attached to the surface of the samples was removed under a flow of nitrogen, and the leak rate was then measured using a helium mass spectrometer leak detector. The results are presented in Table 1.

Table 1. Hermeticity tests ($\times 10^{-4}$ Pa·cm^3/s).

As bonded	Sample 1	Sample 2	Sample 3	Sample 4	Sample 5
	1.53	1.48	1.66	2.47	1.46

For the bonding samples, the leakage rate was in the range of 1.46×10^{-4}–2.47×10^{-4} Pa·cm^3/s. The average value was 1.72×10^{-4} Pa·cm^3/s which is significantly lower than the leak rate rejection limit.

3.4. High-Temperature Reliability

To verify the high-temperature reliability, HTS at 350 °C for 10 h was arranged. A cross-sectional SEM image of the stored interfacial microstructure for Cu–Sn bonding was recorded which is shown in Figure 15a. Some small voids were present, but obvious defects in the bonded interface were absent. This indicates that the quality of the bonding interface is almost unaffected by the HTS test. However, this may influence shear strength and hermeticity. The Au–Au bonding is shown in Figure 15b; the interface included a smooth interface with no void looks like that before HTS.

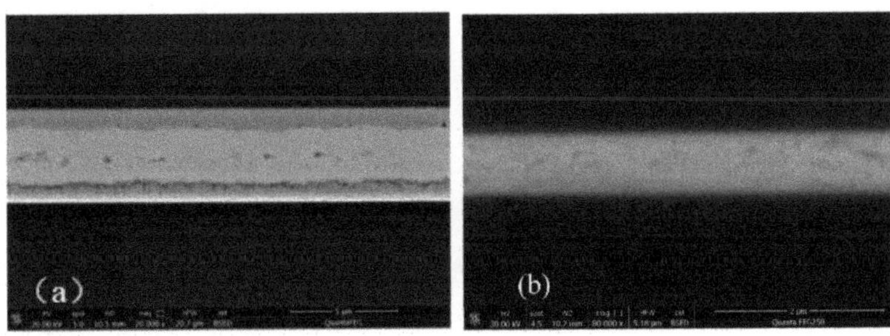

Figure 15. SEM images of the stored interfacial structure. (**a**) Cu–Sn bonding. (**b**) Au–Au bonding.

Then, five dual-bonding samples were tested, and the shear strength was found to be in the range of 11.5–16.3 MPa. The average shear strength was 13.5 MPa. The five samples in Table 1 are still used for hermeticity experiment after HTS. The results are shown in Table 2, compared with the data in Table 1. It is found that the leakage rate of each sample has increased to some extent, which presumably related to the formation of voids observed at the Cu–Sn interface before. The average value was 2.1×10^{-4} Pa cm^3/s, but the overall change of shear strength and hermeticity is small. These results indicate that the dual-bonding structure can be effectively used in high-temperature environments. The bonding interface exhibits excellent performance even after the samples were subjected to conditions of the HTS test.

Table 2. Hermeticity tests ($\times 10^{-4}$ Pa·cm^3/s).

Stored	Sample 1	Sample 2	Sample 3	Sample 4	Sample 5
	1.65	1.87	2.39	2.86	1.73

3.5. Static Test

A multimeter was first used to test whether the resistance of graphene was still present after HTS. The results showed a slight decrease in the resistance of graphene compared with before HTS. This decrease can be attributed to the different thermal matching of graphene and metal electrodes at high temperatures. Then we used a piston manometer to pressurize graphene devices. The experimental arrangement is connected the precision digital pressure gauge to a piston manometer which can record the real time pressure of piston manometer. The change in the resistance value of the graphene sensor was recorded by digital multimeter. The output resistance versus the pressure curve is shown in Figure 16. The graph includes three cycle tests, with each cycle test including pressurization and relief process. The output resistance response showed excellent repeatability and stability. The sensitivity of the pressure sensor is expressed using the following equation:

$$S = \Delta R / \Delta P \tag{1}$$

where ΔR is the output resistance variation value, and ΔP is the pressure variation value. The average sensitivity of the pressure sensor is 3.11 Ω/MPa. Because there is no circuit in this case, the advantage of device sensitivity cannot be exploited. Relevant circuits are then designed to improve the sensitivity of the output response.

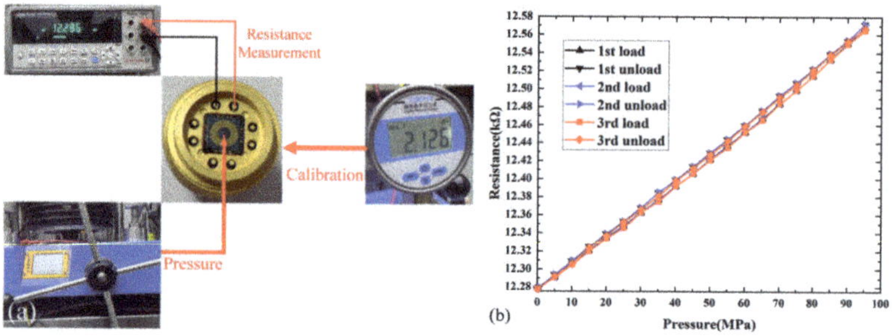

Figure 16. Electromechanical characteristics of the pressure sensor. (**a**) The schematic diagram for graphene pressure sensor measurement. (**b**) Relationships between the pressure and resistance for three cycles.

4. Conclusions

Hermetic packaging for a high-temperature graphene pressure sensor was realized based on the dual bonding of Cu–Sn and Au–Au. The dual-bonding performance was evaluated through various experiments. The bonding interface of Cu–Sn had no Sn overflow and transformed into a stable Cu–Cu_3Sn–Cu structure, and Au–Au bonding included a smooth interface with no void. Shear test showed that shear strength of Cu–Sn bonding was higher than that of Au–Au bonding. Thus, it is reasonable that the final fracture of the dual-bonding samples occurs mostly at the Au–Au bonding. Hermeticity of the dual bonding was one order of magnitude less than standard leakage rate. After HTS test at 350 °C for 10 h, the dual-bonding performance also revealed no obvious change. The output resistance response exhibited considerable sensitivity and outstanding repeatability through three cycles of static pressure test, which was mainly attributed to the excellent sealing protection. In conclusion, dual bonding of Cu–Sn and Au–Au would be suitable for hermetic packaging in a particular condition and further promote the development of high-temperature graphene pressure sensor.

Author Contributions: Conceptualization, methodology, project administration, writing-review and editing, J.W.; data curation and writing-original draft H.Z.; investigation and validation, X.C.; funding acquisition, supervision, M.L. All authors have read and agreed to the published version of the manuscript.

Funding: This research was funded by the "173" Projects of China (Grant No. 2021JCJQJJ0172, 2020JCJQZD043, and 2017JCJQZD006).

Data Availability Statement: Not applicable.

Acknowledgments: The authors would like to appreciate the facility assistance from Suzhou Institute of Nano-Tech and Nano-Bionics, Chinese Academy of Sciences.

Conflicts of Interest: The authors declare no conflict of interest.

References

1. Chun, S.; Kim, Y.; Jung, H.; Park, W. A flexible graphene touch sensor in the general human touch range. *Appl. Phys. Lett.* **2014**, *105*, 041907. [CrossRef]
2. Xu, H.; Wu, J.; Chen, Y.; Zhang, H.; Zhang, J. Substrate Engineering by Hexagonal Boron Nitride/SiO2for Hysteresis-Free Graphene FETs and Large-Scale Graphene p-n Junctions. *Chem. Asian J.* **2013**, *8*, 2446–2452. [CrossRef] [PubMed]
3. Kühne, M.; Paolucci, F.; Popovic, J.; Ostrovsky, P.; Maier, J.; Smet, J.H. Ultrafast lithium diffusion in bilayer graphene. *Nat. Nanotechnol.* **2017**, *12*, 895–900. [CrossRef] [PubMed]
4. Lan, C.; Li, C.; Wang, S.; He, T.; Zhou, Z.; Wei, D.; Guo, H.; Yang, H.; Liu, Y. Highly responsive and broadband photodetectors based on WS2–graphene van der Waals epitaxial heterostructures. *J. Mater. Chem. C* **2017**, *5*, 1494–1500. [CrossRef]
5. Loan, P.T.K.; Wu, D.; Ye, C.; Li, X.; Tra, V.T.; Wei, Q.; Fu, L.; Yu, A.; Li, L.-J.; Lin, C.-T. Hall effect biosensors with ultraclean graphene film for improved sensitivity of label-free DNA detection. *Biosens. Bioelectron.* **2018**, *99*, 85–91. [CrossRef] [PubMed]
6. Balandin, A.A.; Ghosh, S.; Bao, W.; Calizo, I.; Teweldebrhan, D.; Miao, F.; Lau, C.N. Superior Thermal Conductivity of Single-Layer Graphene. *Nano Lett.* **2008**, *8*, 902–907. [CrossRef] [PubMed]
7. Attia, N.F.; Elashery, S.E.; Zakria, A.M.; Eltaweil, A.S.; Oh, H. Recent advances in graphene sheets as new generation of flame retardant materials. *Mater. Sci. Eng. B* **2021**, *274*, 115460. [CrossRef]
8. Elashery, S.E.A.; Attia, N.F.; Mohamed, G.G.; Omar, M.M.; Tayea, H.M.I. Hybrid Nanocomposite Based Graphene Sensor for Ultrasensitive Clomipramine HCl Detection. *Electroanalysis* **2021**, *33*, 2361–2371. [CrossRef]
9. Karimov, K.S.; Sulaiman, K.; Ahmad, Z.; Akhmedov, K.M.; Mateen, A. Novel pressure and displacement sensors based on carbon nanotubes. *Chin. Phys. B* **2015**, *24*, 018801. [CrossRef]
10. Yu, F.; Liu, Q.; Gan, X.; Hu, M.; Zhang, T.; Li, C.; Kang, F.; Terrones, M.; Lv, R. Ultrasensitive Pressure Detection of Few-Layer MoS_2. *Adv. Mater.* **2017**, *29*, 1603266. [CrossRef]
11. Bolotin, K.; Sikes, K.; Jiang, Z.; Klima, M.; Fudenberg, G.; Hone, J.; Kim, P.; Stormer, H. Ultrahigh electron mobility in suspended graphene. *Solid State Commun.* **2008**, *146*, 351–355. [CrossRef]
12. Wang, L.; Meric, I.; Huang, P.Y.; Gao, Q.; Gao, Y.; Tran, H.; Taniguchi, T.; Watanabe, K.; Campos, L.M.; Muller, D.A.; et al. One-Dimensional Electrical Contact to a Two-Dimensional Material. *Science* **2013**, *342*, 614–617. [CrossRef]
13. Yin, Y.; Cheng, Z.; Wang, L.; Jin, K.; Wang, W. Graphene, a material for high temperature devices–intrinsic carrier density, carrier drift velocity and lattice energy. *Electroanalysis* **2014**, *4*, 5758. [CrossRef]
14. Bunch, J.S.; van der Zande, A.M.; Verbridge, S.S.; Frank, I.W.; Tanenbaum, D.M.; Parpia, J.M.; Craighead, H.G.; McEuen, P.L. Electromechanical Resonators from Graphene Sheets. *Science* **2007**, *315*, 490–493. [CrossRef] [PubMed]
15. Sorkin, V.; Zhang, Y.W. Graphene-based pressure nano-sensors. *J. Mol. Model.* **2011**, *17*, 2825–2830. [CrossRef] [PubMed]
16. Smith, A.D.; Niklaus, F.; Paussa, A.; Vaziri, S.; Fischer, A.C.; Sterner, M.; Forsberg, F.; Delin, A.; Esseni, D.; Palestri, P.; et al. Electromechanical Piezoresistive Sensing in Suspended Graphene Membranes. *Nano Lett.* **2013**, *13*, 3237–3242. [CrossRef] [PubMed]
17. Smith, A.; Vaziri, S.; Niklaus, F.; Fischer, A.; Sterner, M.; Delin, A.; Östling, M.; Lemme, M. Pressure sensors based on suspended graphene membranes. *Solid-State Electron.* **2013**, *88*, 89–94. [CrossRef]
18. Wang, Q.; Hong, W.; Dong, L. Graphene "microdrums" on a freestanding perforated thin membrane for high sensitivity MEMS pressure sensors. *Nanoscale* **2016**, *8*, 7663–7671. [CrossRef] [PubMed]
19. Wang, J.; Zhao, S.; Li, M.; Wang, P.; Li, M. Solid–liquid interdiffusion bonding of Cu–Sn–Cu interconnection and sealing for high-temperature pressure sensor based on graphene. *IEEE Trans. Compon. Packag. Manuf. Technol.* **2019**, *10*, 65–71. [CrossRef]
20. Choi, W.-B.; Ju, B.-K.; Jeong, S.-J.; Lee, N.-Y.; Koh, K.-H.; Haskard, M.; Sung, M.-Y.; Oh, M.-H. Anodic bonding technique under low-temperature and low-voltage using evaporated glass. *J. Vac. Sci. Technol. B* **2002**, *15*, 477–481. [CrossRef]
21. Noguchi, K.; Ikeda, M.; Shimizu, I.; Kawamura, Y.; Ohno, Y. Effect of Cooling Rate on Microstructure and Strength Properties of Tin-Silver-Copper Solder Ball Bonding. *Mater. Trans.* **2001**, *42*, 761–768. [CrossRef]
22. Cao, Y.; Ning, W.; Le, L. Wafer-level package with simultaneous TSV connection and cavity hermetic sealing by solder bonding for MEMS device. *IEEE Trans. Electron. Packag. Manuf.* **2009**, *32*, 125–132.
23. Zhang, S.; Yang, M.; Wu, Y.; Du, J.; Lin, T.; He, P.; Huang, M.; Paik, K.W. A study on the optimization of anisotropic conductive films for Sn-3Ag-0.5Cu-based flex-on-board application at a 250 degrees C bonding temperature. *IEEE Trans. Compon. Packag. Manuf. Technol.* **2018**, *8*, 383–391. [CrossRef]
24. Hang, C.; Liu, J.; Wang, J.; Fu, X.; Chen, H.; Li, M. A low-temperature Cu-to-Cu interconnection method by using nanoporous Cu fabricated by dealloying electroplated Cu–Zn. *J. Mater. Sci. Mater. Electron.* **2020**, *31*, 18381–18388. [CrossRef]
25. Jiang, H.; Robertson, S.; Zhou, Z.; Liu, C. Cu–Cu bonding with Cu nanowire arrays for electronics integration. In Proceedings of the 2020 IEEE 8th Electronics System-Integration Technology Conference (ESTC), Tønsberg, Norway, 15–18 September 2020.
26. Hao, F.; Wang, J.; Li, M. Low-temperature anodic bonding for wafer-level Al–Al interconnection in MEMS grating gyroscope. *IEEE Trans. Compon. Packag. Manuf. Technol.* **2020**, *11*, 19–24. [CrossRef]
27. Wang, J.; Wang, Q.; Wu, Z.; Wang, D.; Cai, J. Solid-state-diffusion bonding for wafer-level fine-pitch Cu/Sn/Cu interconnect in 3-D integration. *IEEE Trans. Compon. Packag. Manuf. Technol.* **2017**, *7*, 19–26. [CrossRef]
28. Kurashima, Y.; Matsumae, T.; Takagi, H. Room-temperature Au–Au bonding in atmospheric air using direct transferred atomically smooth Au film on electroplated patterns. *Microelectron. Eng.* **2018**, *189*, 33–38. [CrossRef]

29. Wang, J.; Zhu, Z.; Qi, Y.; Li, M. A Novel Crossbeam Structure with Graphene Sensing Element for N/MEMS Mechanical Sensors. *Nanomaterials* **2022**, *12*, 2101. [CrossRef] [PubMed]
30. Matsumae, T.; Kurashima, Y.; Umezawa, H.; Mokuno, Y.; Takagi, H. Room-temperature bonding of single-crystal diamond and Si using Au/Au atomic diffusion bonding in atmospheric air. *Microelectron. Eng.* **2018**, *195*, 68–73. [CrossRef]
31. Yamamoto, M.; Matsumae, T.; Kurashima, Y.; Takagi, H.; Suga, T.; Takamatsu, S.; Itoh, T.; Higurashi, E. Effect of Au Film Thickness and Surface Roughness on Room-Temperature Wafer Bonding and Wafer-Scale Vacuum Sealing by Au-Au Surface Activated Bonding. *Micromachines* **2020**, *11*, 454. [CrossRef]
32. Yamamoto, M.; Higurashi, E.; Suga, T.; Sawada, R.; Itoh, T. Properties of various plasma surface treatments for low-temperature Au–Au bonding. *Jpn. J. Appl. Phys.* **2018**, *57*, 04FC12. [CrossRef]
33. Wu, D.; Tian, W.; Wang, C.; Huo, R.; Wang, Y. Research of Wafer Level Bonding Process Based on Cu–Sn Eutectic. *Micromachines* **2020**, *11*, 789. [CrossRef] [PubMed]

micromachines

Article

Au Wire Ball Welding and Its Reliability Test for High-Temperature Environment

Chenyang Wu [1,2], Junqiang Wang [1,2,*], Xiaofei Liu [1,2], Mengwei Li [1,2,*], Zehua Zhu [1,2] and Yue Qi [1,2]

1 Notional Key Laboratory of Instrumentation Science & Dynamic Measurement, Taiyuan 030051, China
2 Academy for Advanced Interdisciplinary Research, North University of China, Taiyuan 030051, China
* Correspondence: wangjq210@nuc.edu.cn (J.W.); lmwnuc@163.com (M.L.);
Tel.: +86-158-1133-8056 (J.W.); +86-139-3424-8366 (M.L.)

Abstract: The long-term application of sensors in a high-temperature environment needs to address several challenges, such as stability at high temperatures for a long time, better wiring interconnection of sensors, and reliable and steady connection of the sensor and its external equipment. In order to systematically investigate the reliability of thin coatings at high temperatures for a long time, Au and Cr layers were deposited on silicon substrates by magnetron sputtering. Additionally, samples with different electrode thicknesses were annealed at different temperatures for a varied duration to study the effect of electrode thickness, temperature, and duration on the reliability of samples. The results of tensile and probe tests before and after heat treatment revealed that the mechanical strength and electrical properties have changed after annealing. In addition, the bonding interface was analyzed by a cross-sectional electron microscope. The analysis showed that long-term continuous high-temperature exposure would result in thinning of the electrode, formation of pores, recrystallization, and grain growth, all of which can affect the mechanical strength and electrical properties. In addition, it was observed that increasing the thickness of the gold layer will improve reliability, and the test results show that although the thin metal layer sample is in poor condition, it is still usable. The present study provides theoretical support for the application of thin coatings in high temperatures and harsh environments.

Keywords: Au wire bonding; high-temperature annealing; focused ion beam (FIB); morphology analysis

Citation: Wu, C.; Wang, J.; Liu, X.; Li, M.; Zhu, Z.; Qi, Y. Au Wire Ball Welding and Its Reliability Test for High-Temperature Environment. *Micromachines* **2022**, *13*, 1603. https://doi.org/10.3390/mi13101603

Academic Editor: Wensheng Zhao

Received: 12 August 2022
Accepted: 24 September 2022
Published: 27 September 2022

Publisher's Note: MDPI stays neutral with regard to jurisdictional claims in published maps and institutional affiliations.

Copyright: © 2022 by the authors. Licensee MDPI, Basel, Switzerland. This article is an open access article distributed under the terms and conditions of the Creative Commons Attribution (CC BY) license (https://creativecommons.org/licenses/by/4.0/).

1. Introduction

Transducers, such as temperature sensors and high-temperature pressure sensors, are used in harsh environments, including aerospace engines, supersonic aircraft, automotive engines, and metallurgical industries [1–5]. These sensors are used for quite a long time in harsh environments [6–8]. One of the most critical technologies in sensor packaging is wire bonding, wherein the bonding materials and bonding parameters will affect the quality and operation life of sensors [9]. Therefore, it is necessary to analyze the long-term stability, service life, ability to withstand harsh environments, mechanical strength of wire bonding. Thus, these bottlenecks need to be addressed urgently in developing a suitable long-term stable contact and packaging technology [10,11].

High purity Au wires have been used as an interconnection for most device packages [12,13] as Au has high temperature resistance, excellent conductivity, and good oxidation resistance. Presently, gold wire ball welding has become one of the commonly used welding processes in sensor packaging technology [14,15]. In general, Au wires ball welding is produced by the electric-flame-off (EFO) system, which generates high-voltage discharge to the tail wire and generates electric sparks. The high temperature generated instantly melts the tail end of the gold wire to produce a free air ball (FAB) [16,17]. Later, the ultrasonic power source generates mechanical vibration energy, which causes deformation between the gold ball and the pad, resulting in mutual diffusion of metal atoms. This completes the welding of the first solder joint; generally, the height of the welding ball welded by this process is 2–3 times the wire diameter. The factors that affect the quality of welding points are ultrasonic power, contact time, welding pressure, and hot table temperature [18–22], all of which need to be adjusted to maximize the welding quality.

Generally, the barrier layer will be sputtered under the Au layer to increase the adhesion with the substrate and prevent the Au layer from diffusing between the substrates. The most common metals used for barrier layers are Ti, Cr, Ni, and Ta [23–26]. The traditional pad coating is composed of 3–5 μm nickel layer and 1–1.25 μm thick gold layer. The processing difficulty and high cost of gold, thinner coating such as 0.3–0.5 μm of the gold layer are opted to replace the thick gold layer. Nevertheless, the thin coating of gold leads to some challenges, including poor bonding of lead wire, especially after high-temperature baking, and failure of bond due to the lower annealing temperature of Au (600 °C) [27–31]. In practical application, high-temperature sensors need to work in high-temperature and harsh environments. Therefore, it is necessary to analyze the lead bonding failure at high temperatures and explore the bonding failure mechanism at different thicknesses and temperatures in order to develop better bonding strategies.

In this work, the Au wire ball bonding method is employed. Under the optimal parameters debugged, Au wire ball bonding has been carried out to produce a thin metal coating. The bonding quality has been analyzed by testing the pull force of gold wire on the coatings with different thicknesses. The thin metal coating of 50/300 nm and 50/400 nm of Cr/Au plated on bare silicon by magnetron sputtering, separately. Then the temperature resistance experiments were carried out at 600 °C and 700 °C for 2 h and 5 h, separately. Further, the test samples were analyzed by scanning electron microscopy (SEM) and probe station, and the morphology and electrical conductivity of the samples were observed, and their stability under the harsh environment at high temperature for a long time was assessed.

2. Materials and Methods

The investigated Au wires with 99.99% purity and 25 μm diameter were bonded to the gold electrodes on the silicon substrate by ball bonding in a wiring machine. Figure 1 shows the process of fabricating a gold electrode on a 2-inch diameter bare silicon chip. In Figure 1a, bare silicon wafers are treated in a vacuum with oxygen plasma to remove surface contaminants and enhance the adhesion between the wafer and the sputtered metal. After photolithographic treatment of bare silicon wafers, as shown in Figure 1b, a 50-nm layer of metal Cr is first sputtered on the surface of the wafer using a magnetron sputtering machine (LAB18, K. J. Lesker, Inc., Jefferson Hills, PA, USA) as a seed layer. Afterward, 300-nm and 400-nm thicknesses of Au are sputtered on the surface of the wafer as bonded electrodes. Subsequently, the samples soaked in acetone solution were stripped to remove excess Cr/Au metals. The stripping results are shown in Figure 1c. Finally, the prepared sample was diced as the final experimental slice Figure 1d. The sample is square, the size of the square is 1 cm × 1 cm, and the electrode spacing is 1 mm after dicing according to the above process steps. Each sample contains 9 pairs of electrodes with opposite directions.

The wire bonding process was performed on the two electrodes of the cut experimental sample (Micro Point Pro, iBond5000, Yokneam Elite, Israel). During the bonding process, the ultrasonic and heat energies provided by the cleaver were transferred to the welding wire and pad. The energy absorbed by gold wire is

$$E = k\Delta T^2 + uFAft \tag{1}$$

where k is the influence coefficient, ΔT is the temperature difference before and after wire bonding, u is the friction coefficient, F is the pressure exerted by the cleaver, A is ultrasonic amplitude, f is ultrasonic frequency, and t is the time of ultrasonic action on the gold wire. There are several key factors affecting the quality of solder joints. During the bonding process, we carried out a series of optimizations and finally determined the best bonding pressure as 0.4 MPa, corresponding to the parameter F in the above formula; the ultrasonic power as 100 mV. The ultrasonic power is the ultrasonic (U/s) energy applied to the welding spot. The ultrasonic power is divided into two modes: high power and low power. The parameters of high power correspond to large ultrasonic amplitude. The two parameters in the above formula, ultrasonic amplitude A and ultrasonic frequency f, are reflected by the ultrasonic power; the ultrasonic time as 30 ms, which corresponds to the parameter t in the above formula; and the hot bench temperature as 150 °C, corresponding to the temperature difference ΔT before and after wire bonding in the above formula. In addition, the looping parameters are loop top, reverse length, and kink height, all of which have an impact on the overall bonding quality, and the functional relationship between them is shown in Figure 2. The diameter of the ball is generally 2–5 times of the diameter of the gold wire, the height of the loop top is generally between 200 μm and 800 μm, which can be adjusted according to the actual application, the reverse is to ensure that the gold wire is arcuate. In this paper, the loop top set in the wire bonding machine is 600 μm, the kink height is 300 μm, and the reverse length is 50 μm.

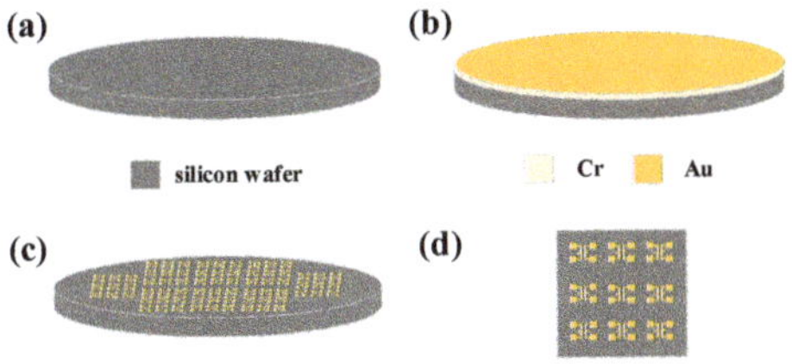

Figure 1. Process flow diagram for fabricating Cr/Au electrodes on silicon wafers. (**a**) Bare silicon pre-treatment. (**b**) Sputtering Cr/Au metal layer. (**c**) Acetone immersion peeling. (**d**) Dicing Silicon Crystals.

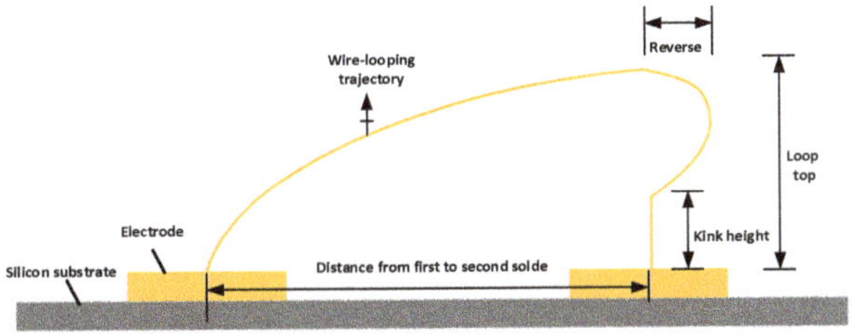

Figure 2. 3D schematic diagram of key parameters of gold wire ball welding bonding.

The sample characteristics and experimental conditions are shown in Table 1. The Cr/Au metal layer with a thickness of 50/300 nm and 50/400 nm is maintained in the high-temperature environment of 500 °C and 700 °C for 2 h and 5 h, respectively. Then the electronic scanning mirror and probe table resistance of the samples are observed after high-temperature annealing, and the energy of the thin metal layer to withstand high temperature, and harsh environment is obtained.

Table 1. Characteristics of experimental samples and experimental environment.

Thickness of Thin Metal Layer Cr/Au (nm)	Sample	Temperature (°C)	Time (h)
50/300	S-1	No	0
50/300	S-2	500	2
50/300	S-3	500	5
50/300	S-4	700	2
50/300	S-5	700	5
50/400	S-6	No	0
50/400	S-7	500	2
50/400	S-8	500	5
50/400	S-9	700	2
50/400	S-10	700	5

3. Results and Discussion

3.1. Bonding Result and Quality Test

After gold wire ball bonding, the bonding samples were observed under an optical microscope and scanning electron microscope to identify the bonding bumps, lead arc, and overall bonding morphology. Figure 3a,b shows the observation results under the optical microscope and the scanning electron microscope. Figure 3c shows the bonding sphere and the position of the electrode surface under a high-power microscope. It can be observed from the SEM image that the shape of the solder ball is three-dimensional, the length of the tail line generated by translation is moderate, and no dragging phenomenon was identified, which ensures the reliability of the bonding results. In addition, the thickness of the electrode Pad was measured with a step meter to ensure the consistency of the experiment. Figure 3d,e show that the thickness of the two electrodes, as expected, was found to be 367.9nm and 482.2 nm. Further, the horizontal curve confirms the uniform thickness of the film. Thus, the quality and uniformity of the metal film are achieved.

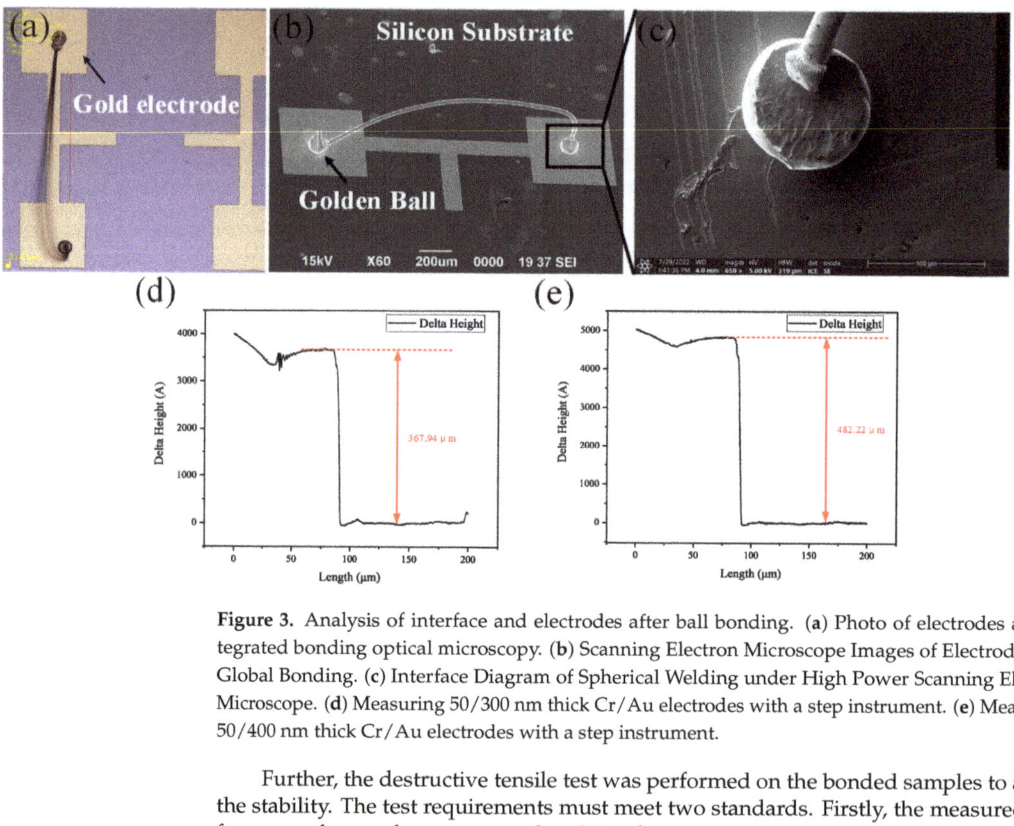

Figure 3. Analysis of interface and electrodes after ball bonding. (**a**) Photo of electrodes and integrated bonding optical microscopy. (**b**) Scanning Electron Microscope Images of Electrodes and Global Bonding. (**c**) Interface Diagram of Spherical Welding under High Power Scanning Electron Microscope. (**d**) Measuring 50/300 nm thick Cr/Au electrodes with a step instrument. (**e**) Measuring 50/400 nm thick Cr/Au electrodes with a step instrument.

Further, the destructive tensile test was performed on the bonded samples to assess the stability. The test requirements must meet two standards. Firstly, the measured pull force must be equal to or greater than 50% of the maximum yield force of the undeformed wire. In addition, the standard deviation of the sample in the laboratory should be less than 15%, and the standard deviation of the sample for manufacturing application should be less than 25%. Another prerequisite is that there is no bonding layer in laboratory applications and less than 10% in manufacturing applications [32]. The pull test results are shown in Table 2, and the data is in agreement with both criteria. The destructive tensile test data suggests that the critical failure force of the sample with a 400 nm thick electrode gold layer is greater than that of the 300 nm thick gold layer. Therefore, the metal layer with an appropriate thickness within the acceptable range can enhance the bonding quality of the sample.

Table 2. Pull test of different bonding points on the same experimental sample.

Sample and Sample Bonding Point	Thickness of Metal Layer Cr/Au (nm)	Breaking Critical Pull Value (g)	Standard Deviation (g)	Average Pull Value (g)
S1–1	50/300	7.080	–	–
S1–2		9.271	1.549	8.176
S1–3		7.566	1.150	7.973
S1–4		10.341	1.512	8.563
S1–5		14.486	3.115	9.829
S1–6		10.515	2.800	9.943
S2–1	50/400	16.171	–	–
S2–2		15.506	0.470	15.839
S2–3		17.208	0.858	16.295
S3–1	50/400	12.497		
S3–2		13.357	0.607	12.927
S3–3		12.591	0.667	12.815

3.2. Testing of Mechanical Properties before High Temperature Annealing

Further, the destructive tensile test was performed on the bonded samples to assess the stability. The test requirements must meet two standards. Firstly, the measured pull force must be equal to or greater than 50% of the maximum yield force of the undeformed wire. In addition, the standard deviation of the sample in the laboratory should be less than 15%, and the standard deviation of the sample for manufacturing application should be less than 25%. Another prerequisite is that there is no bonding layer in laboratory applications and less than 10% in manufacturing applications [32]. The pull test results are shown in Table 2, and the data is in agreement with both criteria. The destructive tensile test data suggests that the critical failure force of the sample with a 400 nm thick electrode gold layer is greater than that of the 300 nm thick gold layer. Therefore, the metal layer with an appropriate thickness within the acceptable range can enhance the bonding quality of the sample.

The hot press ball welding generally involves two solders, the first solder joint is ball welding, and the second solder joint is wedge welding, in short, ball-wedge bonding mode. As the bonding tension of ball welding is much greater than wedge welding bonding, it is necessary to carry out reinforcement treatment in the second solder joint. For materials with small breaking forces, such as gold wire, the ball pressing reinforcement method can be used for repair welding.

The equipment used in the experiment is the Dage4000 push-pull test system. The measurable pull range is 1 g–10,000 g, the system accuracy is ±0.25%, and the maximum test speed can reach 5000 μm/s. The experimental method used in this experiment is to place a pull hook directly below the middle position of the lead and slowly pull it in the z-axis direction until the lead breaks. The moving speed is set at 2500 μm/s, and the accuracy of the z-axis is ± 10μm. As shown in Figure 4, both solder joints are spherical. Figure 4a,b show an optical microscope picture and the SEM image of the fracture position after the tensile failure test, respectively. As can be observed from the figure, the fracture position is concentrated in the spherical tail of the first and second solder joints and does not appear at the second solder joint, indicating that the ball pressing reinforcement has a significant effect on the bonding strength of the second solder joint.

In addition, the shear force test was also carried out. The average shear force of the electrode with a thickness of 300 nm was 7.506 g, and the average shear force of the electrode with a thickness of 400 nm was 12.825 g. Figure 4c shows the shear force test curve of one point. Cracking occurred at the maximum of 8.091 g. The test process was conducted according to JESD22-B117 standard. Figure 4c is a picture of one of the experimental data.

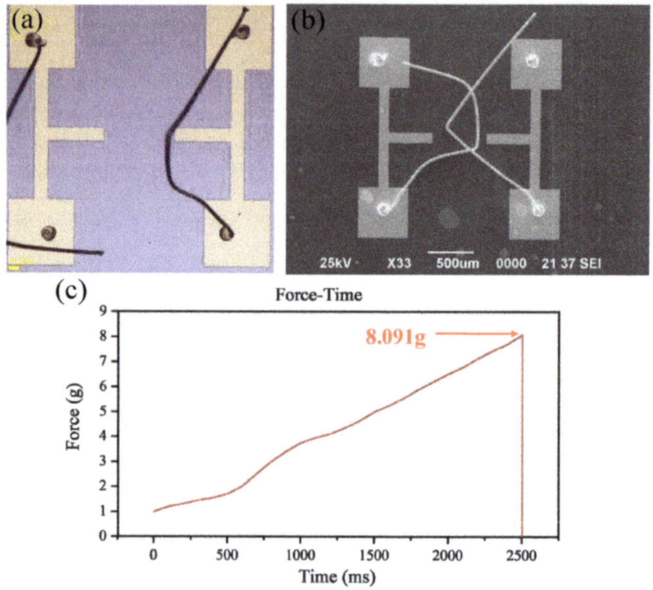

Figure 4. Pull failure experimental results observed under the optical microscope and scanning electron microscope and photo of shear force test. (**a**) Optical microscope image after tensile test. (**b**) SEM image after tensile failure test. (**c**) Time-force diagram of shear force test.

3.3. High-Temperature Reliability Test

In order to test the bonding reliability of samples under high temperature and harsh environments and test the performance of thin metal layer after high-temperature annealing, 10 samples with different metal layer thicknesses were randomly selected for the high-temperature annealing experiment. Similarly, the samples from the same batch were also selected and placed at room temperature for the same time for comparative experiments. The selected samples were heated in a high-temperature oven for 2 and 5 h, and the annealing is carried out in an atmospheric atmosphere using a tubular annealing boiler separately, and heated cross sections were prepared for comparison with samples at room temperature. The samples after high-temperature annealing and the control group samples were observed and compared under the scanning electron microscope. The results are shown in Figure 5. Initially, the surface of the sample was observed, followed by the changes in the morphology of the bonding interface before and after annealing by cutting the bonding interface. In general, the cutting can be carried out by using the grinding prototype to solidify the silicone grease grinding sample, or by using a focused ion beam (FIB). The present work used the FIB method.

The samples before and after annealing are cut, and the profile is observed. It can be clearly observed from Figure 5a that before annealing, the Au and Cr layer of the sample have obvious interfaces, and the section is flat without any cavities and cracks. Moreover, the thickness of the Au/Cr layer (51.37/350.9 nm) measured by a large magnification electron microscope is consistent with the actual measured value. Figure 5b shows the cross-section at the electrode. The situation at the electrode is similar to that at the bond. The Au and Cr layers are flat and smooth. Figure 5c,d are SEM images of 300-nm-thick electrodes annealed at 500 °C for 2 h and 5 h, separately. The joints of the electrodes and the balls are not cracked, while there are many cracks and holes in the electrodes. However, the detailed deeper impact can be studied by observing the cut profile. Figure 5e shows the FIB cut interface between the bonded free gold balloon and the electrodes after annealing for 5 h at 700 °C. The high-power electron microscopy of the cut interface confirms that there is no longer a gap between the bonded free gold balloon and the electrodes. This is because the grain size of the bonded free gold balloon increases with temperature, disappearing the gap. However, some cracks and holes are still observed. This suggests the positive effect of heat treatment on the bonding interface between the gold sphere and the electrodes.

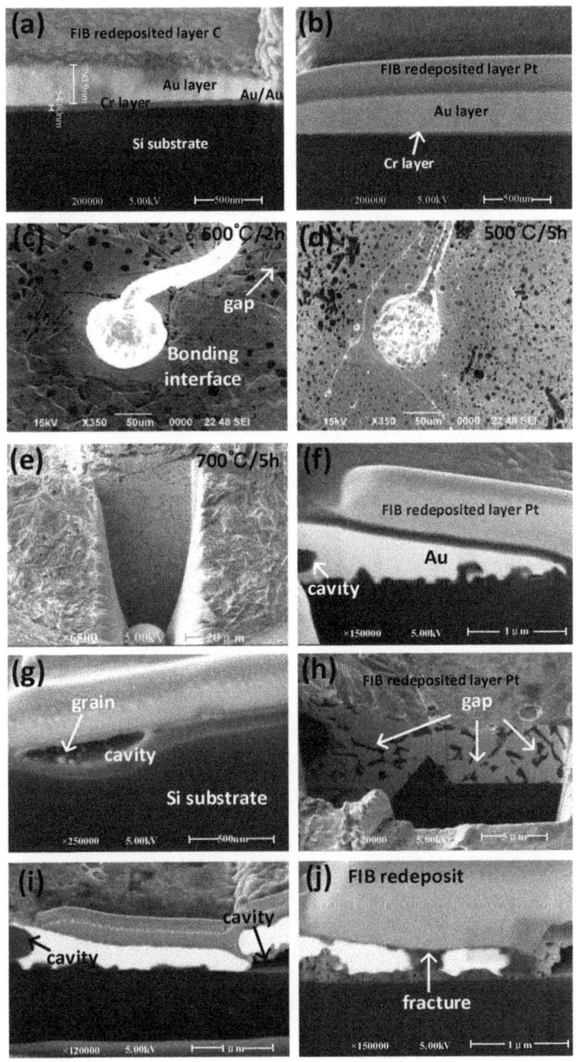

Figure 5. SEM image of FIB cross section cutting of samples covered with a C/Pt (1 μm) electrodeposition layer on gold electrode surface before and after high-temperature annealing as well as the overall SEM image of the uncut bond site after annealing and the SEM image of the FIB cross section cutting of the sample without covered with a C/Pt electrodeposited layer. Overview image of electrode covered with (**a**) C electrodeposition layer and (**b**) Pt electrodeposition layer. Bonding interface image after annealing at (**c**) 500 °C for 2 h and (**d**) 500 °C for 5 h. (**e**) SEM image of bonding interface sectional drawing cut by FIB after annealing at 700 °C for 5 h. SEM image of FIB cut with 300 nm thick electrode annealed at (**f**) 500 °C for 2 h and (**g**) 500 °C for 5 h, SEM image of FIB cut with 400 nm thick electrode annealed at (**h**) 500 °C for 5 h and (**i**) 700 °C for 2 h. (**j**) SEM image of FIB cut with 300 nm thick electrode annealed at 700 °C for 5 h.

Figure 5f shows the image of a 300 nm thick electrode annealed at 500 °C for 2 h. From the figure, it can be observed that a small part of the gold layer evaporated, while most of it is still intact, and only some small holes appeared, without obvious fracture. Figure 5g shows the image of a 300 nm thick electrode annealed at 500 °C for 5 h. The gold layer has become significantly thinner, accompanied by some large extended gaps. Notably, some crystalline grains have also been observed, indicating that recrystallization and grain growth have occurred, and this can well explain the disappearance of gaps in Figure 5e. The 300-nm thick gold electrode samples annealed for 2 and 5 h at 700 °C are similar to those annealed for 2 and 5 h at 500 °C, except that more voids and more severe grain regeneration were observed at 700 °C, but the difference is not significant from that at 500 °C, so it is not discussed here.

Figure 5h is the image of 400-nm thick electrodes annealed at 500 °C for 5 h. The image of 2 h annealing is also similar to that of the 300 nm thick electrode. Unlike 300 nm thick electrodes with voids, there are no obvious voids. However, there are some small cracks and voids, and the remaining thickness of the electrodes is consistently thick. Therefore, it can be concluded that increasing the thickness of electrodes can improve the high-temperature resistance of samples. Figure 5i shows the image of a 400 nm thick sample after annealing at 700 °C for 2 h. Compared with Figure 5f, it can be clearly seen that although there are some voids, the homogeneity of the gold layer has been significantly improved. Moreover, increasing the thickness of the gold layer has a positive impact on the temperature resistance of the sample. Figure 5j is the image of a 400 nm thick sample after annealing at 700 °C for 5 h. In addition to obvious voids and grain growth, there are some similar faults identified in the gold layer. The thickness of the gold layer can be clearly seen, and the resistance is also identified as normal when tested with the probe bench. This suggests that the 400 nm thick sample can work for a long time in the high-temperature environment of 700 °C.

It can be concluded that both 50/300 nm and 50/400 nm thick Cr/Au electrodes can withstand high temperatures and harsh environments for a long time. Within the allowable range, the thicker the electrode, the higher the tolerable temperature, and the longer the temperature resistance time. The high-temperature environment has a significant positive impact on the interface between free air golden ball and electrode bonding.

3.4. Comparative Testing of Mechanical and Electrical Performance

Destructive tensile tests were carried out on several groups of samples after annealing and compared with that before annealing. The results are shown in the histogram in Figure 6. It can be observed that the tensile strength of the sample after high-temperature annealing is biased to a lower value. After annealing for 2 h, the pull value of the samples decreased by nearly 50%, and the samples treated at 700 °C showed a larger decrease than that treated at 500 °C. The critical tensile failure value of the samples annealed at high temperature for 5 h is lower than 1/4 of that of the samples not annealed. Due to the evaporation of the gold wire in the high-temperature environment, the diameter decreases, the mechanical strength of the gold wire decreases, and thus the strength of the connection between the free air gold ball and the electrode also decreases. The reason for the reduction of the tension value may be attributed to these two factors.

The resistance values of samples before and after high-temperature annealing were measured with a probe table. As can be observed from the histogram, the resistance values of 300 nm and 400 nm thick samples after annealing at 500 °C and 700 °C for 2 h have decreased. The range of resistance reduction is as small as 15%, which can be attributed to the evaporation of electrode and gold wire at high temperatures. However, the resistance values of the two samples showed an increase by nearly twice the original value after annealing at 500 °C for 5 h. This increase can be attributed to various factors such as the diffusion of Cr layer metal to the Au layer, the diffusion of oxygen atoms to the Au layer, and the occurrence of recrystallization and grain growth. In addition, an interesting phenomenon is that the resistance of samples annealed at 700 °C for 5 h shows a decreasing trend. This is probably due to the different diffusion degrees of Cr and O atoms at different temperatures, and the elimination of metal dislocation and stress by heat treatment [33].

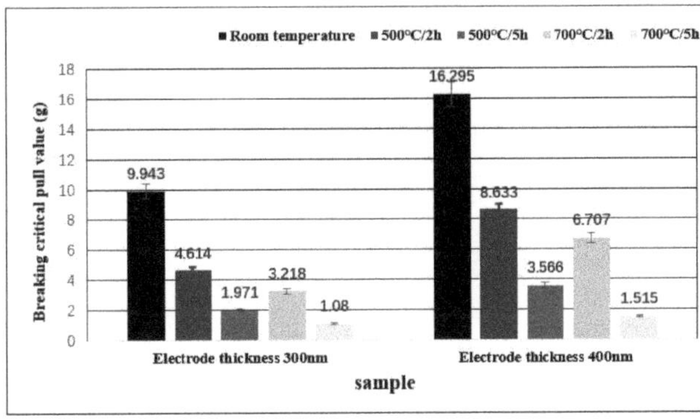

Figure 6. Histogram depicting the critical failure tension of the initial wire bonded samples at room temperature and the heat-treated bonded samples.

In this work, the resistivity of the annealed film can be expressed as [34]:

$$\rho = \rho_{Au} + \rho_{CrI} + \rho_{OI} + \rho_D \quad (2)$$

where ρ_{Au} is the Au bulk resistivity, ρ_{CrI} is the resistivity due to diffusion of Cr atoms into the Au layer, ρ_{OI} is the resistivity due to diffusion of O atoms into Au layer, and ρ_D is the resistivity caused by impurities, point defects, and grain boundaries in the Au layer.

For the Au electrode annealed at 500 °C and 700 °C, as discussed in Figure 7, high temperature will produce grains, recrystallization and impurities, which can be considered that ρ_D in the above formula decreases. The conductivity of Au is better than that of Cr, a higher temperature will cause the Cr layer to diffuse into the Au layer, and the O atoms in the air will also diffuse into the Au layer, which will lead to an increase in $\rho_{CrI} + \rho_{OI}$. There is only slight grain Recrystallization at 500 °C, so the main reason for the increase in resistance at 500 °C is that Cr and O atoms diffuse into the Au layer, resulting in an increase in resistance. The diffusivity of Cr and O atoms in Au is higher at 500 °C than at 700 °C, and recrystallization is more obvious at 700 °C. At the same time, heat treatment can also eliminate stress and dislocation in the metal. As a result, the resistivity at 700 °C is much lower than that at 500 °C [34].

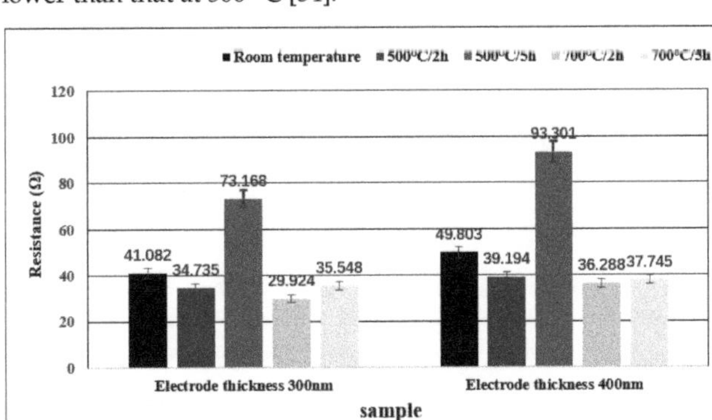

Figure 7. Histogram depicting the resistance change of the initial wire bonding sample at room temperature and after annealing.

4. Conclusions

The present work mainly emphasizes the thermoelectric reliability of a gold wire ball welding system based on a silicon substrate. Samples of Cr/Au electrodes with electrode thickness of 50/300 nm and 50/400 nm were annealed at high temperatures of 500 °C and 700 °C for 2 h and 5 h, respectively, to simulate the high temperature and harsh environment in practical applications. The results showed that after high-temperature annealing, the gap between the gold ball and the electrode bonding interface disappears, and high-temperature heat treatment has a positive effect on the bonding interface. However, as the annealing temperature and time increase, some pores will appear, accompanied by recrystallization and grain growth, affecting the electrical and mechanical properties of the sample.

In a comparison of the annealed samples after the tensile destructive test and the samples at room temperature, it is found that the critical value of tensile force is significantly reduced, indicating that high-temperature heat treatment significantly reduces the mechanical strength of the samples. The resistance values of annealed samples were measured with a probe station. It was found that when the samples were treated at a high temperature for a short time, the resistance values decreased due to the thermal volatilization of gold. However, when the samples were annealed at 500 °C for a long time, the resistance of the samples increased due to the diffusion of Cr and O atoms and the growth of grains. As the activation free energy of gold decreases at 700 °C and heat treatment eliminates metal dislocation and stress, the resistance value of samples after heat treatment for a long time at 700 °C decreases significantly below the initial value. Conclusively, the present work can provide the basis for the selection of thin metal layers for gold wire ball bonding and improve the reliability of application in high temperature and harsh environments to promote its application in the field of high-temperature pressure sensors or high-temperature sensors.

Author Contributions: Writing—review and editing, writing—original draft, C.W.; conceptualization, project administration and methodology, J.W.; investigation and validation, X.L.; funding acquisition, supervision, M.L.; software, Z.Z.; data curation, Y.Q. All authors have read and agreed to the published version of the manuscript.

Funding: This research was funded by the "173" Projects of China (Grant No. 2021JCJQJJ0172, 2020JCJQZD043).

Data Availability Statement: Not applicable.

Acknowledgments: The authors would like to appreciate the facility assistance from Suzhou Institute of Nano-Tech and Nano-Bionics, Chinese Academy of Sciences.

Conflicts of Interest: The authors declare no conflict of interest.

References

1. Zhan, Y.; Xue, S.; Yang, Q.; Xiang, S.; He, H.; Zhu, R. A novel fiber Bragg grating high-temperature sensor. *Opt.-Int. J. Light Electron Opt.* **2008**, *119*, 535–539. [CrossRef]
2. Zhao, Y.; Zhao, L.; Jiang, Z. A novel high temperature pressure sensor on the basis of SOI layers. *Sens. Actuators A Phys.* **2003**, *108*, 108–111. [CrossRef]
3. Jin, X.; Yu, L.; Garcia, D.; Ren, R.X.; Zeng, X. Ionic liquid high-temperature gas sensor array. *Anal. Chem.* **2006**, *78*, 6980–6989. [CrossRef]
4. Lu, C.C.; Liao, K.H.; Udrea, F.; Covington, J.A.; Gardner, J.W. Multi-field simulations and characterization of CMOS-MEMS high-temperature smart gas sensors based on SOI technology. *J. Micromech. Microeng.* **2008**, *18*, 075010. [CrossRef]
5. Zhang, Y. Single-crystal sapphire-based optical high-temperature sensor for harsh environments. *Opt. Eng.* **2004**, *43*, 157–164. [CrossRef]
6. Mihailov, S.J. Fiber Bragg Grating Sensors for Harsh Environments. *Sensors* **2012**, *12*, 1898–1918. [CrossRef]
7. Gregory, O.J.; You, T. Ceramic temperature sensors for harsh environments. *IEEE Sens. J.* **2005**, *5*, 833–838. [CrossRef]
8. Jiang, H.; Yang, R.; Tang, X.; Burnett, A.; Lan, X.; Xiao, H.; Dong, J. Multilayer fiber optic sensors for in situ gas monitoring in harsh environments. *Sens. Actuators B Chem.* **2013**, *177*, 205–212. [CrossRef]
9. Pufall, R. Automatic process control of wire bonding. In Proceedings of the IEEE 43rd Electronic Components and Technology Conference (ECTC '93), Orlando, FL, USA, 1–4 June 1993.

10. Werner, M.R.; Fahrner, W.R. Review on materials, microsensors, systems and devices for high-temperature and harsh-environment applications. *IEEE Trans. Ind. Electron.* **2001**, *48*, 249–257. [CrossRef]
11. Manikam, V.R.; Cheong, K.Y. Die Attach Materials for High Temperature Applications: A Review. *IEEE Trans. Compon. Packag. Manuf. Technol.* **2011**, *1*, 457–478. [CrossRef]
12. Stephan, D.; Chew, Y.H.; Goh, H.M.; Pasamanero, E.; Theint EP, P.; Calpito DR, M.; Ling, J. A Comparison Study of the Bondability and Reliability Performance of Au Bonding Wires with Different Dopant Levels. In Proceedings of the 2007 9th Electronics Packaging Technology Conference, Singapore, 10–12 December 2007.
13. März, B.; Graff, A.; Klengel, R.; Petzold, M. Interface microstructure effects in Au thermosonic ball bonding contacts by high reliability wire materials. *Microelectron. Reliab.* **2014**, *54*, 2000–2005. [CrossRef]
14. Mertens, J.; Finot, E.; Thundat, T.; Fabre, A.; Nadal, M.H.; Eyraud, V.; Bourillot, E. Effects of temperature and pressure on microcantilever resonance response. *Ultramicroscopy* **2003**, *97*, 119–126. [CrossRef]
15. Santos, D.F.; Guerreiro, A.; Baptista, J.M. Simultaneous Plasmonic Measurement of Refractive Index and Temperature Based on a D-Type Fiber Sensor with Gold Wires. *IEEE Sens. J.* **2017**, *17*, 2439–2446. [CrossRef]
16. Hsueh, H.-W.; Hung, F.-Y.; Lui, T.-S.; Chen, L.-H. Microstructure, electric flame-off characteristics and tensile properties of silver bonding wires. *Microelectron. Reliab.* **2011**, *51*, 2243–2249. [CrossRef]
17. Srikanth, N.; Murali, S.; Wong, Y.; Vath, C.J. Critical study of thermosonic copper ball bonding. *Thin Solid Film.* **2004**, *462*, 339–345. [CrossRef]
18. Gonzalez, B.; Knecht, S.; Handy, H.; Ramirez, J. The effect of ultrasonic frequency on fine pitch aluminum wedge wirebond. In Proceedings of the 1996 Proceedings 46th Electronic Components and Technology Conference, Orlando, FL, USA, 28–31 May 1996.
19. Zulkifli, M.N.; Harun, F.; Jalar, A. Effect of surface roughness and hardness of leadframe on the bondability of gold wedge bonds. *Microelectron. Int.* **2019**, *36*, 62–67. [CrossRef]
20. Lum, I.; Mayer, M.; Zhou, Y. Footprint study of ultrasonic wedge-bonding with aluminum wire on copper substrate. *J. Electron. Mater.* **2006**, *35*, 433–442. [CrossRef]
21. Hang, C.; Song, W.; Lum, I.; Mayer, M.; Zhou, Y.; Wang, C.; Moon, J.; Persic, J. Effect of electronic flame off parameters on copper bonding wire: Free-air ball deformability, heat affected zone length, heat affected zone breaking force. *Microelectron. Eng.* **2009**, *86*, 2094–2103. [CrossRef]
22. Lee, J.; Mayer, M.; Zhou, N.; Persic, J. Microelectronic Wire Bonding with Insulated Au Wire: Effects of Process Parameters on Insulation Removal and Crescent Bonding. *Mater. Trans.* **2008**, *49*, 2347–2353. [CrossRef]
23. Xiao, L.; Zhao, Z.; Du, L.; Wu, S.; Liu, Q. Annealing effect on the stability of platinum thin films covered by SiO_2 or SiN_x layer. In Proceedings of the 8th Annual IEEE International Conference on Nano/Micro Engineered and Molecular Systems, Suzhou, China, 7–10 April 2013.
24. Yang, C.-C.; Li, B.; Baumann, F.H.; Li, J.; Edelstein, D.; Rosenberg, R. Microstructure Modulation in Copper Interconnects. *IEEE Electron Device Lett.* **2014**, *35*, 572–574. [CrossRef]
25. Yang, C.-C.; Li, B.; Edelstein, D.; Rosenberg, R. Effects of Metal Capping on Thermal Annealing of Copper Interconnects. *IEEE Electron Device Lett.* **2012**, *33*, 1051–1053. [CrossRef]
26. Trupina, L.; Miclea, C.; Amarande, L.; Cioangher, M. Growth of highly oriented iridium oxide bottom electrode for Pb(Zr,Ti)O3 thin films using titanium oxide seed layer. *J. Mater. Sci.* **2011**, *46*, 6830–6834. [CrossRef]
27. Ruffino, F.; Romano, L.; Pitruzzello, G.; Grimaldi, M.G. High-temperature annealing of thin Au films on Si: Growth of SiO2 nanowires or Au dendritic nanostructures? *Appl. Phys. Lett.* **2012**, *100*, 341. [CrossRef]
28. Duan, C.Y.; Feng, Y.; Zhao, X.X.; Shen, H. Preparation of Self-Assembled Au Nanoparticles Arrays for Silicon Solar Cell Applications. *Adv. Mater. Res.* **2013**, *803*, 257262. [CrossRef]
29. Salas, E.; Riobóo, R.J.J.; Prieto, C.; Every, A. Surface acoustic wave velocity of gold films deposited on silicon substrates at different temperatures. *J. Appl. Phys.* **2011**, *110*, 322. [CrossRef]
30. Purwidyantri, A.; El-Mekki, I.; Lai, C.-S. Tunable Plasmonic SERS "Hotspots" on Au-Film over Nanosphere by Rapid Thermal Annealing. *IEEE Trans. Nanotechnol.* **2017**, *16*, 551–559. [CrossRef]
31. Karakouz, T.; Tesler, A.B.; Sannomiya, T.; Feldman, Y.; Vaskevich, A.; Rubinstein, I. Mechanism of morphology transformation during annealing of nanostructured gold films on glass. *Phys. Chem. Chem. Phys.* **2013**, *15*, 4656–4665. [CrossRef]
32. Brachmann, E.; Seifert, M.; Ernst, D.; Menzel, S.B.; Gemming, T. Pt-wire bonding optimization for electroplated Pt films on γ-Al_2O_3 for high temperature and harsh environment applications. *Sens. Actuators A: Phys.* **2018**, *284*, 129–134. [CrossRef]
33. Lovesey, S.W.; Balcar, E. A Guide to Electronic Multipoles in Photon Scattering and Absorption. *J. Phys. Soc. Jpn.* **2013**, *82*, 021008. [CrossRef]
34. Qian, H.; Qiu, H.; Wu, P.; Chen, X. Effect of annealing on the characteristics of Au layers grown on the high-temperature deposited Ni50Fe50 layers. *Vacuum* **2006**, *80*, 899–903. [CrossRef]

micromachines

Article

Electrical Modeling and Characterization of Graphene-Based On-Chip Spiral Inductors

Da-Wei Wang [1], Meng-Jiao Yuan [1], Jia-Yun Dai [2,*] and Wen-Sheng Zhao [1,*]

[1] Zhejiang Provincial Key Lab of Large-Scale Integrated Circuit Design, School of Electronics and Information, Hangzhou Dianzi University, Hangzhou 310018, China
[2] Science and Technology on Monolithic Integrated Circuits and Modules Laboratory, Nanjing 210016, China
* Correspondence: jydai2016@163.com (J.-Y.D.); wshzhao@hdu.edu.cn (W.-S.Z.)

Abstract: This paper investigates the electrical performance of graphene-based on-chip spiral inductors by virtue of a physics-based equivalent circuit model. The skin and proximity effects, as well as the substrate loss effect, are considered and treated appropriately. The graphene resistance and inductance are combined into the circuit model. It is demonstrated that the electrical characteristics of the on-chip square spiral inductor can be improved by replacing copper with graphene. Moreover, graphene exhibits more effectiveness in improving the inductance in tapered inductors than uniform ones.

Keywords: graphene; on-chip spiral inductor; circuit model; kinetic inductance; quantum resistance

Citation: Wang, D.-W.; Yuan, M.-J.; Dai, J.-Y.; Zhao, W.-S. Electrical Modeling and Characterization of Graphene-Based On-Chip Spiral Inductors. *Micromachines* **2022**, *13*, 1829. https://doi.org/10.3390/mi13111829

Academic Editor: Ha Duong Ngo

Received: 8 October 2022
Accepted: 23 October 2022
Published: 26 October 2022

Publisher's Note: MDPI stays neutral with regard to jurisdictional claims in published maps and institutional affiliations.

Copyright: © 2022 by the authors. Licensee MDPI, Basel, Switzerland. This article is an open access article distributed under the terms and conditions of the Creative Commons Attribution (CC BY) license (https://creativecommons.org/licenses/by/4.0/).

1. Introduction

With the improvement of CMOS technologies, radio-frequency integrated circuits (RF ICs) have become possible and have drawn increasing attention in the past decades. To make RF ICs lightweight, multi-functional, and low-cost, on-chip spiral inductors with high inductance density and low power consumption have been widely utilized in the design of low-noise amplifiers, voltage-controlled oscillators, and filters [1].

In order to improve the inductor performance, it is intuitive to pursue high inductance density and quality factors. There are various factors affecting inductor characteristics, such as substrate resistivity and metal thickness. Increasing metal thickness can improve the quality factor by minimizing Ohmic losses but is counter-productive at high frequencies due to accentuated proximity effect loss [2]. The magnetic field from all turns of a spiral inductor accumulates in the middle area, thereby resulting in severe current crowding in the inner turns. As the current crowding can be mitigated with narrow and widely spaced inner turns, a tapered spiral inductor was proposed in [3] to improve the quality factor without consuming extra area.

To facilitate the miniaturization of RF ICs, the scaling down of on-chip spiral inductors is inevitable [4]. However, there exists an inherent limitation in the scalability of conical spiral inductors, as the inductance value is limited by the laws of electromagnetic induction [5]. This is, the magnetic flux is proportional to the surface area, and the magnetic inductance cannot be scaled to retain desired inductance density.

Carbon nanomaterials, including carbon nanotube (CNT) and graphene, have been proposed as promising alternative candidates for interconnected applications. They have been demonstrated to have unique physical properties such as a long mean free path (MFP), extremely high ampacity, and large thermal conductivity, and they exhibit superior performance and reliability to traditional metal wires [6–9]. More importantly, carbon nanomaterials possess large kinetic inductance, which makes them suitable for building inductors in future scaled RF ICs. It was proven that CNT-based on-chip spiral inductors can provide better quality factors than their Cu counterparts [10]. In comparison with CNT, graphene is more compatible with the traditional CMOS process [11], and there are ways to control graphene chirality and doping levels [12]. Graphene-based on-chip spiral inductors were explored in-depth recently [5,13]. It was experimentally demonstrated that the graphene kinetic inductance is beneficial for improving the inductor performance. Although the graphene-based inductors were investigated numerically in [13], the anomalous skin effect could be neglected as it mainly affects the electrical characteristics beyond 100 GHz, and the modeling methodology could be therefore simplified.

This paper aims to provide a simple circuit model for investigating graphene-based inductors. Firstly, the equivalent circuit model of an on-chip tapered spiral inductor is developed, with the proximity effect and substrate loss taken into account. The quantum contact resistance, scattering resistance, and kinetic inductance of graphene ribbon (GR) are combined into the model to explore the electrical characteristics. By virtue of the circuit model, the electrical performance of graphene-based inductors is investigated, with some guidance provided for future development. The rest of this paper is organized as follows. Section 2 presents the geometry of the on-chip tapered spiral inductor and its physics-based equivalent circuit model. The graphene kinetic inductance is discussed in Section 3, and it is combined into the model to explore the inductor characteristics. The performance analysis of graphene-based spiral inductors is carried out in Section 4. Some conclusions are finally drawn in Section 5.

2. Circuit Model of On-Chip Square Spiral Inductor

On-chip spiral inductors can be divided into rectangular, square, circular, hexagon, and octagonal types according to their geometries. Due to their simple configuration, square spiral inductors are widely used in the design of RF ICs, as shown in Figure 1. In the figure, D_{out} is the outer diameter, and w_1 and s_1 denote the width and spacing of the outermost turn, respectively. For the tapered spiral inductor, the difference in the width between adjacent turns is defined as *taper*, with the pitch kept constant. Accordingly, the ith turn has a width of $w_i = w_1 - (i-1) \cdot taper$, and its spacing with an adjacent inner turn is $s_i = s_1 + (i-1) \cdot taper$.

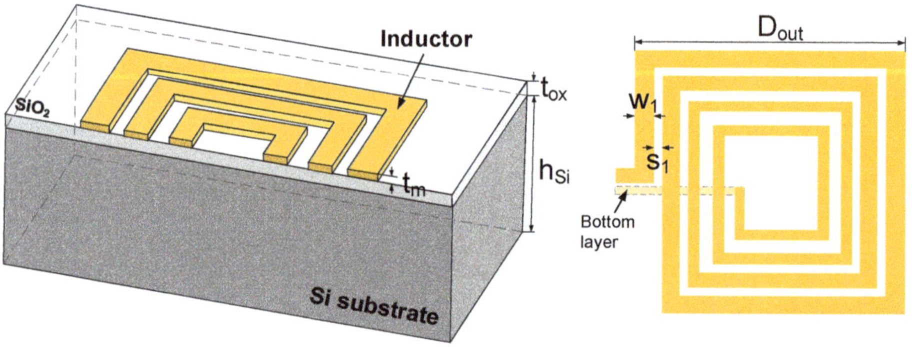

Figure 1. Schematic of on-chip square spiral inductor.

Figure 2 shows the π-equivalent circuit model of the on-chip tapered spiral inductor, which is composed of oxide capacitance C_{ox}, substrate capacitance C_{Si}, and substrate resistance R_{Si} [14]. To obtain the inductance value accurately, the widely applied Greenhouse formulas of a planar spiral inductor are adopted, and the total inductance of the spiral inductor is calculated by

$$L_{\text{dc}} = L_{\text{self}} + \sum M_{+} + \sum M_{-} \quad (1)$$

where L_{self} is the self-inductance of a metal line, and M_{+} and M_{-} are mutual inductances between two lines.

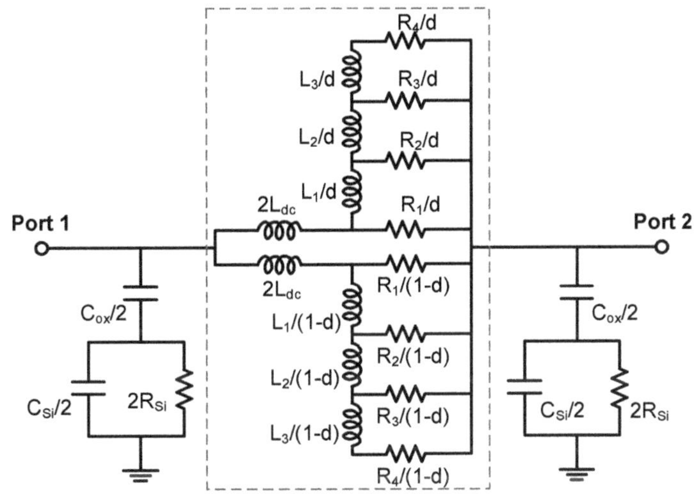

Figure 2. Equivalent circuit model.

The dc resistance of a spiral inductor is calculated by

$$R_{\mathrm{dc}} = \frac{1}{\sigma_m t_m} \sum_{i=1}^{N} \frac{l_i}{w_i} \tag{2}$$

where σ_m is the metal conductivity. To accurately evaluate the inductor characteristics, the frequency-dependent effect should be treated appropriately. A ladder network marked by the dashed box in Figure 2 is added for modeling skin and proximity effects [15]. At low frequencies, the current is uniformly distributed inside the inductor, while the current accumulation occurs at the surface layer of the inductor at high frequencies. The high-frequency resistance R_1 is calculated by

$$R_1 = \frac{1}{2\sigma_m t_m \delta_{\mathrm{max}}} \sum_{i=1}^{N} l_i \tag{3}$$

where δ_{max} denotes the skin depth at the maximum operating frequency f_{max}, and it is given as $\delta_{\mathrm{max}} = 1/\sqrt{\pi\mu\sigma_m f_{\mathrm{max}}}$. The calculations of other resistances and inductances can refer to [15]. Further, 2-branch networks are employed to improve the model accuracy, with the proximity factor defined as [15]

$$d = L_{\mathrm{dc}} \cdot \left[2L_{\mathrm{dc}} - \frac{1}{2I} \sum_{i=1}^{N} w_i l_i (B_{i,i} + B_{i,\mathrm{others}}) \right]^{-1} \tag{4}$$

where I is the excitation current, and $B_{i,i}$ and $B_{i,\ \mathrm{others}}$ represent the magnetic fields due to the ith turn and the other turns except ith turn, respectively. The minimum and maximum values of d are 0.5 and 1, which correspond to the cases of no proximity effect and maximum proximity effect [16].

The oxide and silicon capacitance are calculated by

$$C_{\mathrm{ox}} = \frac{\varepsilon_{\mathrm{ox}}}{t_{\mathrm{ox}}} \sum_{i=1}^{N} w_i l_i \tag{5}$$

$$C_{\mathrm{Si}} = 2\varepsilon_{\mathrm{Si}} \sum_{i=1}^{N} \frac{w_i l_i}{2h_{\mathrm{Si}} + \sqrt{\frac{w_i l_i}{\pi}} - \sqrt{4h_{\mathrm{Si}}^2 + \frac{w_i l_i}{\pi}}} \tag{6}$$

$$R_{\mathrm{Si}} = \varepsilon_{\mathrm{Si}} / (\sigma_{\mathrm{Si}} C_{\mathrm{Si}}) \tag{7}$$

where $\varepsilon_{\mathrm{ox}}$ is the permittivity of the oxide layer, and $\varepsilon_{\mathrm{Si}}$ and σ_{Si} are the permittivity and conductivity of the silicon substrate, respectively.

To evaluate the circuit model, a set of square spiral inductors are simulated using a full-wave electromagnetic simulator, i.e., ANSYS HFSS (2021 version, Ansys, Inc., Canonsburg, PA, USA). In the simulation; the silicon conductivity is 10 S/m, and the geometrical parameters are as follows: $D_{\mathrm{out}} = 200$ μm; $w_1 = 13$ μm; and $s_1 = 7$ μm. For the tapered spiral inductor, $taper = 2$ μm/turn. The effective inductance and quality factor are obtained from the simulated Y-parameters and plotted in Figure 3:

$$Q = \frac{\mathrm{Im}(Y_{11})}{\mathrm{Re}(Y_{11})} \tag{8}$$

$$L_{\mathrm{eff}} = \frac{1}{2\pi f} \mathrm{Im}\left(\frac{1}{Y_{11}}\right) \tag{9}$$

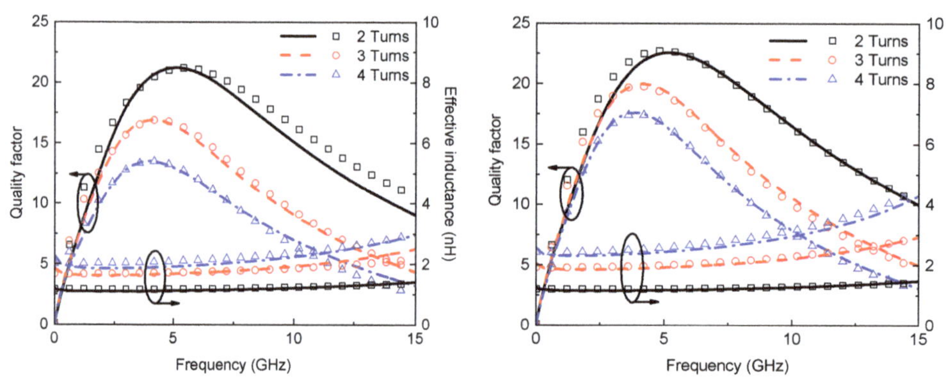

(**a**) (**b**)

Figure 3. Effective inductance and quality factor of on-chip (**a**) uniform and (**b**) tapered spiral inductors (lines: simulation; symbols: model).

It is evident that the results obtained by the circuit model agree well with the simulated results. The uniform spiral inductor possesses smaller dc resistance than the tapered one but has larger high-frequency resistance due to skin and proximity effects. The effective inductance of the spiral inductor can be increased by introducing tapering, and therefore, the quality factor can be improved.

3. Modeling of Graphene-Based Inductors

According to Faraday's law of electromagnetic induction, the current change in the metal turns of a spiral inductor produces time-varying magnetic flux. As the magnetic flux is proportional to the inductor area, the decreased size degrades the inductance density, thereby limiting the scaling of RF ICs. Kinetic inductance, which originates in the kinetic energy required by mobile charge carriers in alternative electromotive force, is usually ignored in conventional metals due to their small relaxation time and large conducting channel number [5]. However, the momentum relaxation time of graphene is on the order of picoseconds, and therefore, graphene possesses large kinetic inductance, which makes it suitable for building on-chip spiral inductors in future scaled RF ICs.

3.1. GR Impedance

Figure 4a shows the structure of a multilayer GR interconnect with side contacts [12]. In the figure, w_g and t_g are the width and thickness, δ is the spacing between adjacent graphene layers, and the number of graphene layers is calculated as $n = 1 + \text{Inter}(t_g/\delta)$, where "Inter(·)" denotes that only the integer part is considered [10]. The corresponding equivalent circuit model of a multilayer GR is depicted in Figure 4b. The number of conducting channels per layer of graphene sheet is given as [7]

$$N_{\text{ch}} = \sum_{i=0}^{n_C} \left(1 + e^{\frac{E_i - E_F}{k_B T}}\right)^{-1} + \sum_{i=0}^{n_V} \left(1 + e^{\frac{E_i - E_F}{k_B T}}\right)^{-1} \tag{10}$$

where the first and second summations on the right-hand side of (10) represent the contributions of the conduction subbands and valence subbands, respectively, k_B is the Boltzmann constant, T is the temperature, E_F is the Fermi energy, and E_i denotes the ith conduction (valence) subbands with the lowest (highest) energy. For GR with $w_g > 10$ nm and $E_F > 0.1$ eV, N_{ch} has a linear relationship with w_g and E_F, i.e., $N_{\text{ch}} = \alpha w_g E_F$, where $\alpha = 1.2\ \text{eV}^{-1} \cdot \text{nm}^{-1}$ is the fitting coefficient [17].

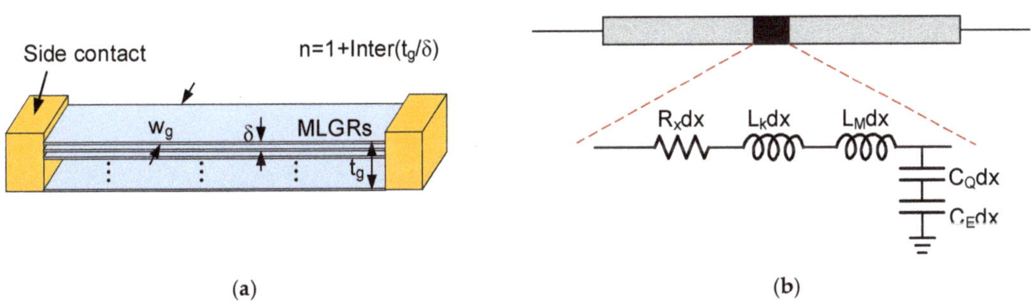

Figure 4. (**a**) Schematic of multilayer GR interconnect and its (**b**) equivalent distributed circuit model.

The kinetic inductance of a single-layer GR is given by [17]

$$L_K \approx \frac{8\ \text{nH}/\mu\text{m}}{N_{\text{ch}}} = \frac{8\ \text{nH}/\mu\text{m}}{\alpha w_g E_F} \tag{11}$$

It is worth noting that graphene tends to graphite as the layer number increases [18]. However, decoupled graphene layers were experimentally demonstrated in [19], and it is expected that a multilayer GR with a certain thickness can be realized in the future. Therefore, the multilayer GR is regarded as a stack of single-layer GRs in this study, and its inductance is calculated as $L_K \approx 8\ \text{nH}/\mu\text{m}/(n\alpha w_g E_F)$. Figure 5 shows the kinetic inductances of single- and multilayer GRs. It is evident that the GR inductance decreases with E_F, and single-layer GR exhibits much larger kinetic inductance than multilayer GR. However, the single-layer GR is not suitable for building inductors due to its ultrahigh resistive loss.

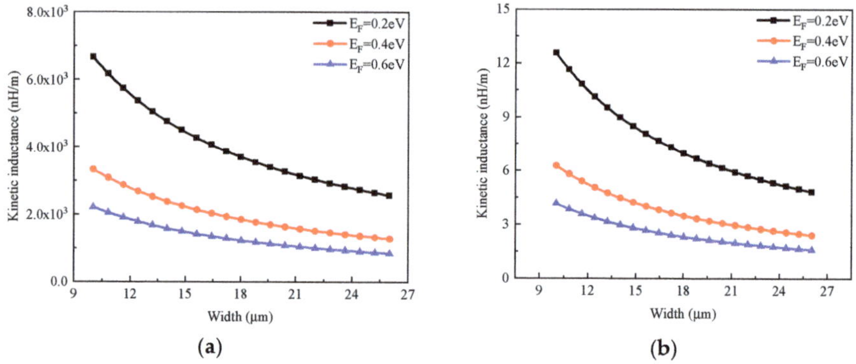

Figure 5. Kinetic inductances of (**a**) single-layer and (**b**) multilayer GRs.

As shown in Figure 6a, the kinetic inductance of multilayer GR is in inverse proportion to the layer number and thereby decreases with increasing thickness. For building an on-chip spiral inductor, the GR length is usually larger than the effective MFP λ_{eff}, and therefore, the scattering resistance of GR can be approximated as [17]

$$R_S \approx \frac{12.9 \text{ k}\Omega}{nN_{\text{ch}}\lambda_{\text{eff}}} = \frac{12.9 \text{ k}\Omega}{n\alpha w_g E_F \lambda_{\text{eff}}} \tag{12}$$

where λ_{eff} is related to various scattering mechanisms, and it has a dominating effect in determining the signal transmission performance [7]. As shown in Figure 6b, the GR resistance decreases with the thickness and Fermi level. To improve the graphene conduction, it is necessary to increase the Fermi level and MFP by appropriate doping techniques [20]. It is worth noting that the fabricated GRs usually cannot come up to theoretical predictions due to manufacturing errors and defects.

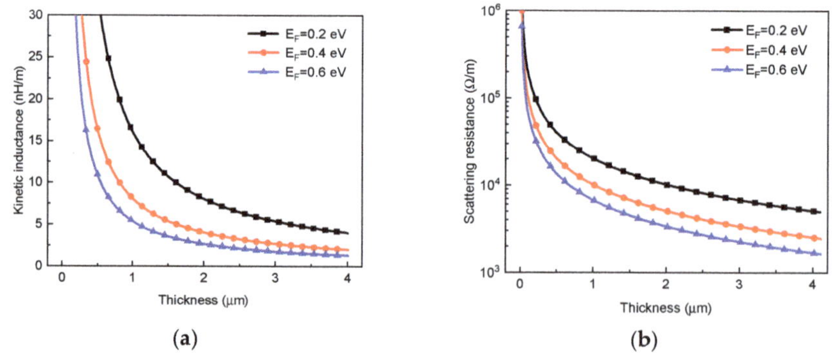

Figure 6. (**a**) Kinetic inductances and (**b**) resistance of GR interconnect versus thickness for different Fermi levels.

3.2. Spiral Inductors

Although an anomalous skin effect exists in GR due to the large ratio of in-plane to out-plane MFP, it mainly appears as the operating frequency exceeds several tens of gigahertz [13]. Therefore, the anomalous skin effect can be neglected, as it has little influence on the inductor characteristics in the frequency range of interests of this study. Considering the kinetic inductance, the dc inductance of the graphene-based spiral inductor is given by

$$L_{\text{dc}} = L_{\text{self}} + \sum M_+ + \sum M_- + L_{K,t} \tag{13}$$

where $L_{K,t}$ represents the total kinetic inductance, and it can be calculated by

$$L_{K,t} = \left(\frac{1}{n\alpha E_F} \sum_{i=1}^{N} \frac{l_i}{w_{\text{eff},i}} \right) \times 8\text{nH}/\mu\text{m} \tag{14}$$

The effective width $w_{\text{eff},\,i}$ is equal to two times the skin depth [5]. The effective conductivity of multilayer GR is given by

$$\sigma_{\text{eff}} = \frac{1}{R_S w t} = \frac{n\alpha E_F \lambda_{\text{eff}}}{t} \times \frac{1}{12.9\ \text{k}\Omega} \tag{15}$$

Note that the width in both the numerator and denominator cancel out, and the effective conductivity would be irrelevant to the width. Therefore, the conductivity can be treated as a constant for a specific GR line in the modeling of graphene-based inductors. By utilizing the above equations and circuit model shown in Figure 2, the electrical characteristics of graphene-based on-chip spiral inductors can be investigated. The quantum contact resistances are incorporated into the ends of the model, but it has little influence on the inductor performance. The metallic vias in graphene-based inductors could be made of Co [12]. Note that the modeling methodology can also be applied to other spiral inductor structures, such as circular, hexagon, and octagonal types. To verify the model, the quality factor of the graphene-based spiral inductor is obtained and compared with the experimental results in [5]. The geometrical parameters are as follows: $D_{out} = 200$ µm, $w = 25$ µm, and $s = 5$ µm. It can be seen from Figure 7 that the modeling results basically agree with the experimental results, and the deviation may be attributed to incorrect geometrical parameters, which will be investigated in the next study.

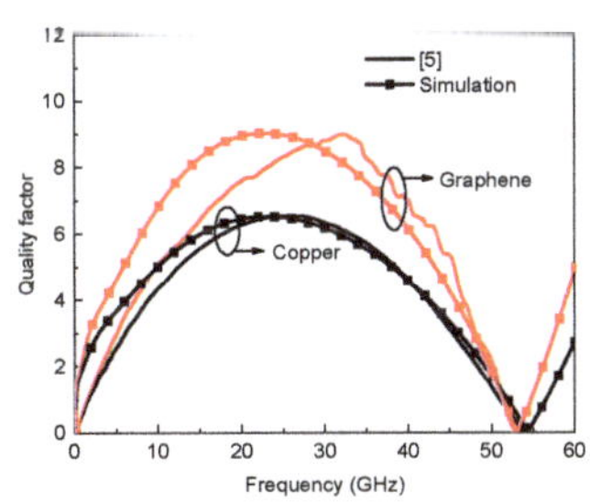

Figure 7. Quality factors of Cu- and graphene-based inductors.

4. Results and Discussion

For graphene-based on-chip spiral inductors, the inductance mainly comes from the magnetic inductance and the kinetic inductance, which originates in the kinetic energy required by mobile electrons, and its value depends on the number of conduction channels. In order to characterize the influence of kinetic inductance, the inductances of copper and graphene-based on-chip spiral inductors made of copper and graphene are plotted in Figure 8. It is evident that graphene becomes superior to copper for building on-chip inductors with a decreasing geometric size (e.g., thickness and width), implying that graphene is more suitable for the applications of future scaled RF ICs. Moreover, as shown in Figure 8b, the advantage of a graphene-based inductor over its copper counterpart can be strengthened by increasing *taper*. This is mainly because the inner ring of a graphene-based tapered spiral inductor could provide larger kinetic inductance due to its smaller width than that of a uniform inductor.

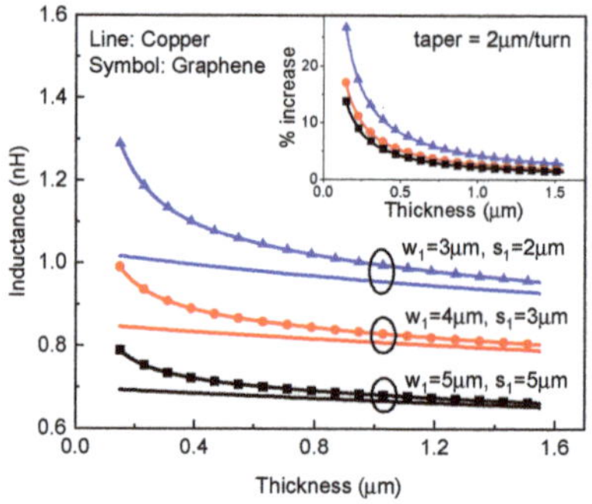

(a)

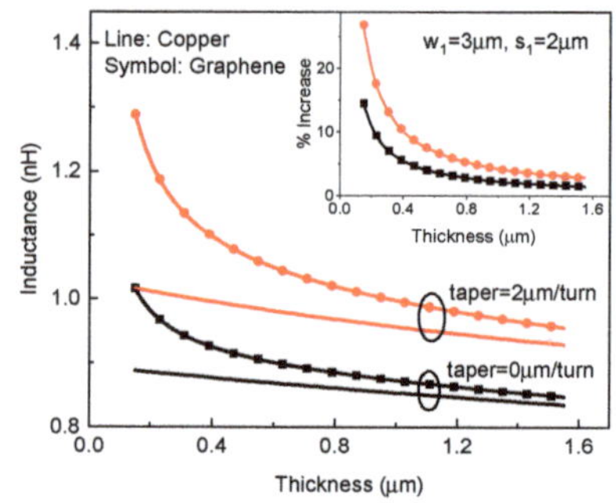

(b)

Figure 8. Inductance versus thickness for copper and graphene-based on-chip spiral inductors with different (**a**) width, spacing, and (**b**) taper. The inset plots the percentage increase in inductance of graphene inductor to copper inductor.

In general, graphene doping can increase the Fermi level and thereby decrease the resistive loss. Moreover, the coupling effect between adjacent graphene layers can be alleviated and finally cancelled by intercalation doping. Therefore, alternate dopants such as AsF_5, Br_2, $FeCl_3$, and KI are continually being investigated [21,22]. Among these dopants, Br is easy to diffuse into graphene layers, and its doping process is relatively simple and efficient. The average thickness increment of Br-doped multilayer graphene is about 6.7% of the original thickness [5], i.e., the average layer spacing between adjacent graphene layers is about 0.3628 nm. As doping is a process of charge transfer, the carrier density and conductivity vary with the doping time. As the doping time exceeds 70 min, the resistivity of Br_2-doped graphene becomes lower than that of bulk copper, and the Fermi level reaches 0.5 eV [23]. Here, two cases of Br_2-doped graphene are considered, i.e., the conductivities are set as 5×10^7 S/m and 10^8 S/m, respectively.

By virtue of the circuit model in Figure 2, the effective inductance and quality factor of on-chip tapered spiral inductors made of copper and Br_2-doped graphene are plotted in Figure 9. It is evident that the effective inductance can be improved by replacing copper with graphene. For graphene with a conductivity of 5×10^7 S/m, the quality factor of a graphene-based spiral inductor is slightly lower than that of its copper counterpart. However, as the graphene conductivity exceeds 10^8 S/m, the quality factor can be significantly improved due to the reduction of metal resistive loss. Figure 10 shows the effective inductance and quality factor of an on-chip spiral inductor with different values of $taper$. It can be seen from Figure 10a that the effective inductance can be increased by increasing $taper$. However, as the inductor characteristics are affected by various factors, such as substrate loss, the quality factor of a tapered spiral inductor is slightly larger than that of a uniform one. Further, the frequency-dependent impedances and scattering parameters of copper and graphene-based on-chip uniform spiral inductors are plotted in Figure 11. It is evident that GR could provide smaller resistance and a slightly larger inductance than its copper counterpart, but their scattering parameters are comparable due to other influences, such as the substrate loss effect.

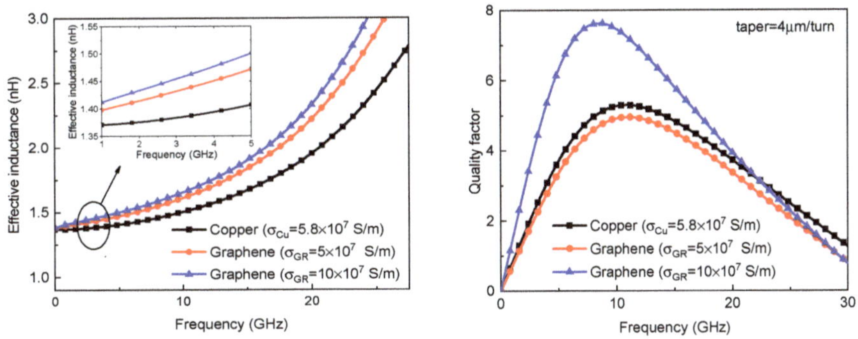

(a) (b)

Figure 9. (**a**) Effective inductance and (**b**) quality factors of on-chip tapered spiral inductors made of copper and graphene.

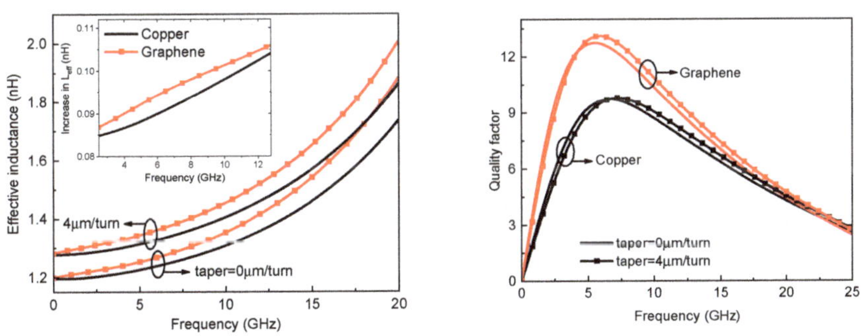

(a) (b)

Figure 10. (**a**) Effective inductance and (**b**) quality factor of on-chip uniform and tapered spiral inductors.

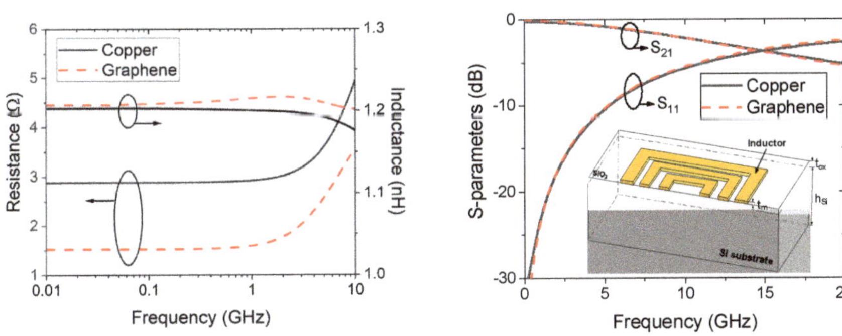

(a) (b)

Figure 11. (**a**) Impedances (the model in the dashed box of Figure 2 and (**b**) scattering parameters of on-chip uniform spiral inductors.

Further, the parametric study of graphene-based on-chip spiral tapered inductors was conducted. The influence of thickness on the inductor characteristics is shown in Figure 12a. With the increasing thickness, the effective inductance decreases, while the quality factor increases significantly due to the reduction of metal resistive loss. As shown in Figure 12b, the increase in turn number increases the effective inductance and decreases the quality factor. Similarly, as shown in Figure 12c, the increasing outermost diameter increases the effective inductance. However, the magnetic field penetrating the inductor coil generates current in the substrate, thereby leading to an increase in eddy current loss [24]. The larger the outermost diameter, the higher the eddy current loss. Thus, due to the increase in both effective inductance and eddy current loss, the maximum quality factor is slightly changed with the outermost diameter. Moreover, the self-resonant frequency decreases dramatically with the increasing outermost diameter. Figure 12d shows the characteristics of graphene-based inductors with different widths and spacings. With a decreasing width, the magnetic inductance increases, and the resistive loss is reduced due to the suppressed proximity effect, thereby increasing the maximum quality factor. The influence of oxide layer thickness on the quality factor of a graphene-based on-chip spiral inductor is explored, as shown in Figure 13a. The thicker the oxide layer is, the greater the quality factor is. The self-resonant frequency increases with the increasing oxide layer thickness. Moreover, as shown in Figure 13b, the quality factor increases with the decreasing substrate conductivity due to the reduced substrate loss. By virtue of the circuit model, some guidance is given for the design and fabrication of graphene-based inductors, which is anticipated to be validated in a future experimental study. It is worth noting that although the simulation could give insight into the device's operation, real-world implementation is inevitable, as the simulation might be too idealistic. Moreover, much effort should be spent in all fields related to the fabrication of graphene-based inductors, including graphene growth and doping.

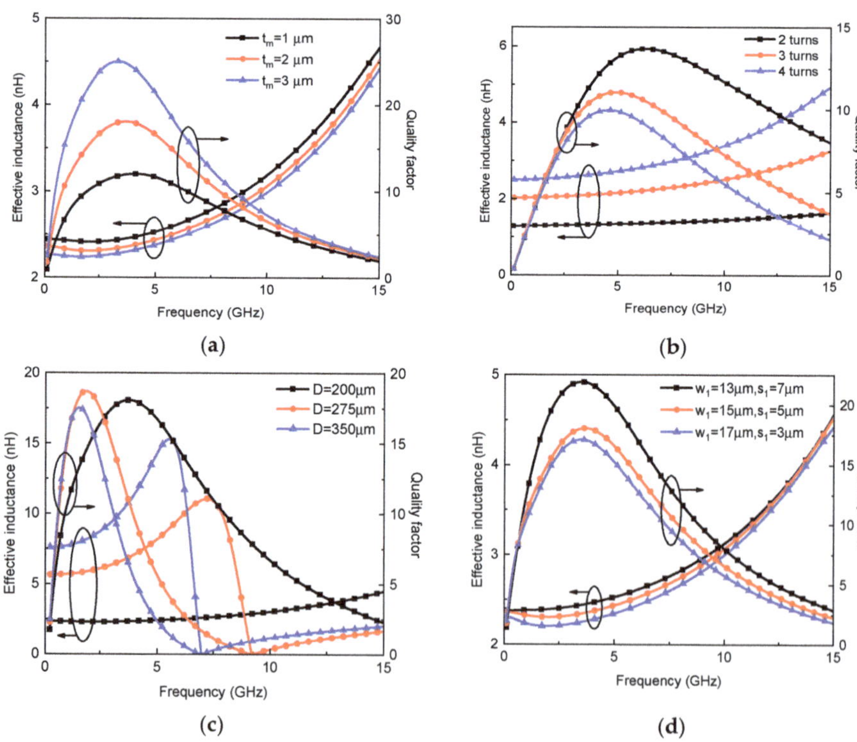

Figure 12. Effective inductance and quality factor of graphene-based on-chip tapered spiral inductors with different (**a**) thicknesses, (**b**) turn numbers, (**c**) outermost diameter, (**d**) width and spacing.

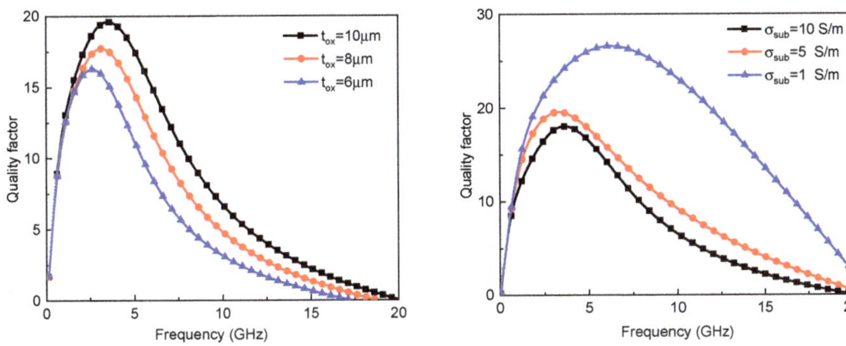

(a) (b)

Figure 13. Quality factor of graphene-based on-chip tapered spiral inductors with different (**a**) oxide layer thicknesses and (**b**) substrate conductivities.

In summary, the graphene-based on-chip square spiral inductors are investigated by virtue of the circuit model. Although graphene possesses large kinetic inductance, its advantage over copper becomes significant only when the inductor dimensions are smaller than a certain value. This is, the area of the nanoscale inductor can be reduced by replacing copper with GR. On the other hand, the conductivity can be increased by employing doping techniques, thereby improving the inductor quality factor. However, the simulation is realistic, and much effort should be devoted to the fabrication of high-quality GRs.

5. Conclusions

In this paper, the ultimate electrical performance of graphene-based spiral inductors was investigated theoretically. The equivalent circuit model of a traditional on-chip spiral inductor was developed and verified. The graphene kinetic inductance was combined into the model, and the electrical characteristics of on-chip spiral inductors made of copper and graphene were captured and compared. In the modeling procedure, it was found that the quantum contact resistance has little influence on the inductor performance. Graphene could be superior to its copper counterpart for building inductors when the geometry is very small, indicating that graphene is more suitable for future nanoscale RFICs. Moreover, a tapered spiral inductor could better play the advantage of graphene than a uniform one, as the inner ring has larger kinetic inductance. Although graphene possesses high kinetic inductance, the quality factor of the graphene-based inductor may be limited by its resistive loss, and therefore, the intercalation doping technique should be pursued. By virtue of the circuit model, the influences of geometrical parameters on the performance of graphene-based on-chip spiral inductors were finally explored, with several design guidelines provided.

Author Contributions: Conceptualization, W.-S.Z.; methodology, D.-W.W., writing—original draft preparation, M.-J.Y.; writing—review and editing, J.-Y.D. and W.-S.Z. All authors have read and agreed to the published version of the manuscript.

Funding: This research was funded by the Natural Science Foundation of China (NSFC) under Grants 62222401 and 61874038 and by the Zhejiang Provincial Natural Science Foundation under Grants LXR22F040001 and LD22F040003.

Data Availability Statement: Not applicable.

Conflicts of Interest: The authors declare no conflict of interest.

References

1. Yue, C.P.; Wong, S.S. On-chip spiral inductors with patterned ground shields for Si-based RF ICs. *IEEE J. Solid-State Circuits* **1998**, *33*, 743–752. [CrossRef]
2. Vanukuru, V.N.R. High *Q* inductors utilizing thick metals and dense-tapered spirals. *IEEE Trans. Electron Devices* **2015**, *62*, 3095–3099. [CrossRef]
3. Lopez-Villegas, J.M.; Samitier, J.; Cané, C.; Losantos, P.; Bausells, J. Improvement of the quality factor of RF integrated inductors by layout optimization. *IEEE Trans. Microw. Theory Techn.* **2000**, *48*, 76–83. [CrossRef]
4. Yue, C.P.; Wong, S.S. Scalability of RF CMOS. In Proceedings of the 2005 IEEE Radio Frequency Integrated Circuits (RFIC) Symposium, Long Beach, CA, USA, 12–14 June 2005.
5. Kang, J.; Matsumoto, Y.; Li, X.; Jiang, J.; Xie, X.; Kawamoto, K.; Kenmoku, M.; Chu, J.H.; Liu, W.; Mao, J.; et al. On-chip intercalated-graphene inductors for next-generation radio frequency electronics. *Nat. Electron.* **2018**, *1*, 46–51. [CrossRef]
6. Chiariello, A.G.; Maffucci, A.; Miano, G. Circuit models of carbon-based interconnects for nanopackaging. *IEEE Trans. Compon. Packag. Manufac. Technol.* **2013**, *3*, 1926–1938. [CrossRef]
7. Naeemi, A.; Meindl, J.D. Compact physics-based circuit models for graphene nanoribbon interconnects. *IEEE Trans. Electron Devices* **2009**, *56*, 1822–1833. [CrossRef]
8. Kumar, A.; Kumar, V.R.; Kaushik, B.K. Transient analysis of crosstalk induced effects in mixed CNT bundle interconnects using FDTD technique. *IEEE Trans. Electromagn. Compat.* **2019**, *61*, 1621–1629. [CrossRef]
9. Kumar, A.; Kaushik, B.K. Edge-roughness aware EM-RA model for signal integrity analysis of MLGNR interconnects. *IEEE Trans. Compon. Packag. Manuf. Technol.* **2021**, *11*, 273–283. [CrossRef]
10. Li, H.; Banerjee, K. High-frequency analysis of carbon nanotube interconnects and implications for on-chip inductor design. *IEEE Trans. Electron Devices* **2009**, *56*, 2202–2214. [CrossRef]
11. Nishad, V.K.; Nishad, A.K.; Kaushik, B.K.; Sharma, R. First-principle analysis of transition metal edge-passivated armchair graphene nanoribbons for nanoscale interconnects. *IEEE Trans. Nanotechnol.* **2021**, *20*, 92–98. [CrossRef]
12. Agashiwala, K.; Jiang, J.; Parto, K.; Zhang, D.; Yeh, C.-H.; Banerjee, K. Demonstration of CMOS-compatible multi-level graphene interconnects with metal vias. *IEEE Trans. Electron Devices* **2021**, *68*, 2083–2091. [CrossRef]
13. Sarkar, D.; Xu, C.; Li, H.; Banerjee, K. High-frequency behavior of graphene-based interconnects—Part II: Impedance analysis and implications for inductor design. *IEEE Trans. Electron Devices* **2011**, *58*, 853–859. [CrossRef]
14. Sathyasree, J.; Vanukuru, V.; Nair, D.R.; Chakravorty, A. A substrate model for on-chip tapered spiral inductors with forward and reverse excitations. *IEEE Trans. Electron Devices* **2019**, *66*, 802–805. [CrossRef]
15. Sathyasree, J.; Vanukuru, V.; Nair, D.; Chakravorty, A. Compact modeling of proximity effect in high-*Q* tapered spiral inductors. *IEEE Electron Device Lett.* **2018**, *39*, 588–590. [CrossRef]
16. Jayaraman, S.S.; Vanukuru, V.; Nair, D.; Chakravorty, A. A scalable, broadband, and physics-based model for on-chip rectangular spiral inductors. *IEEE Trans. Magn.* **2019**, *55*, 1–6. [CrossRef]
17. Zhao, W.-S.; Yin, W.-Y. Comparative study on multilayer graphene nanoribbon (MLGNR) interconnects. *IEEE Trans. Electrom. Compat.* **2014**, *56*, 638–645. [CrossRef]
18. Wang, Z.; Zhang, Q.; Peng, P.; Tian, Z.; Ren, L.; Zhang, X.; Huang, R.; Wen, J.; Fu, Y. Q-factor of CVD monolayer graphene and graphite inductors. *J. Phys. D Appl. Phys.* **2017**, *50*, 345103. [CrossRef]
19. Reina, A.; Jia, X.; Ho, J.; Nezich, D.; Son, H.; Bulovic, V.; Dresselhaus, M.S.; Kong, J. Large area, few-layer graphene films on arbitrary substrates by chemical vapor deposition. *Nano Lett.* **2008**, *9*, 30–35. [CrossRef]
20. Jiang, J.; Kang, J.; Cao, W.; Xie, X.; Zhang, H.; Chu, J.H.; Liu, W.; Banerjee, K. Intercalation doped multi-layer-graphene-nanoribbons for next-generation interconnects. *Nano Lett.* **2017**, *17*, 1482–1488. [CrossRef] [PubMed]
21. Das, S.; Bhattacharya, S.; Das, D.; Rahaman, H. Modeling and analysis of electro-thermal impact of crosstalk induced gate oxide reliability in pristine and intercalation doped MLGNR interconnects. *IEEE Trans. Device Mater. Reliab.* **2019**, *19*, 543–550. [CrossRef]
22. Liu, W.; Kang, J.; Banerjee, K. Characterization of $FeCl_3$ intercalation doped cvd few-layer graphene. *IEEE Electron Device Lett.* **2016**, *37*, 1246–1249. [CrossRef]
23. Tongay, S.; Hwang, J.; Tanner, D.B.; Pal, H.K.; Maslov, D.; Hebard, A.F. Supermetallic conductivity in bromine-intercalated graphite. *Phys. Rev. B* **2010**, *81*, 115428. [CrossRef]
24. Ibrahim, N.M.; Kuhn, W.B. An approach for the calculation of magnetic field within square spiral inductors at low frequency. *Int. J. Num. Model. Electron. Net. Devices Fields* **2022**, *15*, 339–354. [CrossRef]

micromachines

MDPI

Article

A Novel Interposer Channel Structure with Vertical Tabbed Vias to Reduce Far-End Crosstalk for Next-Generation High-Bandwidth Memory

Hyunwoong Kim [1], Seonghi Lee [1], Kyunghwan Song [1], Yujun Shin [1], Dongyrul Park [1], Jongcheol Park [2], Jaeyong Cho [3] and Seungyoung Ahn [1,*]

1 Korea Advanced Institute of Science and Technology, Cho Chun Shik Graduate School of Mobility, Deajeon 34051, Korea; khw3399@kaist.ac.kr (H.K.); ijhu12@kaist.ac.kr (S.L.); kyunghwan.song@kaist.ac.kr (K.S.); yujun.shin@kaist.ac.kr (Y.S.); dongryulpark@kaist.ac.kr (D.P.)
2 Department of System IC Development, National NanoFab Center, Daejeon 34141, Korea; jcpark@nnfc.re.kr
3 Huwin, Seongnam 13558, Korea; jycho@huwin.co.kr
* Correspondence: sahn@kaist.ac.kr; Tel.: +82-42-350-1263

Abstract: In this paper, we propose and analyze a novel interposer channel structure with vertical tabbed vias to achieve high-speed signaling and low-power consumption in high-bandwidth memory (HBM). An analytical model of the self- and mutual capacitance of the proposed interposer channel is suggested and verified based on a 3D electromagnetic (EM) simulation. We thoroughly analyzed the electrical characteristics of the novel interposer channel considering various design parameters, such as the height and pitch of the vertical tabbed via and the gap of the vertical channel. Based on the frequency-dependent lumped circuit resistance, inductance, and capacitance, we analyzed the channel characteristics of the proposed interposer channel. In terms of impedance, insertion loss, and far-end crosstalk, we analyzed how much the proposed interposer channel improved the signal integrity characteristics compared to a conventional structure consisting of micro-strip and strip lines together. Compared to the conventional worst case, which is the strip line, the eye-width, the eye-height, and eye-jitter of the proposed interposer channel were improved by 17.6%, 29%, and 9.56%, respectively, at 8 Gbps. The proposed interposer channel can reduce dynamic power consumption by about 28% compared with the conventional interposer channel by minimizing the self-capacitance of the off-chip channel.

Keywords: broadside structure; far-end crosstalk; impedance; interposer channel; silicon interposer; vertical tabbed via

Citation: Kim, H.; Lee, S.; Song, K.; Shin, Y.; Park, D.; Park, J.; Cho, J.; Ahn, S. A Novel Interposer Channel Structure with Vertical Tabbed Vias to Reduce Far-End Crosstalk for Next-Generation High-Bandwidth Memory. *Micromachines* **2022**, *13*, 1070. https://doi.org/10.3390/mi13071070

Academic Editor: Wensheng Zhao

Received: 13 June 2022
Accepted: 3 July 2022
Published: 5 July 2022

Publisher's Note: MDPI stays neutral with regard to jurisdictional claims in published maps and institutional affiliations.

Copyright: © 2022 by the authors. Licensee MDPI, Basel, Switzerland. This article is an open access article distributed under the terms and conditions of the Creative Commons Attribution (CC BY) license (https://creativecommons.org/licenses/by/4.0/).

1. Introduction

Recently, memory usage has steadily increased as technology has developed in the era of the fourth industrial revolution, with artificial intelligence, big data centers, robots, autonomous vehicles, augmented reality (AR), and virtual reality (VR). Representative memories are double data rate (DDR), graphics double data rate (GDDR), and high-bandwidth memory (HBM). In order to store and process these large amounts of data, demand for high-bandwidth memory has been rapidly increasing. HBM is the essential memory structure for industries in the fourth industrial revolution because it can implement high bandwidth more effectively than any other memory structure. HBM has a 3D stacked structure based on through-silicon vias (TSVs). The development of silicon interposer interconnection technologies, such as interposer channel and TSV, are essential for the development of next-generation HBM and chip-to-chip interfaces [1,2]. HBM is designed as a parallel interface structure with 1024 IOs [3].

To meet the needs of the next generation of HBMs, it is essential to address the issue of signal integrity. Signal integrity needs to be improved to resolve issues such as channel loss, crosstalk, impedance mismatching, and inter-symbol interference (ISI).

The HBM interconnection consists of the TSV, interposer channel, pad, package, etc. Because the interposer channel has the longest physical channel length in the off-chip interface, it must be considered and analyzed in terms of signaling within the target frequency band. Various channel components such as the interposer channel effectively have low-pass filter characteristics, so signal degradation occurs as the frequency increases.

The conventional HBM interposer channel has three problems, as shown in Figure 1. Firstly, the HBM interposer channel has a lower impedance than the transmitter (TX) and receiver (RX) driver. In general, although the dielectric can be manufactured as thin as about 1 μm, it is possible to manufacture a relatively wide channel with a width of at least 2 μm or more due to the limitations of the semiconductor manufacturing process. Since the channel is relatively wide, the impedance formed is inevitably very small. In [4], in order to increase the impedance of the interposer channel, the ground slot under the signal channel was presented. Secondly, from the paper [5], the power consumption of an off-chip interconnection was found to be generally 100 times larger than that on-chip. Thus, when designing an off-chip interconnection, it is essential to consider the low power consumption. The third HBM interface problem is the very high channel density, due to the 1024 IO channels. If the clock and address line are combined, a very large number of channels need to be placed. There is not enough free space. Therefore, a solution to the HBM interposer channel structure is required. The new interposer channel solution should alleviate the influence of impedance discontinuity and minimize the issue of off-chip power consumption, while not causing a routing issue.

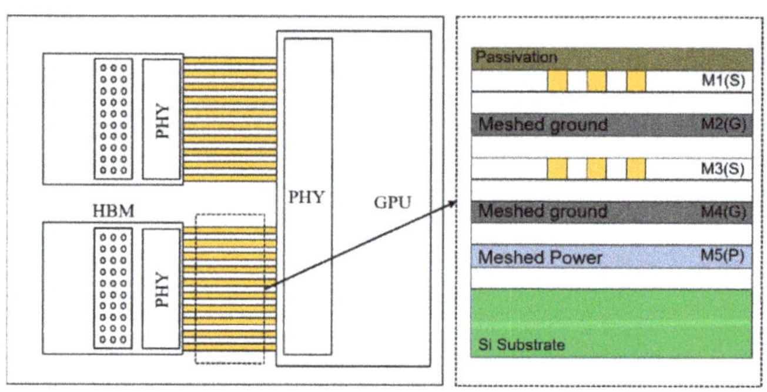

Figure 1. Conventional HBM interposer channel.

Regarding signal integrity, previous research on the HBM interface has been carried out in various ways, which can be divided into several categories [4,6–10]. Refs. [4,6] present the signal integrity design method and the analyses of conventional interposer channels such as the micro-strip and strip line. There has been an insufficient number of suggestions for methods to improve the signal integrity characteristics of the interposer channel itself. Refs. [7,8] are related studies, proposing and analyzing passive equalizers. The passive equalizer can minimize ISI and flatten channel loss by intentionally leaking a low-frequency band signal. However, it requires additional chip area. Because the HBM interposer channel is already very dense, it is not realistic for application to the HBM. Refs. [9,10] report the on-chip solutions. The representative on-chip solutions are the crosstalk reduction circuit method and data bus inversion (DBI) coding. These methods can effectively minimize the influence of far-end crosstalk noise. However, these methods require an additional circuit and can cause additional power consumption. Moreover, it is not possible to fundamentally improve the cause of channel loss in the case of impedance mismatching, reflection, etc. With DBI coding, a DBI channel is required, so issues with space also occur. Ref. [11] studies the structure of inserting the shielding channel in the interposer. Shielding channels can reduce crosstalk between channels. However, it can generate unnecessary self-capacitance, causing the impedance mismatching issue. Furthermore, inserting shielding channels incurs additional costs and requires more space. Thus, it is not a suitable method for the HBM interface.

In this paper, a novel broadside interposer channel structure with a vertical tabbed via for high-speed signaling and low-power consumption is proposed for next-generation HBM. The broadside structure solves the problem of low impedance and the issue of off-chip interconnection power consumption. The proposed structure uses the vertical tabbed via to mitigate the far-end crosstalk of the broadside structure. The proposed channel structure is presented and analyzed based on the modeling of self-capacitance and mutual capacitance. We also take into account the effect of design parameters such as the physical dimensions of the vertical tabbed via and evaluate it from a signal integrity perspective. The signaling performance of the proposed channel structure was analyzed for impedance, channel loss, far-end crosstalk, and eye-diagram compared to a conventional channel structure that includes the micro-strip line and strip line. We also evaluated the dynamic power consumption and compared it with a that of a conventional channel structure.

2. Proposal and Modeling of the Broadside Interposer Channel Structure with a Vertical Tabbed Via for Next Generation HBM

2.1. The Structure of the Proposed Novel Broadside Interposer Channel Structure

The proposed channel structure is shown in Figure 2. The proposed channel structure has two features. First, the proposed channel structure is a broadside strip line channel structure. This broadside structure can minimize channel loss by reducing impedance mismatching. In addition, if a multi-channel system is required that is more than three signal layers, the proposed broadside channel structure can be designed with a high channel density structure because it can minimize the exposure of E-field and H-field to the outside. This will be detailed in Section 4.

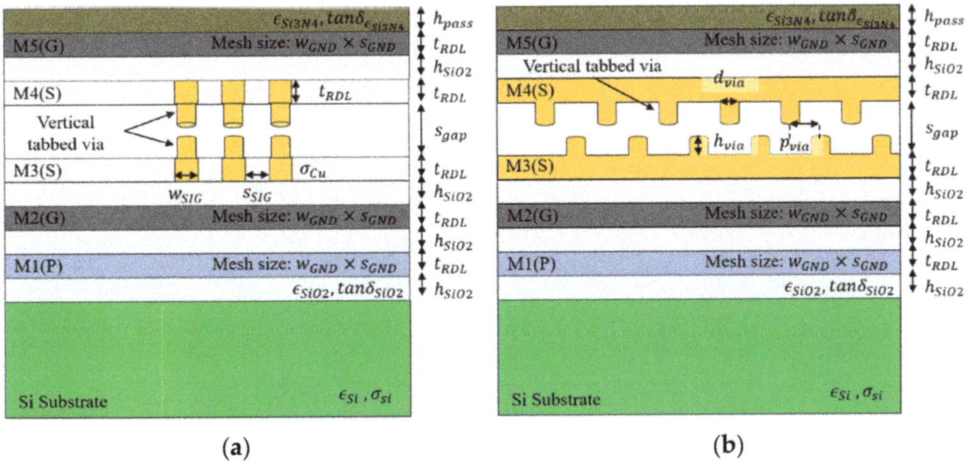

Figure 2. The proposed interposer channel structure: (**a**) front view and (**b**) side view.

Second, the proposed channel structure has a vertical tabbed via. The vertical tabbed via can mitigate the influence of far-end crosstalk between vertical channels. This effect is similar to the effect of reducing far-end crosstalk by increasing mutual capacitance, such as tabbed routing mainly used in the breakout zone of PCBs [12,13]. The physical dimension and electrical properties are summarized in Table 1. The interposer channel width and space are assumed to be 3 μm and 3 μm, respectively, and the silicon conductivity is 10 σ/m.

Table 1. Physical dimensions and material properties of the silicon interposer.

Symbol	Parameter	Value	Symbol	Parameter	Value
w_{SIG}	Width of channel	3 µm	w_{GND}	Width of a meshed ground layer	3 µm
s_{SIG}	Space of channel	3 µm	s_{GND}	Space of a meshed ground layer	3 µm
h_{sub}	Height of Si substrate	100 µm	$l_{channel}$	Length of channel	5 mm
t_{RDL}	Height of RDL layer	1 µm	ε_{Si3N4}	Relative permittivity of Si_3N_4	6.5
h_{SiO_2}	Height of SiO_2 layer	1 µm	ε_{SiO_2}	Relative permittivity of SiO_2	4.1
h_{pass}	Height of passivation layer	1 µm	$tan\ \delta_{Si3N4}$	Loss tan gent of Si_3N_4	0.001
s_{gap}	Gap of the vertical channel	Design parameter (4 µm /6 µm/8 µm/10 µm)	$tan\ \delta_{SiO_2}$	Loss tan gent of SiO_2	0.001
h_{via}	Height of vertical tabbed via	Design parameter (0.5 µm/1 µm/2 µm)	ε_{Si}	Relative permittivity of Si	11.9
d_{via}	Diameter of vertical tabbed via	1.5 µm	σ_{Si}	Conductivity of Si substrate	10 σ/m
p_{via}	Pitch of vertical tabbed via	Design parameter (3 µm/9 µm/18 µm)	σ_{Cu}	Conductivity of copper	5.8×10^7 σ/m

2.2. Analytical Modeling of Self and Mutual Capacitance for the Proposed Interposer Channel Structure

The electrical characteristics of the proposed channel structure were analyzed depending on the design parameters by building an analytical model of self-capacitance and mutual capacitance.

The self-capacitance of the proposed channel structure can be calculated by dividing it into four sections based on the cross-section, as shown in Figure 3. The four sections were determined according to the ground slots and the vertical tabbed via. The self-capacitance consists of various fringing capacitances between the bottom ground layer and the top ground layer. The equations of fringing capacitances for self-capacitance are expressed through (1)–(17) [14–18].

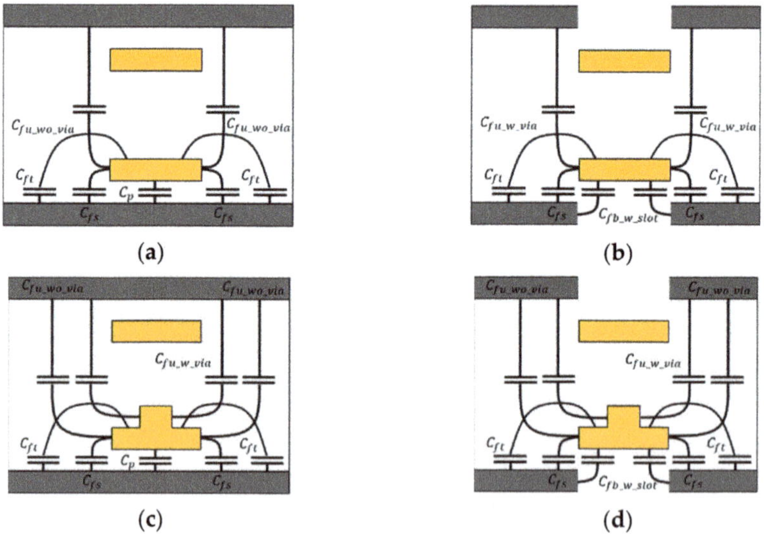

Figure 3. Cross-section view of self-capacitance of the proposed channel structure. (**a**) section 1: without the vertical tabbed via and without the ground slot, (**b**) section 2: without the vertical tabbed via and with the ground slot, (**c**) section 3: with the vertical tabbed via and without the ground slot, (**d**) section 4: with the vertical tabbed via and with the ground slot.

The C_p is simply determined using parallel-plate capacitor formulas. The C_{fs} is the analytical expression of external fringing capacitance. The effective thickness of the substrate is determined depending on the height of the s_{gap}. The C_{ft} is calculated by

(3) valid at the range $0.5 \leq h_{SiO_2}/t_{RDL} \leq 4$ [15]. From the equations in [15], the factor of A_0, B_0, C_0 for C_{ft} can be calculated. The $C_{fu\_wo\_via}$ and $C_{fu\_w\_via}$ are also the external fringing capacitances between channel and ground capacitance [16]. The $C_{fu\_w\_slot}$ is the internal fringing capacitance, which is unlike external fringing capacitance, since the electric field can be distorted into a non-perfect parabola shape [16].

For section 2 and section 4 with a ground slot due to the meshed ground layer, the total capacitance of those sections was calculated, including the $C_{fb\_w\_slot}$ generated from the four sides, as shown in Figure 3b,d. For section 3 and section 4 with the vertical tabbed via, the total capacitance of those sections was calculated, including the $C_{fu\_w\_via}$ from two sides, as shown in Figure 3c,d.

$$C_p = \varepsilon_{SiO_2}\varepsilon_0 \frac{w_{SIG}}{h_{SiO_2}} \tag{1}$$

$$C_{fs} = \frac{\varepsilon_{SiO_2}\varepsilon_0}{\pi}\left[2t_{b1}\ln(t_{b1}+1) - (t_{b1}-1)\ln\left(t_{b1}^2 - 1\right)\right] \tag{2}$$

$$C_{ft} = \frac{\varepsilon_{SiO_2}\varepsilon_0 e^{\left(-\frac{A_0}{B_0}\right)}\left[\ln\left(1+\frac{2w_{SIG}}{S_{SIG}}\right) + e^{\left(-\frac{C_0}{-3S_{SIG}}\right)}\right]}{\varepsilon_{SiO_2}\varepsilon_0 e^{\left(-\frac{A_0}{B_0}\right)} + A\left[\ln\left(1+\frac{2w_{SIG}}{S_{SIG}}\right) + e^{\left(-\frac{C_0}{-3S_{SIG}}\right)}\right]} \tag{3}$$

$$C_{fu\_wo\_via} = \frac{\varepsilon_{SiO_2}\varepsilon_0}{\pi}\left[2t_{b\_wo\_via}\ln(t_{b\_wo\_via}+1) - (t_{b\_wo\_via}-1)\ln\left(t_{b\_wo\_via}^2 - 1\right)\right] \tag{4}$$

$$C_{fb\_w\_slot} = \frac{\varepsilon_{SiO_2}\varepsilon_0}{\pi} A_1\left[\frac{s_{SIG}}{2h_{SiO_2}+t_{RDL}} - \frac{2}{\pi}\ln\left(\frac{\pi}{2}\frac{s_{SIG}}{h_{SiO_2}+t_{RDL}}\right)\right] + B_1 \tag{5}$$

$$C_{fu\_w\_via} = \frac{\varepsilon_{SiO_2}\varepsilon_0}{\pi}\left[2t_{b\_w\_via}\ln(t_{b\_w\_via}+1) - (t_{b\_w\_via}-1)\ln\left(t_{b\_w\_via}^2 - 1\right)\right] \tag{6}$$

$$C_{section1} = C_p + 2C_{fs} + 2C_{ft} + 2C_{fu\_wo\_via} \tag{7}$$

$$C_{section2} = 4C_{fb\_w\_slot} + 2C_{fs} + 2C_{ft} + 2C_{fu\_wo\_via} \tag{8}$$

$$C_{section3} = C_p + 2C_{fs} + 2C_{ft} + 2C_{fu\_w\_via} + 2C_{fu\_wo\_via} \tag{9}$$

$$C_{section4} = 4C_{fb\_w\_slot} + 2C_{fs} + 2C_{ft} + 2C_{fu\_w\_via} + 2C_{fu\_wo\_via} \tag{10}$$

$$C_{self} = C_{section1} + C_{section2} + C_{section3} + C_{section4} \tag{11}$$

where

$$t_{SIG\_eff} = \frac{t_{RDL} + s_{gap}}{6} \tag{12}$$

$$t_{b1} = \frac{2h_{SiO_2} + t_{SIG\_eff}}{2h_{SiO_2}} \tag{13}$$

$$h_{wo\_via} = s_{gap} + 2h_{SiO_2} \tag{14}$$

$$h_{w\_via} = s_{gap} + 2h_{SiO_2} - h_{via} \tag{15}$$

$$t_{b\_wo\_via} = \frac{2h_{wo\_via} + t_{SIG}}{2h_{wo\_via}} \tag{16}$$

$$t_{b\_w\_via} = \frac{2h_{w\_via} + t_{RDL}}{2h_{w\_via}} \tag{17}$$

$$C_{s\_wo\_via} = \varepsilon_{SiO_2}\varepsilon_0 \frac{w_{SIG}}{s_{gap}} \tag{18}$$

$$C_{s\_w\_via} = \varepsilon_{SiO_2}\varepsilon_0 \frac{w_{SIG}}{s_{gap} - h_{via}} \tag{19}$$

$$C_{f\_wo\_via} = 0.5\frac{\varepsilon_r}{60c_0\pi}\left[\frac{K(k_{oi1})}{K(k'_{oi1})} - \frac{K(k_{ei1})}{K(k'_{ei1})}\right] \tag{20}$$

$$C_{f\_w\_via} = 0.5\frac{\varepsilon_r}{60c_0\pi}\left[\frac{K(k_{oi2})}{K(k'_{oi2})} - \frac{K(k_{ei2})}{K(k'_{ei2})}\right] \tag{21}$$

$$C_{section\_A} = C_{s\_wo\_via} + 2C_{f\_wo\_via} \tag{22}$$

$$C_{section\_B} = C_{s\_w\_via} + 2C_{f\_w\_via} \tag{23}$$

$$C_{mutual} = C_{section\_A} + C_{section\_B} \tag{24}$$

where

$$k_{oi1} = tanh\left(\frac{\pi}{4}\frac{w_{SIG}}{s_{SIG}}\right)/tanh\left(\frac{\pi}{4}\frac{s_{gap} + w_{SIG}}{s_{SIG}}\right) \tag{25}$$

$$k_{oi2} = tanh\left(\frac{\pi}{4}\frac{w_{SIG}}{s_{SIG}}\right)/tanh\left(\frac{\pi}{4}\frac{s_{gap} - h_{via} + w_{SIG}}{s_{SIG}}\right) \tag{26}$$

$$k_{ei1} = tanh\left(\frac{\pi}{4}\frac{w_{SIG}}{s_{SIG}}\right)tanh\left(\frac{\pi}{4}\frac{s_{gap} + w_{SIG}}{s_{SIG}}\right) \tag{27}$$

$$k_{ei2} = tanh\left(\frac{\pi}{4}\frac{w_{SIG}}{s_{SIG}}\right)tanh\left(\frac{\pi}{4}\frac{s_{gap} - h_{via} + w_{SIG}}{s_{SIG}}\right) \tag{28}$$

$$k\prime^2_{oi1} = 1 - k^2_{oi} \tag{29}$$

$$k\prime^2_{oi2} = 1 - k^2_{o2} \tag{30}$$

$$k\prime^2_{ei1} = 1 - k^2_{ei} \tag{31}$$

$$k\prime^2_{ei2} = 1 - k^2_{e2} \tag{32}$$

The mutual capacitance of the proposed structure can be calculated by dividing it into two sections based on the cross-section, as shown in Figure 4. In mutual capacitance, unlike in self-capacitance, the effect of the ground slot is negligible. The equation of fringing capacitance for mutual capacitance is expressed through (18)–(32) [17]. The $C_{s\_wo\_via}$ and $C_{s\_w\_via}$ are simply determined by the parallel-plate capacitor formula. The $C_{f\_wo\_via}$ and $C_{f\_w\_via}$ are an empirical gap capacitance in the dielectric [19]. The $K(k)$ and $K(k')$ are the complete elliptic integral of the first type and its complement.

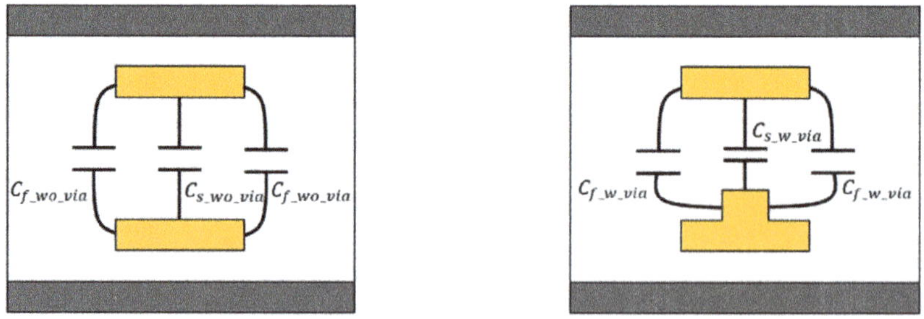

(a) (b)

Figure 4. Modeling of mutual capacitance of the proposed channel structure. (**a**) Section A: without the vertical tabbed via, (**b**) Section B: with the vertical tabbed via.

For the design parameters for s_{gap} and h_{via}, the analytical modeling of the self-capacitance and mutual capacitance was verified based on 3D electromagnetic (EM) simulation, using the ANSYS Q3D 3D field simulator, as shown in Figure 5. With the variation in s_{gap} and h_{via}, the trend in the capacitance model for the analytical modeling and 3D EM simulation is overall very similar, and the error is within 5%. As the s_{gap} increases, the electric field is more strongly coupled to the ground layer of M2, which is close to the target channel of M3. However, the electric field between the M5 ground layer is relatively weak. Therefore, the total self-capacitance value eventually increases because the increase in the fringing capacitance between the ground layer of M2 close to the target channel of M3 is larger than the decrease of the fringing capacitance between the ground layer of M5 farther away from the target channel of M3. As the height of the vertical tabbed via increases with the same vertical distance of channel, it can be seen that the mutual capacitance significantly increases.

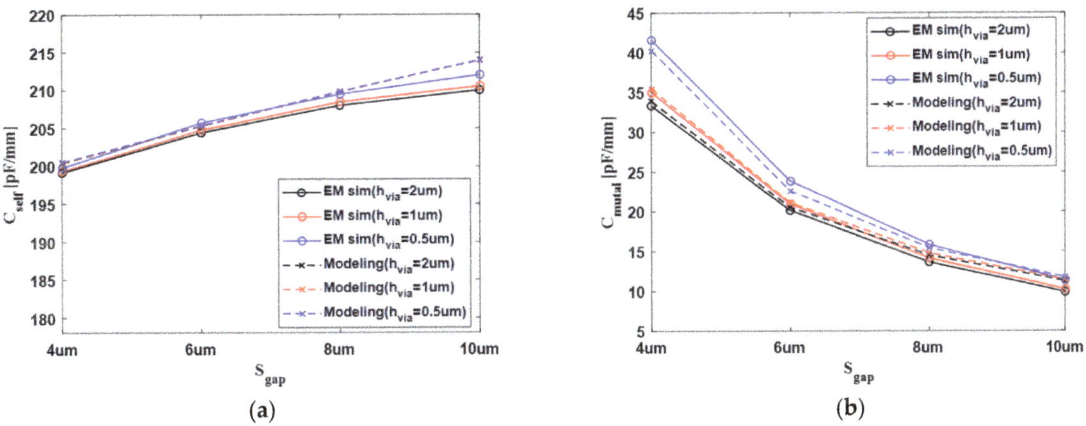

Figure 5. Verification of capacitance modeling: (**a**) self-capacitance and (**b**) mutual capacitance.

When the s_{gap} has a value of about 10 μm, the self-capacitance has a saturated constant value. The 10 μm is a sufficiently wide gap of the channel because it is 10 times the thickness of the channel. Thus, this modeling of capacitance is presented with sufficient height.

3. Signal Integrity Analysis of the Proposed Interposer Channel Structure Depending on the Design Parameters

In terms of signal integrity, the characteristics of the proposed channel structure were analyzed depending on the design parameters that can affect the channel performance of the proposed structure. The influence of far-end crosstalk was evaluated through the power-sum far-end crosstalk (PSFEXT) [20]. The PSFEXT can consider all the effects of surrounding channels that include top and bottom aggressor channels for one victim channel. The PSFEXT equation is expressed in (33). The channel length is set to 5 mm. This channel length assumes the longest interposer channel length of the HBM interface.

First, the proposed channel structure was analyzed depending on the presence of a vertical tabbed via. Based on the analytical capacitance modeling, the self-capacitance of the proposed channel with the vertical tabbed via was not dramatically different from the proposed channel structure without the vertical tabbed via. Therefore, there was no significant change in terms of channel loss, as shown in Figure 6a.

On the other hand, when the structure had the vertical tabbed via, the amount of mutual capacitance could increase significantly, based on the analytical modeling. The FEXT coefficient can be expressed in (34). When the input signal transitions from low to high, the FEXT is a negative signal that has a 180-degree phase with the input signal. The FEXT coefficient is negative using (34). Therefore, in order to reduce FEXT, it is necessary to increase the mutual capacitance or reduce the self-capacitance.

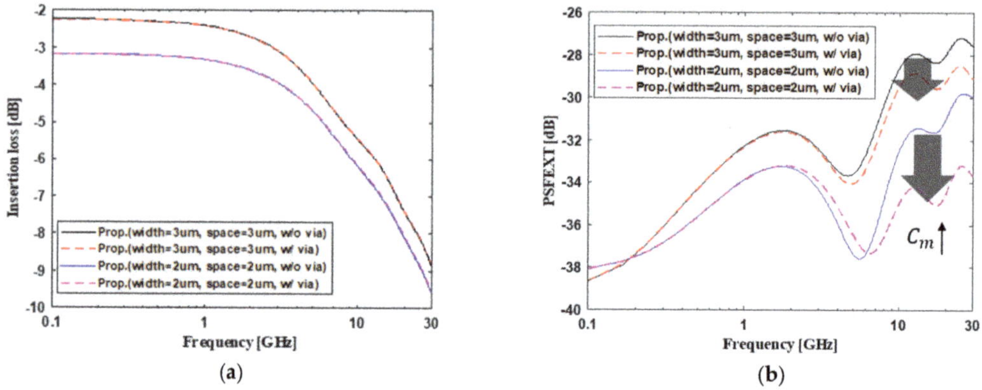

Figure 6. Signal integrity analysis of the effect of the proposed channel structure on the vertical tabbed via in the frequency domain. (**a**) Insertion loss and (**b**) power sum far-end crosstalk (PSFEXT).

The mutual capacitance is increased by inserting the vertical tabbed via, so the PSFEXT can be greatly mitigated, as shown in Figure 6b. In addition, looking at the PSFEXT, capacitive coupling is dominant above 1 GHz, and inductive coupling is dominant below 1 GHz [6]. The increase in mutual capacitance produced by inserting the vertical tabbed via, confirms that the FEXT is greatly reduced only above 1 GHz. In the case of width and space of 3 μm and 2 μm, respectively, there is a frequency band with large FEXT when the vertical tabbed via is inserted. This is because the resonance frequency occurs due to the cable length. If the cable is long, the channel should be designed considering the cable resonance.

$$Powersum\ FEXT\ (PSFEXT) = 10log \sum_{j \in \Omega_{FEXT}} |S_{i,j}|^2 \tag{33}$$

$$FEXT = \frac{len_{coupling}}{2vT_r}\left(\frac{C_m}{C_s} - \frac{L_m}{L_s}\right) \tag{34}$$

Three design parameters of the proposed channel structure were analyzed, the height and the pitch of the vertical tabbed via and the gap of the vertical channel. Next, we analyzed the far-end crosstalk. As mentioned before, the insertion loss did not significantly change depending on the presence of the vertical tabbed via.

As shown in Figure 7a, as the height of the vertical tabbed via increases, the mutual capacitance that occurs between the channels increases. Thus, the far-end crosstalk can be reduced. If the pitch of the vertical tabbed via becomes narrow, the far-end crosstalk can be reduced in the same way, as shown in Figure 7b. As the gap of the vertical channel increases, the far-end crosstalk decreases because the physical channel distance increases, as shown in Figure 7c. In this case, since mutual inductance also decreases, the far-end crosstalk is reduced even below 1 GHz. In the above, at 1 GHz, the value of the FEXT coefficient decreases because the mutual capacitance decreases while the self-capacitance and self-inductance increase. It is important to note that, as the gap in the vertical channel increases, the difference in far-end crosstalk improvement also decreases. That is, when the gap of the vertical channel becomes larger than a certain value, the far-end crosstalk can be saturated.

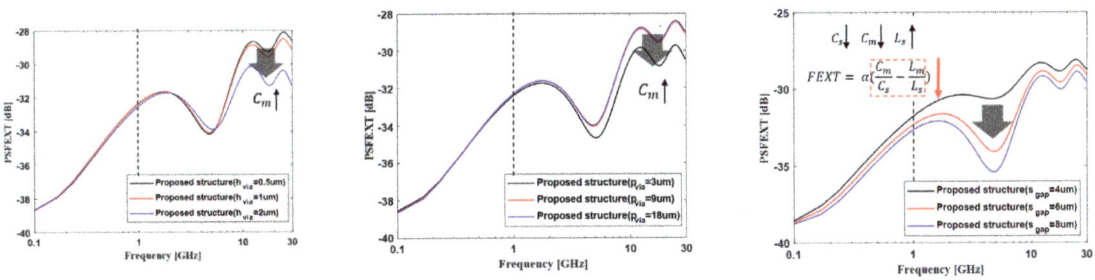

(**a**) (**b**) (**c**)

Figure 7. The PSFEXT analysis of proposed channel structure depending on the design parameters. (**a**) The height of the vertical tabbed via, (**b**) the pitch of the vertical tabbed via, and (**c**) the gap of the vertical channel.

4. Signal Integrity Analysis of the Proposed Interposer Channel Structure Compared with the Conventional Interposer Channel Structure for the HBM

4.1. Signal Integrity Analysis of the Proposed Novel Interposer Channel Structure

The signaling performance of the proposed channel structure and conventional channel structures, that is the micro-strip line and strip line, can be compared in terms of signal integrity. The interposer channel was selected as a five-layer structure, including two signal layers, two ground layers, and one power layer, as shown in Figure 8 [6]. The design parameters of the vertical tabbed via were selected to be 6 µm for s_{gap}, 1 µm for h_{via}, and 9 µm for p_{via} based on the analysis results in the previous chapter. Both the conventional channel structure and the proposed channel structure were placed in a perfectly misaligned arrangement with a meshed ground layer. Since the impedance mismatching is minimized at the position where the signal channel and the meshed ground are perfectly misaligned, the channel loss is less degraded, and the eye margin can be greatly secured [6].

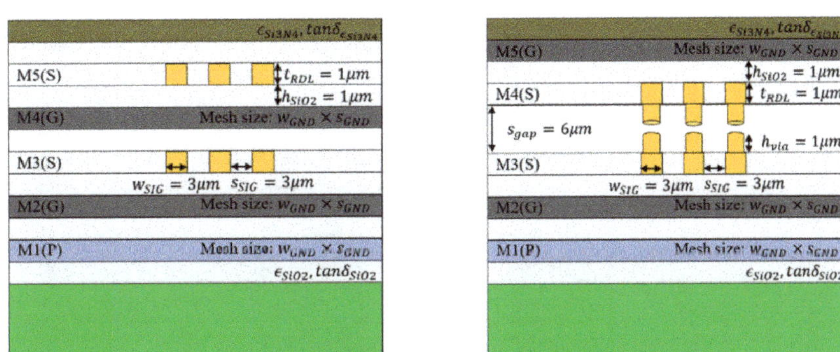

(**a**) (**b**)

Figure 8. Interposer structure: (**a**) conventional channel structure and (**b**) proposed channel structure.

The E-field and H-field distribution are shown in Figure 9. It can be seen that the E-field and H-field are widely exposed because the micro-strip line does not have a ground layer above the channel. On the other hand, the proposed channel structure can minimize field distribution to the outside by arranging the ground layer above and below. In this case, if an additional signal layer is placed, one signal layer and one ground layer are required in the conventional channel structure. However, since the proposed channel structure has a ground layer on the top and bottom to minimize exposure to the outside, only the one signal layer is required. After all, in a system that requires many signal layers, if one layer can be reduced, an advantage in channel density can be obtained.

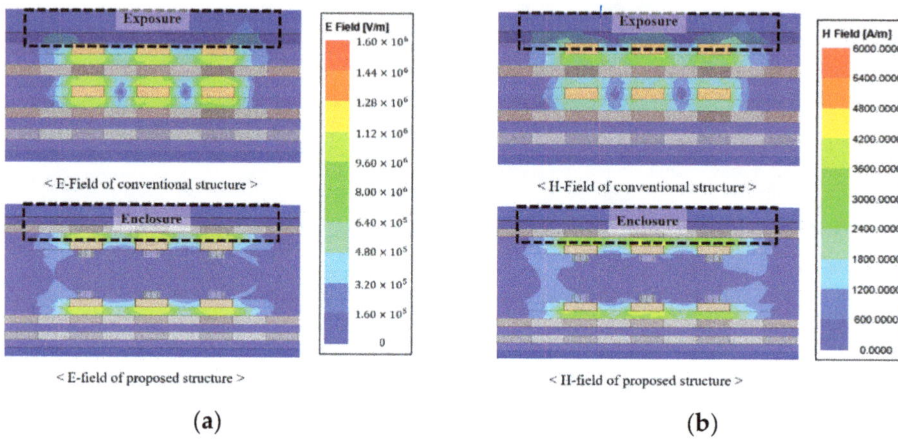

Figure 9. Field distribution of the proposed channel structure and conventional channel structure: (**a**) E-field and (**b**) H-field.

Figure 10 shows an RLC component of the proposed channel structure and conventional channel structure based on the 3D EM simulation, using the ANSYS HFFF 3D field simulator. Since the cross-sectional area of the conductor through which current flows is as large as the area of the vertical tabbed via compared to the conventional channel structure, the conductor loss of the proposed channel structure is smaller than that of the conventional channel structure below 1 GHz, as shown in Figure 10a. In general, above 1 GHz, the AC resistance is determined by the skin effect and proximity effect. The proposed channel structure has a larger AC resistance compared to the conventional channel structure because the proximity effect increases as the distance between the adjacent channels decreases, due to the vertical tabbed via. Additionally, due to chemical mechanical polishing (CMP), the interposer channel uses meshed ground layer, so the AC resistance of the interposer is relatively high. Since the meshed layer has many slots, the return path is relatively longer than that in a solid ground layer in the printed circuit board (PCB). As shown in Figure 10b, the capacitance of the proposed channel structure is significantly reduced compared to that for the strip line. Since the proposed channel structure is arranged in the structure of the broadside, this structure can achieve the effect of minimizing the self-capacitance. As shown in Figure 10c, the self-inductance tends to be greater as there are more regions in which the magnetic field can be generated. Since the strip line is a closed structure due to the ground layer, the self-inductance is relatively small. Below 1 GHz, the proposed channel structure is smaller than that for the micro-strip line because it is a half-closed structure. However, as the frequency increases, the self-inductance becomes similar to that for the micro-strip line.

In terms of signal transmission, the impedance matching has a greater effect on signal transmission characteristics than the signal loss due to AC resistance at a high-frequency band. This is because the signal loss reduces the level of the signal, but the severe reflection due to impedance mismatching makes it impossible to transmit the signal normally. In the low-frequency band, the proposed structure can obtain good performance by reducing the resistance. The reason for this is that signal loss is more important than reflection in the low-frequency bands.

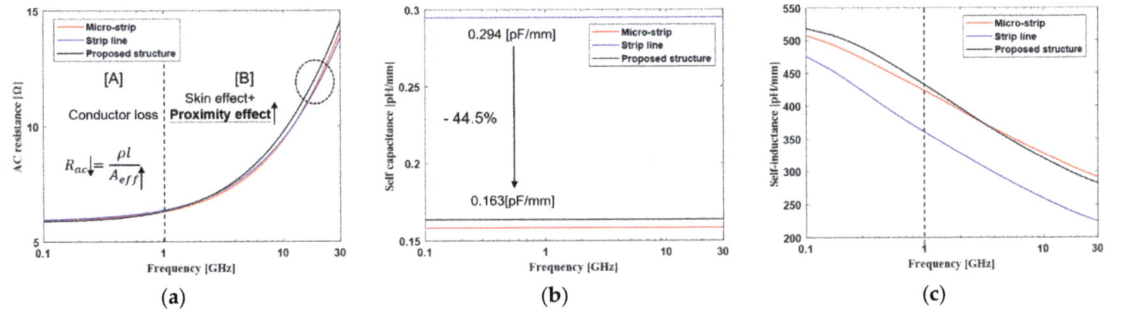

Figure 10. RLC component of the proposed channel structure: (**a**) AC resistance, (**b**) self-capacitance, and (**c**) self-inductance.

The results in Figure 11 confirm the impedance of the proposed channel structure and the conventional channel structure. We can estimate the characteristic impedance using (35) [21]. Based on the analysis of the RLC component, the proposed channel structure has a characteristic impedance similar to that for the micro-strip line, as shown in Figure 11a. Impedance matching was also checked through reflection loss, as shown in Figure 11b. If the reflection loss is less than 12 dB, it means that the reflection of the signal is sufficiently small due to the impedance mismatching for a driver impedance of 50 ohms. The strip line is not significantly impedance matched for some frequency bands under the same width and space channel conditions. On the other hand, it was confirmed that the proposed channel structure provides good impedance matching over a wide frequency band similar to that for a micro-strip line.

$$Z_0 = Z_{sys}\sqrt{\frac{(1+S_{11})^2 - S_{21}^2}{(1-S_{11})^2 - S_{21}^2}} \tag{35}$$

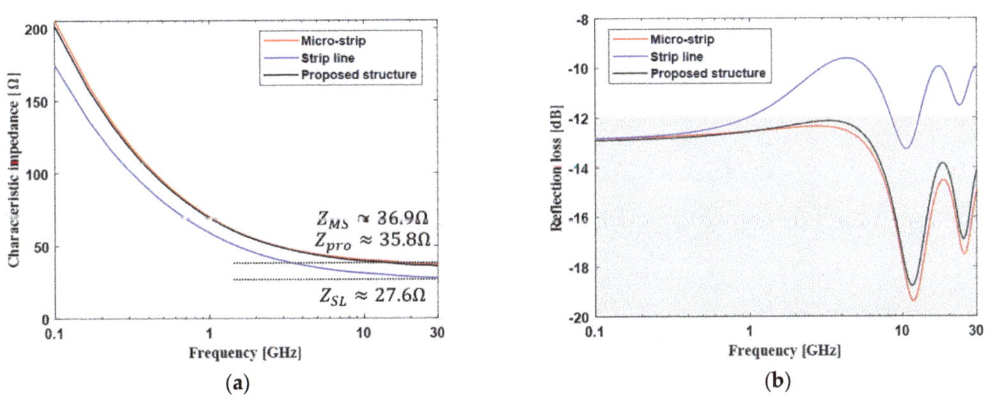

Figure 11. Impedance analysis of the proposed channel structure and conventional channel structure: (**a**) characteristic impedance and (**b**) reflection loss.

The channel loss and far-end crosstalk of the proposed channel structure and conventional channel structure were compared, as shown in Figure 12a. First of all, it was confirmed that the proposed channel structure significantly improves the channel loss in broadband compared to the strip line, which is the worst case for a conventional structure. This effect comes from the improvement obtained by reducing the impedance mismatching.

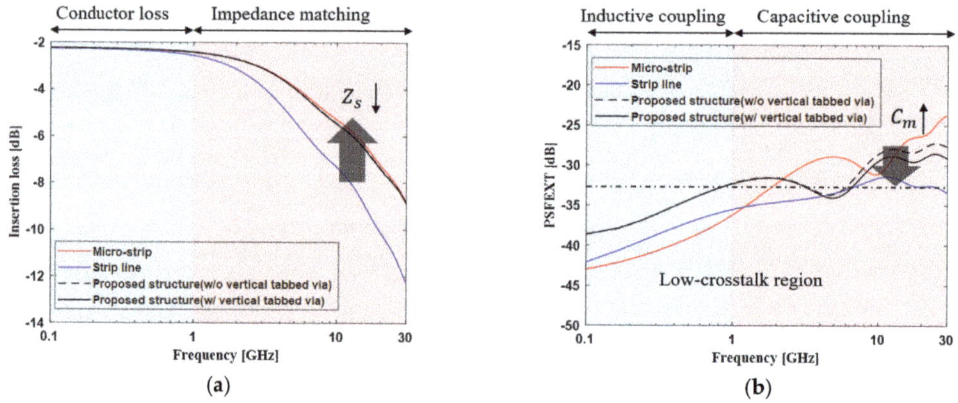

Figure 12. Frequency domain analysis of the proposed channel structure and conventional channel structure: (**a**) insertion loss and (**b**) PSFEXT.

Since the proposed channel structure has vertical symmetry, it shows almost the same performance regardless of the position of the interposer channel. Below 1 GHz, the conductor loss is dominant. The conductor loss is formed according to the AC resistance. The proposed channel structure has the smallest conductor loss because the effective conductor area is the largest. Above 1 GHz, the characteristic impedance is dominant according to the self-capacitance and self-inductance. Since the proposed channel structure and micro-strip line have similar values, it can be confirmed that, above 1 GHz, the channel loss is similar. On the other hand, the strip line has a larger capacitance than the others. It can be seen that the signal is greatly degraded above 1 GHz.

As shown in Figure 12b, in terms of far-end crosstalk, the PSFEXT of the strip line is the smallest of the channels in the overall frequency range. Since the strip line is located between the ground layers, it can be easily predicted that the far-end crosstalk noise is small. For the micro-strip line and the proposed channel structure, it can be confirmed that the far-end crosstalk is relatively large because the ground layer is located only on one side.

The proposed channel structure has structural advantages in terms of far-end crosstalk. This is because the proposed channel structure can generate more mutual capacitance than the micro-strip line due to the vertical tabbed via. In addition, by adding the vertical tabbed via, the influence of far-end crosstalk can be mitigated. The comprehensive evaluation including the channel loss and far-end crosstalk can be checked using the eye diagram in the next section.

The proposed channel structure has an advantage in terms of timing. The effect of LC delay is also determined by the self-inductance and self-capacitance of the off-chip interconnection. Here, LC delay means a delay in the signal generated through the off-chip interconnection. As the LC delay of the channel increases, the ISI can also increase. The LC delay is defined as the difference in time at which the input waveform and output waveform have transitioned to about 50% of their maximum value, respectively. As shown in Figure 13, by minimizing the parasitic self-capacitance, the proposed channel structure can reduce the LC delay by about 20 ps compared to the strip-line, which is the worst case for the conventional structure.

4.2. Eye-Diagram and Dynamic Power Consumption Analysis of the Proposed Novel Interposer Channel Structure Compared with the Conventional Channel Structure

Using an eye diagram, it is possible to compare and analyze the signaling performance of the channel by comprehensively evaluating various signal integrity indicators. As shown in Figure 14, the eye-diagram setup refers to the JEDEC standard [22]. The data rate was evaluated for 4 and 8 Gbps, which are the performance levels of HBM Gen 3 and HBM Gen 4. The load of the TX driver includes the effects of electro-static discharge (ESD) and pad.

The load of the RX driver does not have resistor termination due to the power consumption issue. Referring to the JEDEC standard, if the data rate is doubled, the load tends to be reduced by half. Therefore, the load value for 8 Gbps is assumed to be 0.2 pF. The selected rising time was as 10% of the Nyquist frequency.

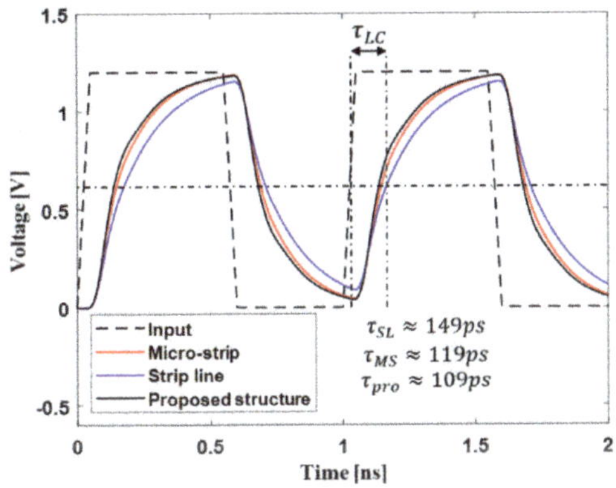

Figure 13. Timing analysis of proposed channel structure and conventional channel structure.

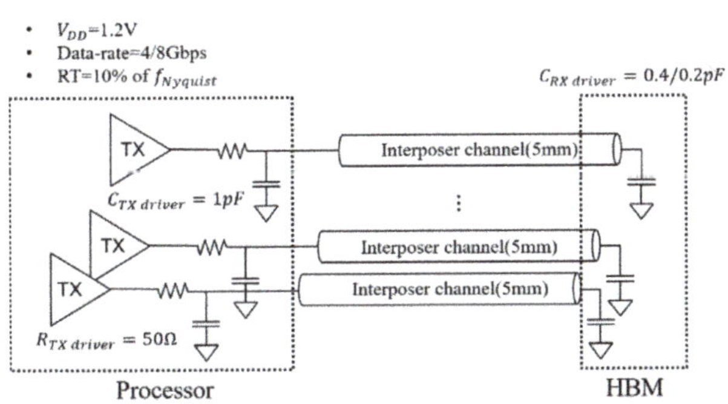

Figure 14. Eye-diagram simulation setup.

The results of the eye diagram for 4 and 8 Gbps are shown in Figure 15. It can be seen that the results for impedance, insertion loss, far-end crosstalk, and eye diagram have the same tendency. In the eye diagram, the proposed channel structure shows similar or better performance to that for the micro-strip line. The strip line significantly degrades the signaling performance. In general, the system performance is determined based on the worst case. By using the proposed channel structure, the overall performance of the system can be surely improved. Since the proposed channel structure is a vertical symmetric structure, all channels have uniform performance. This result can be confirmed more clearly when the higher data rate 8 Gbps signal is transmitted. In the case of 8 Gbps, it can be seen that the strip line is almost closed in the eye-diagram. As the data-rate increases, it means that the strip line cannot transmit the signal.

At 4 Gbps, the proposed channel structure can improve eye width, eye height, and eye jitter by up to 12.9%, 9.9%, and 4.11%, respectively, compared to the worst-case, which is that for the strip line. At 8 Gbps, the channel loss of the worst-case can mean more degradation. The proposed channel structure can improve eye-width, eye-height, eye-jitter by up to 17.6%, 29%, and 9.56%, respectively, compared to the worst case.

The eye margins are summarized, depending on the gap of the vertical channel, in Figure 16. We obtained additional eye-margin because the far-end crosstalk is improved as the gap in the vertical channel increases. However, as mentioned in the previous section, when the gap of the vertical channel is increased above a certain value, the amount of far-end crosstalk improvement can be limited.

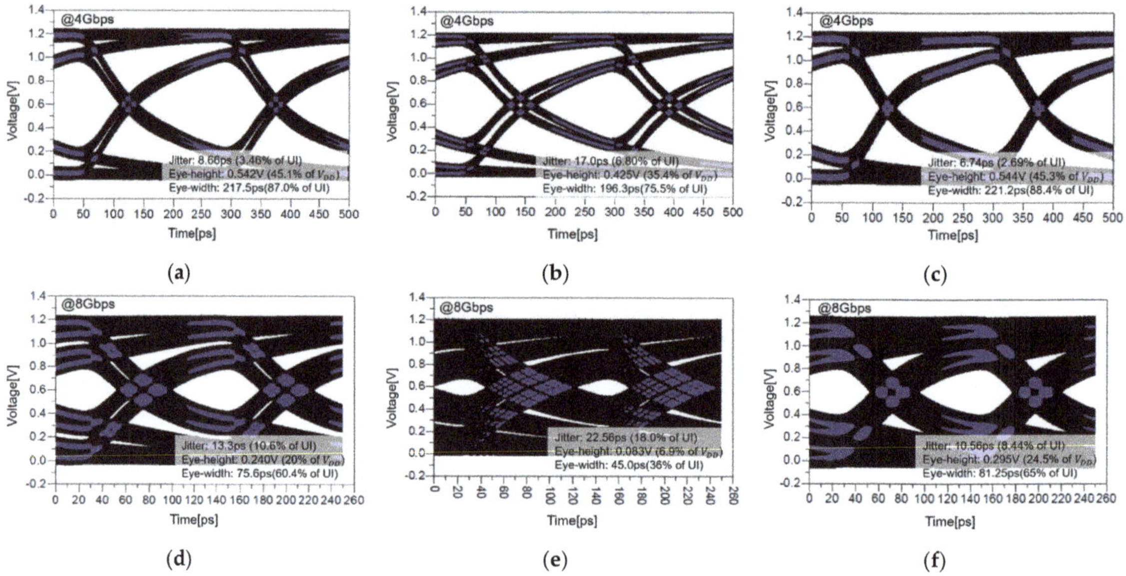

Figure 15. Eye-diagram of the conventional micro-strip line, strip line, and the proposed structure: (**a**) micro-strip line at 4 Gbps, (**b**) strip line at 4 Gbps, (**c**) proposed structure at 4 Gbps, (**d**) micro-strip line at 8 Gbps, (**e**) strip line at 8 Gbps, and (**f**) proposed structure at 8 Gbps.

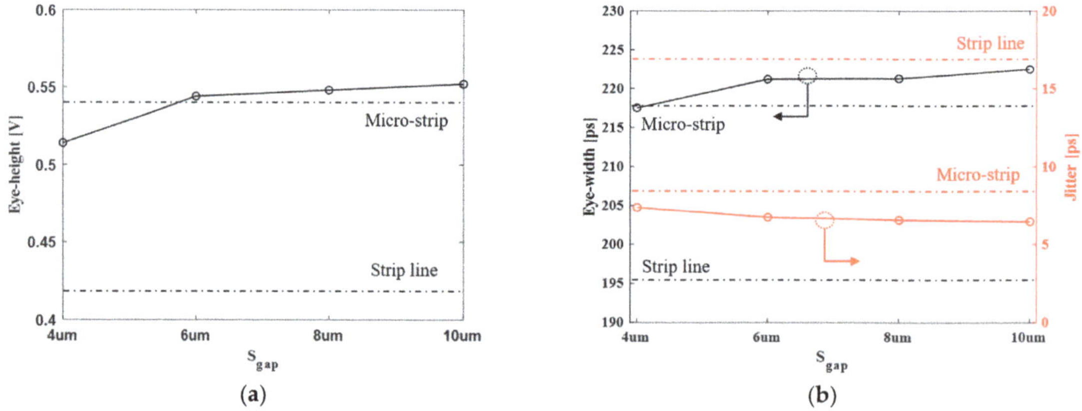

Figure 16. Summary of the eye-diagram data: (**a**) eye-height and (**b**) eye-width and jitter.

Figure 16 indicates how much of a gap of the vertical channel needs to be selected to show a better eye margin than the conventional channel structure. The proposed channel structure shows better signaling performance than the micro-strip line when the vertical channel gap is more than 6 μm.

Additionally, the proposed channel structure has an advantage in terms of power consumption. As mentioned before, the power issue that occurs off-chip should be minimized. Power consumption can be calculated using (36) [23]. Based on the dynamic power consumption formula, the way to reduce power consumption in off-chip is to minimize parasitic capacitance [24]. The proposed channel structure forms a capacitance similar to that of micro-strip line regardless of the channel position. However, since the strip line forms a very large capacitance, the conventional channel structure consumes a lot of power. Compared with the average power consumption, it can be reduced up to about 28%. Table 2 summarizes the amount of power consumption depending on the channel structure and data rate.

$$P_{dynamic} = \frac{1}{2} V_{DD}^2 f_{Nyquist} C_{total} \quad (36)$$

Table 2. Dynamic power consumption of proposed structure and conventional channel structure.

Parameter	Conventional Micro-Strip	Conventional Strip Line	Proposed Structure
$C_{channel}$ [pF/mm]	0.158	0.294	0.163
$P_{channel}$ [mW/mm] at 4 Gbps	0.227	0.423	0.234
$P_{channel}$ [mW/mm] at 8 Gbps	0.455	0.846	0.469
$P_{average\ channel}$ [mW/mm] at 4 Gbps	0.325		0.234 (−28%)

Table 3 summarizes of the evaluation factors in terms of the signal integrity, and power consumption analyzed for the conventional structure and the proposed structure. The proposed structure obtains similar or better signaling performance than the micro-strip line in most evaluation factors, and the proposed structure is significantly improved compared to the strip line. Conventional structures should always be used together with the micro-strip and strip line. Thus, the strip line determines the worst performance of the conventional structure. However, since the proposed structure is vertically symmetrical, the signaling performance of all channels is more uniform than in the conventional structure. Therefore, the proposed structure is the advanced channel structure for the high-speed and low-power channels for the next-generation HBM.

Table 3. Summary of system performance of the proposed and conventional structure.

Parameter	Conventional Micro-Strip	Conventional Strip Line	Proposed Structure
Reflection loss [dB] at 4 GHz	−12.5	−9.62	−12.19
Inseriton loss [dB] at 4 GHz	−3.50	−4.68	−3.55
FEXT [dB] at 4 GHz	−29.09	−33.83	−33.72
Characteristic impedance [Ω] at 4 GHz	45.6	36.4	45.2
LC delay [ps]	119	149	109
Eye-width [ps]	75.6	45.0	84.3
Eye-height [V]	0.240	0.083	0.295
Power consumption [mW/mm] at 8 Gbps	0.455	0.846	0.469

5. Conclusions

In order to realize a next-generation HBM, a structural change to a broadside interposer channel structure is needed to alleviate the impedance mismatching problem. In this paper, to improve high-speed signaling and low-power consumption, a broadside interposer channel with vertical tabbed via is proposed and analyzed for the first time. Using the 3D EM simulation, we suggested and verified the analytical modeling of the self-capacitance and mutual capacitance for the proposed interposer channel depending on various design parameters. Based on the proposed analytical modeling, the insertion loss and far-end crosstalk of the proposed interposer channel were analyzed depending on the design parameters.

The proposed interposer channel structure has four major advantages. First, since the impedance mismatching can be minimized by arranging the broadside structure, the proposed interposer channel can dramatically improve channel loss. Second, by inserting the vertical tabbed via, the shortcoming of the far-end crosstalk issue in the broadside structure can be reduced. Third, the proposed interposer channel can minimize off-chip capacitance by about 28% compared to the conventional structure in terms of dynamic power consumption. Finally, the proposed interposer channel can prevent E-field and H-field from being emitted to the outside of the interposer by arranging a ground layer above and below. It is possible to prevent far-end crosstalk noise occurring between the interposer channel and other components such as TSV, via, etc. In addition, this effect can be achieved in terms of channel density because the number of the signal layer can be reduced in a situation that requires an additional signal layer. Although there is a difference in performance depending on s_{gap}, eye-width, height, and jitter were improved by 17.6%, 29%, and 9.56%, respectively, at 8 Gbps compared to a strip line based on 6 μm.

The proposed interposer channel is expected to be used as a high-speed interconnection channel structure to stably transmit a higher data rate in the future. In addition, it can solve the issue of power consumption.

Author Contributions: Conceptualization, H.K., S.L. and J.P.; methodology, K.S.; validation, D.P. and J.C.; writing, Y.S.; review and editing, S.A. All authors have read and agreed to the published version of the manuscript.

Funding: This research was supported by the National R&D Program through the National Research Foundation of Korea (NRF), funded by the Ministry of Science and ICT (2020M3F3A2A01081585).

Institutional Review Board Statement: Not applicable.

Informed Consent Statement: Not applicable.

Data Availability Statement: The data presented in this study are available on request from the corresponding author.

Acknowledgments: The author would like to thank ANSYS (ANSYS HFSS), KESIGHT KOREA (ADS), and Huwin (Advanced Channel Verification System, ACVS) for the technical support. The EDA tool was supported by the IC Design Education.

Conflicts of Interest: The authors declare no conflict of interest.

References

1. Shan, G.; Zheng, Y.; Xing, C.; Chen, D.; Li, G.; Yang, Y. Architecture of Computing System based on Chiplet. *Micromachines* **2022**, *13*, 205. [CrossRef] [PubMed]
2. Wang, M.; Ma, S.; Jin, Y.; Wang, W.; Chen, J.; Hu, L.; He, S. A RF Redundant TSV Interconnection for High Resistance Si Interposer. *Micromachines* **2021**, *12*, 169. [CrossRef] [PubMed]
3. Jun, H.; Nam, S.; Jin, H.; Lee, J.-C.; Park, Y.J.; Lee, J.J. High-Bandwidth Memory (HBM) Test Challenges and Solutions. *IEEE Des. Test* **2016**, *34*, 16–25. [CrossRef]
4. Cho, K.; Kim, Y.; Lee, H.; Song, J.; Park, J.; Lee, S.; Kim, S.; Park, G.; Son, K.; Kim, J. Signal Integrity Design and Analysis of Differential High-Speed Serial Links in Silicon Interposer with Through-Silicon Via. *IEEE Trans. Compon. Packag. Manuf. Technol.* **2019**, *9*, 107–121. [CrossRef]
5. Kogge, P.; Shalf, J. Exascale computing trends: Adjusting to the "new normal" for computer architecture. *Comp. Sci. Eng.* **2013**, *15*, 16–26. [CrossRef]
6. Cho, K.; Kim, Y.; Lee, H.; Kim, H.; Choi, S.; Song, J.; Kim, S.; Park, J.; Lee, S.; Kim, J. Signal Integrity Design and Analysis of Silicon Interposer for GPU-Memory Channels in High-Bandwidth Memory Interface. *IEEE Trans. Compon. Packag. Manuf. Technol.* **2018**, *8*, 1658–1671. [CrossRef]
7. Kim, H.; Cho, J.; Kim, J.; Choi, S.; Kim, K.; Lee, J.; Park, K.; Pak, J.S.; Kim, J. A Wideband On-Interposer Passive Equalizer Design for Chip-to-Chip 30-Gb/s Serial Data Transmission. *IEEE Trans. Compon. Packag. Manuf. Technol.* **2015**, *5*, 28–39. [CrossRef]
8. Jeon, Y.; Kim, H.; Kim, J.; Je, M. Design of an On-Silicon-Interposer Passive Equalizer for Next Generation High Bandwidth Memory with Data Rate Up To 8 Gb/s. *IEEE Trans. Circuits Syst. I Regul. Pap.* **2018**, *65*, 2293–2303. [CrossRef]
9. Ko, H.G.; Shin, S.; Oh, J.; Park, K.; Jeong, D.K. An 8 Gb/s/μm FFE-combined crosstalk-cancellation scheme for HBM on silicon interposer with 3D-staggered channels. In Proceedings of the 2020 IEEE International Solid-State Circuits Conference-(ISSCC), San Francisco, CA, USA, 16–20 February 2020; pp. 128–130.
10. Bae, S.-J.; Park, K.-I.; Ihm, J.-D.; Song, H.-Y.; Lee, W.-J.; Kim, H.-J.; Kim, K.-H.; Park, Y.-S.; Park, M.-S.; Lee, H.-K.; et al. An 80 nm 4 Gb/s/pin 32 bit 512 Mb GDDR4 Graphics DRAM With Low Power and Low Noise Data Bus Inversion. *IEEE J. Solid-State Circuits* **2008**, *43*, 121–131. [CrossRef]

11. Moon, S.; Jo, C.; Nam, S. PI/SI Analysis and Design Approach for HPC Platform Applications. In Proceedings of the 2019 IEEE 69th Electronic Components and Technology Conference (ECTC), Las Vegas, NV, USA, 28–31 May 2019.
12. Song, K.; Kim, J.; Kim, H.; Lee, S.; Ahn, J.; Brito, A.; Kim, H.; Park, M.; Ahn, S. Modeling, Verification, and Signal Integrity Analysis of High-Speed Signaling Channel with Tabbed Routing in Hig Performance Computing Server Board. *Electronics* **2021**, *10*, 1590. [CrossRef]
13. Kunze, R.K.; Chu, Y.; Yu, Z.; Chhay, S.K.; Lai, M.; Zhu, Y. *Crosstalk Mitigation and Impedance Management Using Tabbed Lines*; Intel Corporation: Santa Clara, CA, USA, 2015.
14. Kim, H.; Kim, J.; Song, K.; Lee, S.; Kim, K.; Kim, S.; Lho, D.; Kim, H.; Park, M.; Ahn, S. Modeling and Signal Integrity Analysis of Mounting Pad with Layer-cutting to reduce Impedance Mismatch for Dual-In-Line Memory Module (DIMM). In Proceedings of the2021 IEEE 30th Conference on Electrical Performance of Electronic Packaging and Systems (EPEPS), San Jose, CA, USA, 17–20 October 2021.
15. Bansal, A.; Paul, B.C.; Roy, K. An Analytical Fringe Capacitance Model for Interconnects Using Conformal Mapping. *IEEE Trans. Comput.-Aided Des. Integr. Circuits Syst.* **2006**, *25*, 2765–2774. [CrossRef]
16. Perlow, S. Analysis of Edge-Coupled Shielded Strip and Slabline Structures. *IEEE Trans. Microw. Theory Tech.* **1987**, *35*, 522–529. [CrossRef]
17. Jiang, W.; Cai, X.-D.; Sen, B.; Wang, G. Equation-Based Solutions to Coupled, Asymmetrical, Lossy, and Nonuniform Microstrip Lines for Tab-Routing Applications. *IEEE Trans. Electromagn. Compat.* **2018**, *61*, 548–557. [CrossRef]
18. Cho, K.; Kim, Y.; Kim, S.; Park, H.; Park, J.; Lee, S.; Shim, D.; Lee, K.; Oh, S.; Kim, J. Fast and accurate power distribution network (PDN) modeling of a silicon interposer for 2.5-D/3-D ICs with multi-array TSVs. *IEEE Trans. Compon. Packag. Manuf. Technol.* **2019**, *9*, 1835–1846. [CrossRef]
19. Bedair, S.S. Characteristics of some asymmetrical coupled transmission lines. *IEEE Trans. Microw. Theory Tech.* **1984**, *MTT-32*, 108–110. [CrossRef]
20. Chen, B.; Pan, S.; Wang, J.; Yong, S.; Ouyang, M.; Fan, J. Differential Crosstalk Mitigation in the Pin Field Area of SerDes Channel With Trace Routing Guidance. *IEEE Trans. Electromagn. Compat.* **2019**, *61*, 1385–1394. [CrossRef]
21. Eisenstadt, W.; Eo, Y. S-parameter-based IC interconnect transmission line characterization. *IEEE Trans. Compon. Hybrids Manuf. Technol.* **1992**, *15*, 483–490. [CrossRef]
22. JEDEC. *High Bandwidth Memory DRAM (HBM1, HBM2) JESD235C*; JEDEC: Arlington, VA, USA, 2020.
23. Kim, D.; Kim, J.; Pak, J.; Lee, H.; Lee, J.; Park, K.; Kim, J. Vertical tree 3-dimensional TSV clock distribution network in 3D IC. In Proceedings of the 2012 IEEE 62nd Electronic Components and Technology Conference, San Diego, CA, USA, 9 May–1 June 2012; pp. 1945–1950. [CrossRef]
24. Kim, J.; Cho, J.; Pak, J.S.; Song, T.; Kim, J.; Lee, H.; Lee, J.; Park, K. I/O power estimation and analysis of high-speed channels in through-silicon via (TSV)-based 3D IC. In Proceedings of the 19th Topical Meeting on Electrical Performance of Electronic Packaging and Systems, Austin, TX, USA, 25–27 October 2010; pp. 41–44.

micromachines

MDPI

Article

Design of Power/Ground Noise Suppression Structures Based on a Dispersion Analysis for Packages and Interposers with Low-Loss Substrates

Youngwoo Kim

Nara Institute of Science and Technology, Division of Information Science, Ikoma-shi 630-0192, Japan; youngwoo@is.naist.jp; Tel.: +81+743-72-5953

Abstract: In this study, power/ground noise suppression structures were designed based on a proposed dispersion analysis for packages and interposers with low-loss substrates. Low-loss substrates are suitable for maintaining signal integrity (SI) of high-speed channels operating at high data rates. However, when the power/ground noise is generated in the power delivery network (PDN), low-loss substrates cannot suppress the power/ground noise, thereby causing PDN-induced crosstalk and various power integrity (PI) issues. To solve these issues, noise suppression structures generating electromagnetic bandgap were proposed and designed. The mechanism of the proposed structures was examined based on a proposed dispersion analysis. The proposed structures were designed and fabricated in glass interposer test vehicles, and the effectiveness of the structures on power/ground noise suppression was experimentally validated by measuring the noise suppression band. The proposed dispersion analysis was also verified by comparing the derived noise stopband edges (f_L and f_U) with electromagnetic (EM) simulation and experimental results, and they all showed good agreement. Compared to EM simulation, the proposed method required smaller computational resources but showed good accuracy. Using the proposed dispersion analysis, various power/ground noise suppression bands were designed considering the applications and design rules of packages and interposers. With measurements and EM/circuit simulations, the effectiveness of the designed structure in maintaining SI/PI was verified. By adopting the designed structures, the noise transfer properties in the PDN were suppressed in the target suppression frequency band, which is key for PI design. Finally, it was verified that the proposed structures were capable of suppressing power/ground noise propagation in the PDN by analyzing PDN-induced crosstalk in the high-speed channel.

Keywords: electromagnetic bandgap (EBG); interposers; low-loss substrates; noise suppression structures; packages; power delivery network (PDN); power/ground noise

Citation: Kim, Y. Design of Power/Ground Noise Suppression Structures Based on a Dispersion Analysis for Packages and Interposers with Low-Loss Substrates. *Micromachines* **2022**, *13*, 1433. https://doi.org/10.3390/mi13091433

Academic Editor: Wensheng Zhao

Received: 30 July 2022
Accepted: 26 August 2022
Published: 30 August 2022

Publisher's Note: MDPI stays neutral with regard to jurisdictional claims in published maps and institutional affiliations.

Copyright: © 2022 by the author. Licensee MDPI, Basel, Switzerland. This article is an open access article distributed under the terms and conditions of the Creative Commons Attribution (CC BY) license (https://creativecommons.org/licenses/by/4.0/).

1. Introduction

Transistor scaling based on Moore's law is facing a limit. At the same time, the realization of electrical systems with wide bandwidth, superior performances, small form factor, low power consumption, and reduced manufacturing cost has been a continuous challenge. System scaling based on through-silicon via and interposer technologies is a promising solution for current industrial challenges [1–5]. Recently, silicon interposers have been widely used to integrate graphic processing unit (GPU) and high bandwidth memory (HBM) to form 2.5-dimensional (2.5-D) systems to realize over terabyte per second (TB/s) system bandwidths for extremely high-performance computing [6,7]. However, reducing the manufacturing cost of silicon interposers is still difficult due to the throughput issues associated with limited wafer dimension. Even though silicon interposer-based integration and packaging provides promising solutions, reducing the manufacturing cost is difficult due to the following reasons: throughput issues associated with limited wafer dimension and additional fabrication steps to isolate metal layers from the conductive silicon substrate.

Moreover, the finite conductivity of the silicon substrate can generate signal integrity (SI) issues in the high-frequency range, which may limit high-speed digital signaling, radio frequency (RF) applications, and mixed-signal integrations [8]. Because of these limitations, packages and interposers with low-loss substrates have gained attention as a long-term alternative to silicon interposers.

Packages and interposers with low-loss substrates have been continuously developed. Glass substrates have several advantages, such as dimensional stability, coefficient of thermal expansion (CTE) similar to silicon, smooth surface, submicron metallization, double-sided build-up lamination, and superior electrical resistivity enabling low insertion loss up to millimeter (mm)-wave range [9–11]. Recently, glass substrates have been widely adopted for various packaging applications, such as bandpass filters for 5G communication [12], MEMS [13], millimeter-wave radar for autonomous driving [14], and antennas [15,16]. Fan-out packages with low-loss materials have also been widely adopted [17]. Ultrathin, high-permittivity, and low-loss substrates have been released in the market for printed circuit boards (PCBs), which can also be embedded in packages to realize embedded decoupling capacitors [18,19]. These substrate materials are fabricated from panels that are much larger in dimension compared to silicon wafers. Therefore, these substrates also have the potential to reduce the fabrication cost due to increased yield. The advantages are depicted in Figure 1.

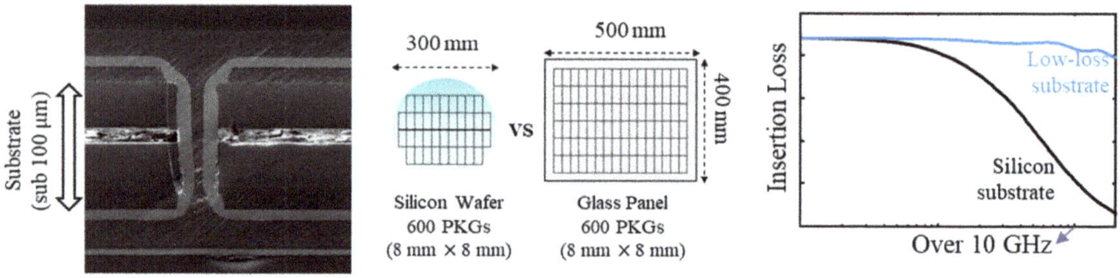

(a) (b) (c)

Figure 1. Advantages of low-loss substrate for packages and interposers. (**a**) Ultrathin (sub 100 μm) and fine-pitch metallization; (**b**) high fabrication yield, which has great potential for cost reduction; (**c**) low signal loss, enabling high-speed signaling.

Despite these advantages and potentials, low-loss substrates are vulnerable to power/ground noise generated in the power delivery network (PDN) [20]. Low-loss substrates are excellent for high-frequency signaling, but at the same time, it is difficult to suppress power/ground noise induced in the PDN. The noise can be induced in the PDN due to various reasons, such as simultaneous switching noise (SSN), return current discontinuity of the signal via penetration of the PDN, and coupling from other signal/power/ground interconnections. When the power/ground noise is generated, it propagates along the PDN without suppression, causing various SI/PI (power integrity) and electromagnetic interference (EMI) issues. These issues are shown in Figure 2. When designing a PDN, the allowed power/ground noise margin is becoming tighter because the operating voltages of electrical systems are continuously decreasing to realize low power consumption [21]. Moreover, due to recent trends requiring system-in-package, heterogeneous integration, and mixed modes, various noises are generated in the broadband frequency range. To maximize the advantages of low-loss substrates, power/ground noise must be suppressed and effective suppression bands should be analyzed in advance.

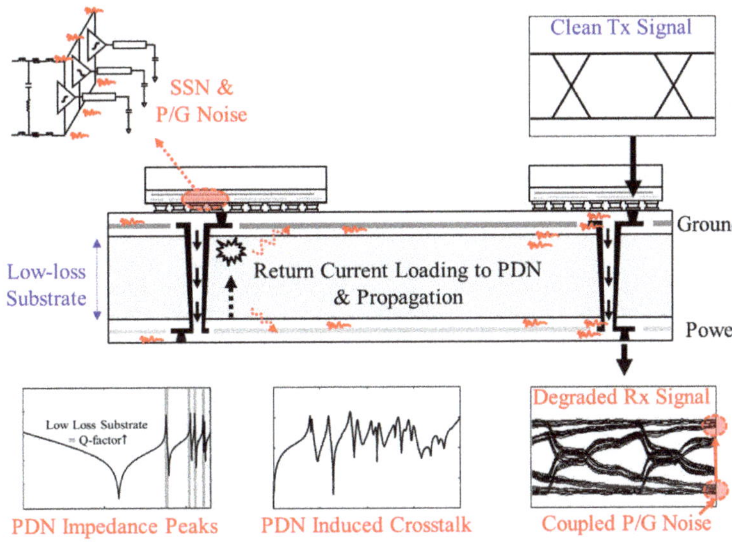

Figure 2. Issues related to low-loss substrate. When the noise is induced in the PDN due to SSN, P/G noise, or return current loading, low substrate loss cannot suppress it. The noise propagates along the PDN and causes SI/PI problems.

Decoupling capacitor arrays and schemes may be insufficient for achieving broadband power/ground noise suppression. Moreover, assembling various decoupling capacitors to achieve broadband suppression can increase the lateral dimensions of packages and interposers, which directly affects the fabrication yield. For security applications, exposed decoupling capacitors attached to the cryptographic core PDN cause electromagnetic (EM) information leakages and security issues [22,23]. Various electromagnetic bandgap (EBG) structures that are mostly embedded inside the PDN have been proposed and validated to achieve wideband power/ground noise suppression [24–31]. Considering recent trends requiring broadband noise suppression without dramatically affecting lateral dimensions of the package/interposer and unexposed areas for some applications, adopting the EBG structure embedded in the PDN is one of the most promising solutions to solve the power/ground noise issues associated with low substrate loss. However, such structures have not been widely developed and applied for packages and interposers with low-loss substrates. Therefore, an efficient design method considering design parameters and material properties is desired. Using the method, noise suppression structures should be developed and verified.

In this study, power/ground noise suppression structures were designed based on a proposed dispersion analysis for packages and interposers with low-loss substrates. The mechanism of noise suppression was thoroughly examined based on the proposed dispersion analysis. The structures were designed and fabricated in glass interposer test vehicles, and the effectiveness of the structures was experimentally validated by measuring the noise suppression band. The proposed dispersion analysis was also verified by comparing derived noise stopband edges (f_L and f_U) with electromagnetic (EM) simulation. It was confirmed that f_L and f_U estimations based on the proposed analysis method showed good agreement with those acquired from experiments and simulations. Compared to EM simulation, the proposed method required smaller computational resources but showed good accuracy, which is suitable for early design stages. Using the proposed dispersion analysis, various power/ground noise suppression bands were designed considering the applications and design rules of packages and interposers. With measurements and EM simulations, the effectiveness of the designed structure in maintaining SI/PI was verified. Finally, it was shown that the proposed structures were capable of suppressing power/ground noise propagation in the PDN by analyzing PDN-induced crosstalk in the high-speed channel.

2. Proposed Dispersion Analysis: Mechanism of Noise Suppression Band Formation and Stopband Edge Estimation

In this section, a dispersion analysis is proposed to explain the mechanism of noise suppression band formulation in the PDN. The proposed dispersion analysis is also capable of deriving noise suppression (stopband) edges. The proposed dispersion analysis is based on a transmission line (TL) theory and mathematics. Compared to full EM simulations, which require heavy computational resources, the proposed method can efficiently estimate the suppression band. In addition, the impacts of material properties and design rules on the suppression band can be easily understood.

Compared to mesh or grid-type PDN structures in silicon interposers, plane-type PDN can be fabricated in packages and interposers with low-loss substrates, such as glass, ceramic, and organic materials [9,18]. In the plane-type PDN, the power/ground noise propagates in forms of transverse electromagnetic (TEM) and quasi-TEM modes. The PDN becomes a transmission line (TL) for the noise wave. To suppress noise propagation, a noise suppression band can be formed by designing certain repetitive structures generating lumped capacitance (C) and inductance (L). The band must be analyzed and engineered to cover the target noise band. In this study, power/ground noise propagation is analyzed in the +x-direction ($\mathbf{N}(x) = e^{-jk_x x}\mathbf{a}_x$). k_x is an effective phase constant defined as $k_x = \alpha_x + j\beta_x$, where α_x becomes an attenuation constant and β_x is a propagation constant of the power/ground noise wave propagating in the +x-direction. Let the size of the repetitive structure formed in the PDN be W_u. A two-dimensional periodic structure can be reduced into a one-dimensional array of unit cells (size of W_u) by placing perfect magnetic conductor (PMC) walls at $y = \pm W_u/2$, as depicted in Figure 3. This assumption and dimension reduction can be applied as the PMC wall can be located anywhere where there is a zero tangential magnetic (H) field [31–33]. By adopting the PMC boundary condition, the noise suppression structure and the PDN can be modeled as TL with lumped C and L. In the following sections, detailed equations for C and L are provided considering the design, physical dimensions, and material properties.

In typical advanced packages and interposers, build-up layers exist in between the substrate and metal layers (ML), as shown in Figure 3. The power/ground noise propagating in the form of the quasi-TEM mode can be modeled into a TL. The TL has a thickness of t_{PDN}, which is summation of the substrate thickness (t_{sub}) and the thickness of two build-up layers (t_{bu}). The dielectric layers in between power and ground planes can be modeled as a single dielectric mixture layer represented with the effective complex dielectric constant ($\varepsilon_{mix}(\omega)$) shown in the following equation:

$$\varepsilon_{mix}(\omega) = \frac{\varepsilon_{sub}(\omega)\varepsilon_{bu}(\omega)}{q\varepsilon_{sub}(\omega) + (1-q)\varepsilon_{bu}(\omega)} \quad (1)$$

where q is the volume fraction of the dielectric layers; $\varepsilon_{sub}(\omega)$ and $\varepsilon_{bu}(\omega)$ are complex permittivity of the substrate and build-up layer, respectively [34]; and ω is the angular frequency ($2\pi f$). Real parts of the $\varepsilon_{mix}(\omega)$, $\varepsilon_{sub}(\omega)$, and $\varepsilon_{bu}(\omega)$ are defined as $\varepsilon_{r,mix}$, $\varepsilon_{r,sub}$, and $\varepsilon_{r,bu}$, respectively.

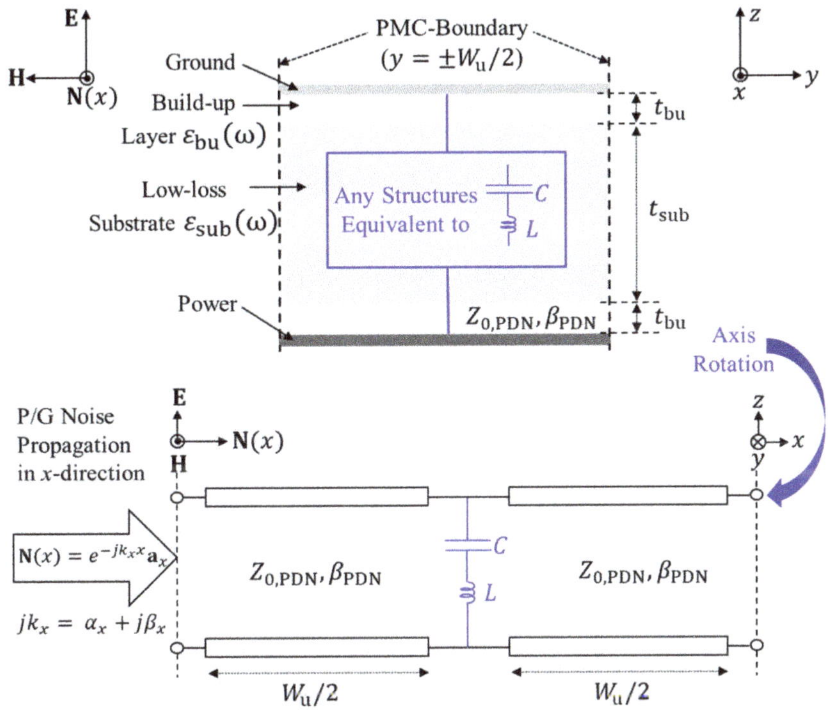

Figure 3. One-dimensional equivalent circuit model of the PDN with structure generating C and L. Considering the H-field direction, PMC boundary condition is adopted to reduce the two-dimensional PDN into one-dimensional domain in the x-direction. $\mathbf{a}_x$ is a unit vector.

Characteristic impedance ($Z_{0,\,\mathrm{PDN}}$) and phase constant (β_{PDN}) of the TL without C and L can be expressed as follows:

$$Z_{0,\mathrm{PDN}} = \frac{\eta_0}{\sqrt{\varepsilon_{\mathrm{r,mix}}}} \frac{t_{\mathrm{PDN}}}{W_{\mathrm{u}}},\ t_{\mathrm{PDN}} = t_{\mathrm{sub}} + 2t_{\mathrm{ub}} \tag{2}$$

and

$$\beta_{\mathrm{PDN}} = \sqrt{\varepsilon_{\mathrm{r,mix}}}\,\omega / c \tag{3}$$

where η_0 is the wave impedance of free space, and c is the speed of light in a vacuum.

To derive the dispersion equations to estimate the noise stopband, *ABCD* parameters of the TL shown in Figure 3 is analyzed. The unit cell of the TL has an effective phase constant k_x. Compared to the silicon, the substrate and build-up layer of the target study have much lower loss factors. Therefore, lossless condition is adopted for an efficient calculation during the dispersion analysis. The *ABCD* parameters of the TL's unit cell can be expressed as follows:

$$\begin{aligned}\begin{bmatrix} A_{\mathrm{u}} & B_{\mathrm{u}} \\ C_{\mathrm{u}} & D_{\mathrm{u}} \end{bmatrix} &= \begin{bmatrix} \cos(k_x W_{\mathrm{u}}) & jZ_{0,\mathrm{u}} \sin(k_x W_{\mathrm{u}}) \\ jZ_{0,\mathrm{u}}{}^{-1} \sin(k_x W_{\mathrm{u}}) & \cos(k_x W_{\mathrm{u}}) \end{bmatrix} \\ &= \begin{bmatrix} \cos\left(\beta_{\mathrm{PDN}} \frac{W_{\mathrm{u}}}{2}\right) & jZ_{0,\mathrm{PDN}} \sin\left(\beta_{\mathrm{PDN}} \frac{W_{\mathrm{u}}}{2}\right) \\ jZ_{0,\mathrm{PDN}}{}^{-1} \sin\left(\beta_{\mathrm{PDN}} \frac{W_{\mathrm{u}}}{2}\right) & \cos\left(\beta_{\mathrm{PDN}} \frac{W_{\mathrm{u}}}{2}\right) \end{bmatrix} \times \begin{bmatrix} 1 & 0 \\ Y & 1 \end{bmatrix} \times \begin{bmatrix} \cos\left(\beta_{\mathrm{PDN}} \frac{W_{\mathrm{u}}}{2}\right) & jZ_{0,\mathrm{PDN}} \sin\left(\beta_{\mathrm{PDN}} \frac{W_{\mathrm{u}}}{2}\right) \\ jZ_{0,\mathrm{PDN}}{}^{-1} \sin\left(\beta_{\mathrm{PDN}} \frac{W_{\mathrm{u}}}{2}\right) & \cos\left(\beta_{\mathrm{PDN}} \frac{W_{\mathrm{u}}}{2}\right) \end{bmatrix}\end{aligned} \tag{4}$$

where $Y = j\omega C/(1 - \omega^2 LC)$, and $Z_{0,\mathrm{u}}$ is the characteristic impedance of TL with the structure inducing L and C. Among the four parameters, A_u is the simplest. By analyzing A_u, a dispersion equation is derived as follows:

$$A_\mathrm{u} = \cos(k_x W_\mathrm{u}) = \cos(\beta_\mathrm{PDN} W_\mathrm{u}) + j\frac{Z_{0,\mathrm{PDN}} Y}{2}\sin(\beta_\mathrm{PDN} W_\mathrm{u}) \tag{5}$$

From (5), the effective phase constant k_x is derived as follows:

$$k_x = \frac{1}{W_\mathrm{u}}\cos^{-1}\left[\cos(\beta_\mathrm{PDN} W_\mathrm{u}) - \frac{\omega\, Z_{0,\mathrm{PDN}}\, C}{2(1 - \omega^2 LC)}\sin(\beta_\mathrm{PDN} W_\mathrm{u})\right] \tag{6}$$

In this study, the lossless condition has been assumed. However, the effective phase constant k_x becomes a complex number. The effective phase constant has an imaginary part when the argument of the inverse cosine function in (6) is outside the interval $[-1, 1]$. The noise stopband is formed at the frequency range where the imaginary part of (6) is nonzero and changes dramatically. The wave becomes evanescent at this frequency range.

To graphically explain the noise propagation characteristics, dispersion diagrams are estimated from (6) and plotted in Figure 4 as an example. The design parameters and values to derive dispersion diagrams shown in Figure 4 are summarized in Table 1. Values are carefully chosen to derive diagrams that can graphically support the method and concept. Without noise suppression structures in the PDN, the propagation constant of the power/ground noise has a constant slope ($2\pi\sqrt{\varepsilon_{r,mix}}/c$) in the frequency domain, which is nondispersive. When the structure forming the L and C is inserted in the PDN, the power/ground noise propagating in the +x-direction ($e^{-jk_x x}$) will experience sudden attenuation at certain frequency band. Let $-jk_x = -\alpha_x - j\beta_x$, and α_x becomes an attenuation constant and β_x represents propagation constant of the power/ground noise wave. In this case, β_x is periodic due to periodicity of the inverse cosine function.

Table 1. Summary of important design parameters and values used to derive dispersion diagrams shown in Figure 4.

$\varepsilon_{r,mix}$	$Z_{0,\mathrm{PDN}}$	W_u	L	C
4.25	9.14 Ω	3.2 mm	125.49 pH	22.831 pF

As can be seen from Figure 4, there are frequency bands where the slope of β_x becomes closer to zero (or β_x is at the Brillouin zone boundaries) or the attenuation constant α_x becomes nonzero. These frequency bands are theoretical noise bandgap or electromagnetic bandgap (EBG) where power/ground noise is suppressed and cannot propagate. However, except for the fundamental stopband, the attenuation constant for other bands are too small to suppress noise propagation.

In this study, the dispersion equations are calculated and plotted under the assumption of lossless substrate and build-up layer. In reality, there always exists attenuation of the wave associated with dielectric loss. Usually, noise suppression bands are valid if they can achieve −40 dB or lower isolation characteristics. Therefore, theoretical upper bands marked in Figure 4 cannot play a role as a suppression band. A band that can suppress power/ground noise will exist inside the fundamental stopband marked in Figure 4.

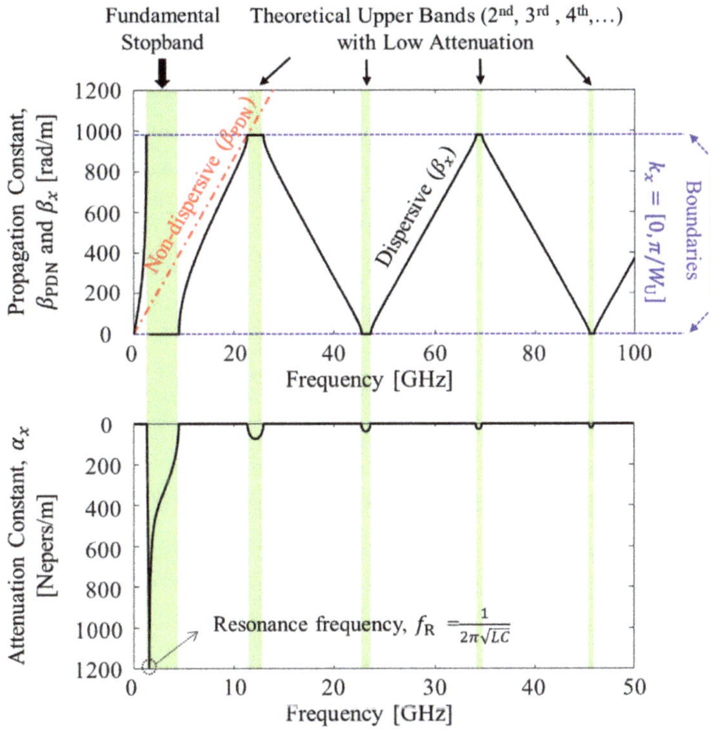

Figure 4. Dispersion diagrams. Without noise suppression structures in the PDN, the power/ground noise is nondispersive. With noise suppression structures, theoretical stopbands are generated.

The lower edge of the noise suppression band (f_L) can be derived by finding the condition satisfying "$A_u = \cos(k_x W_u) = -1$". Among various cases, "$k_x W_u = \pi$" is selected in this study. Because $\beta_{PDN} W_u$ is located far from the Brillouin zone boundary at the lower cut-off frequency, the small-angle approximation "$\sin(\beta_{PDN} W_u) \approx \beta_{PDN} W_u$" and "$\cos(\beta_{PDN} W_u) \approx 1$" can be used. Under these conditions, (6) can be rewritten as follows:

$$-1 = 1 - \frac{\omega Z_{0,PDN} C}{2(1 - \omega^2 LC)} \beta_{PDN} W_u. \tag{7}$$

Because the wave impedance of free space (η_0) can be expressed as $\mu_0 c$, the lower edge of the noise suppression band (f_L) can be derived from (7) and is shown in the following equation:

$$f_L = \frac{1}{\pi \sqrt{C\,(\mu_0 t_{PDN} + 4L)}} \tag{8}$$

By setting different boundary condition, it is possible to estimate the upper edge of the noise suppression band (f_U). It can be obtained by adopting conditions that satisfy "$A_u = \cos(k_x\, W_u) = +1$". Under this condition, (6) becomes the following:

$$\tan\left(\frac{\beta_{PDN} W_u}{2}\right) = \frac{-\omega\, C\, Z_{0,PDN}}{2(1 - \omega^2 LC)} \tag{9}$$

If f_U and the resonant frequency ($f_R = \frac{1}{2\pi\sqrt{LC}}$) are far away from each other, (9) can be approximated as follows:

$$\tan\left(\frac{\pi\sqrt{\varepsilon_{r,mix}}W_u}{c}f_U\right) \approx \pi c f_R{}^2 Z_{0,PDN}\frac{1}{f_U} = \frac{Z_{0,PDN}}{4\pi L}\frac{1}{f_U} \quad (10)$$

From (10), it is difficult to directly estimate f_U. A graphical (numerical) approach is applied to estimate f_U, and an example is shown in Figure 5. In Figure 5, the left-hand side (LHS) and right-hand side (RHS) of (10) are plotted in the frequency domain. The frequency where the two graphs intersect is the upper edge of the noise suppression band (f_U).

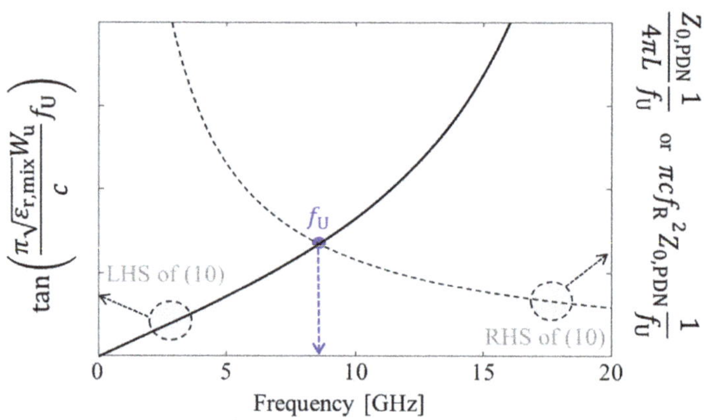

Figure 5. Estimation of f_U based on a graphical approach shown as an example. The LHS and RHS of (10) are plotted in the frequency domain. The intersection frequency of LHS and RHS is f_U.

The following section outlines the design and fabrication of test vehicles with thin and low-loss glass substrate and low-loss polymer as a build-up material. The test vehicles were measured to validate the proposed dispersion analysis and effectiveness of the proposed structures on noise suppression.

3. Verification of the Proposed Dispersion Analysis and Noise Suppression Structures

3.1. Design and Fabricated Test Vehicles

To validate the proposed dispersion analysis and noise suppression structures, two structures were designed and fabricated. In Figure 6, cross-sections/top views of the structures in the glass interposer test vehicles are shown. As shown in Figure 6a, this structure (Type A) had double patches to increase capacitance. Four metal layers (MLs) were needed to form this structure. In Figure 6b, a simpler structure (Type B) is shown, which only had one patch and three MLs were required. Because the copper used to form metal layers (MLs) does not adhere to the glass substrate directly, low-loss polymer was used between the substrate and MLs. In glass packages and interposers, using a low-loss polymer provides various advantages, such as additional mechanical strength, prevention of substrate cracking, prevention of moisture contact, and CTE control [35]. In each test vehicle, 25 (5 in x-direction and 5 in y-direction) unit structures, shown in Figure 6, were embedded in the PDN. The top view of the test vehicle with 25 Type A unit structures is shown in Figure 6c. Similarly, the top view of the test vehicle with 25 Type B unit structures is shown in Figure 6d. In Table 2, the physical dimensions and material properties of the test vehicles are summarized. More detailed process design rules and explanations of the structures have been described in previous works [36,37].

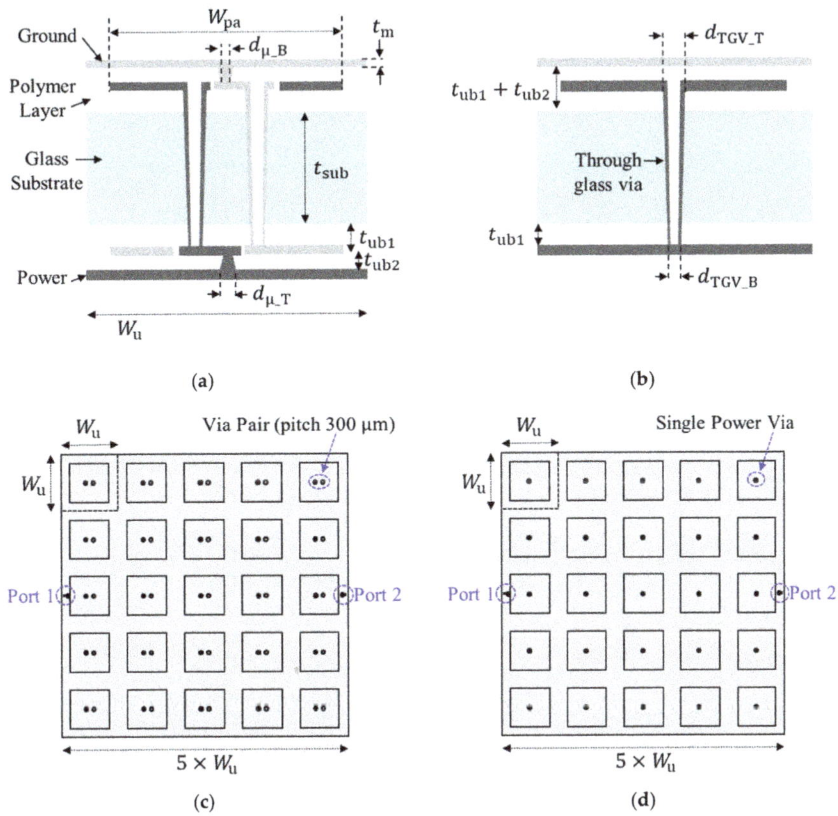

Figure 6. Designed noise suppression structures in the glass interposer test vehicles. (**a**) Type A: double-patch structure; (**b**) Type B: single-patch structure. Top view of the fabricated test vehicles composed with (**c**) Type A and (**d**) Type B. Each test vehicle has 25 unit structures.

Table 2. Physical dimensions and material properties of the glass interposer test vehicles.

	Symbol	**Type A**	**Type B**
Physical Dimensions	t_{sub}	100 μm	
	t_{bu1}	15 μm	17.5 μm
	t_{bu2}	15 μm	35 μm
	t_m	3~5 μm	4~10 μm
	d_{TGV_T}	60 μm	100 μm
	d_{TGV_B}	40 μm	60 μm
	$d\mu_{\_T}$	35 μm	45 μm
	$d\mu_{\_B}$	30 μm	45 μm
	W_u	3.2 mm	
	W_{pa}	2.2 mm	
	p_{TGV}	300 μm	NA
Material Properties	$\varepsilon_{r,sub}$	5.3 @ 2.4 GHz	5.3 @ 2.4 GHz
	$\varepsilon_{r,bu}$	3.2 @ 5.8 GHz	3 @ 10 GHz
	$tan\ \delta_{sub}$	0.004 @ 2.4 GHz	0.004 @ 2.4 GHz
	$tan\ \delta_{bu}$	0.0042 @ 5.8 GHz	0.005 @ 10 GHz
	σ_m	$5.8 \times 10^7 \sigma/\text{m}$	

Each structure has different lumped capacitance (C) and inductance (L). First, the structure of Type A was analyzed to derive lumped C and L. The lumped capacitance (C_A) can be derived by adding capacitance between patches and planes (C_{pa}) and capacitance between the power/ground through-glass via (TGV) pair (C_{TGV}) [38]. They can be summarized as follows:

$$C_A = C_{pa} + C_{TGV} \tag{11}$$

$$C_{pa} = \frac{\varepsilon_0 \varepsilon_{r,bu}\left(2(W_{pa})^2 - \pi(d_{TGV\_T}/2)^2 - \pi(d_{TGV\_B}/2)^2\right)}{t_{bu2} - t_m} \tag{12}$$

$$C_{TGV} = \int_{z=0}^{t_{bu1}} \frac{\pi\varepsilon_{r,bu}}{\cos h^{-1}(p_{TGV}/2r(z))}dz + \int_{z=t_{bu1}}^{t_{bu1}+t_{sub}} \frac{\pi\varepsilon_{r,sub}}{\cos h^{-1}(p_{TGV}/2r(z))}dz + \int_{z=t_{bu1}+t_{sub}}^{2t_{bu1}+t_{sub}} \frac{\pi\varepsilon_{r,bu}}{\cos h^{-1}(p_{TGV}/2r(z))}dz \tag{13}$$

where

$$r(z) = \frac{d_{TGV\_B}}{2} + \frac{d_{TGV\_A} - d_{TGV\_B}}{2(2t_{ub1} + t_{sub})}z \tag{14}$$

The lumped inductance (L_A) can be modeled as follows:

$$L_A = L_{TGV} + 2L_{\mu via} \tag{15}$$

$$L_{TGV} = \int_{z=0}^{2t_{bu1}+t_{sub}} \frac{\mu_0}{\pi}\cos h^{-1}(p_{TGV}/2r(z))\,dz \tag{16}$$

$$L_{\mu via} = \frac{\mu_0(t_{bu2} - t_m)}{4\pi}\left[\ln\left(\frac{4\,W_u{}^2}{\pi d_{\mu via}{}^2}\right) + \frac{\pi d_{\mu via}{}^2}{4\,W_u{}^2} - 1\right] \tag{17}$$

Equation (16) has a close relationship with (13), which dominates L_A. Derivation of (16) is shown in [38]. Because microvia is not paired, the derivation of (17) is a bit different from (16). $L_{\mu via}$ can be derived from the magnetic energy in the unit structure, which is known as $U_m = \frac{1}{2}\int \mathbf{B}\cdot\mathbf{H}\,dv$. By adopting boundary condition $U_m = \frac{1}{2}I^2L$, it is possible to derive (17) [24,39]. This relationship can be used to derive the inductance of the single TGV in the Type B structure. Because microvia is relatively shorter than the TGV pair, d_{μ_T} and d_{μ_B} was averaged to derive $d_{\mu via}$ in (17).

The Type B structure shown in Figure 6b is simple compared to Figure 6a. The lumped capacitance (C_B) can be expressed as follows:

$$C_B = \frac{\varepsilon_0 \varepsilon_{r,bu}\left(2(W_{pa})^2 - \pi(d_{TGV\_T}/2)^2\right)}{t_{bu2} - t_m} \tag{18}$$

The inductance of the single TGV can be obtained by modifying (17). It can be summarized as follows:

$$L_B = L_{single-TGV} = \int_{z=0}^{t_{bu1}+t_{sub}} \left[\ln\left(\frac{W_u{}^2}{dS(z)}\right) + \frac{dS(z)}{W_u{}^2} - 1\right] dz \tag{19}$$

where

$$dS(z) = \pi\left(\frac{d_{TGV\_B}}{2} + \frac{d_{TGV\_A} - d_{TGV\_B}}{2(2t_{ub1} + t_{sub})}z\right)^2. \tag{20}$$

Basically, the derivation process of (17) and (19) is identical because they are both single via confined in the unit cell. Therefore, the two equations are similar. In (19) and (20), the tapered structure of the TGV is reflected. If the length of the TGV is very short and it is not tapered, the integral calculation in (19) can be simplified as in (17). The derived parameters can be inserted into the proposed dispersion analysis explained in the previous section to analyze the noise suppression band.

In Figure 7, fabricated glass interposer test vehicles are shown. Metal patches generating C, through-glass via (TGV), and planes are shown. Optical microscope and scanning electron microscope (SEM) were used to take images of various structures inside the glass interposer test vehicles. As can be seen from Figure 7, measurements were conducted on the probe station. Various measurements were conducted in both frequency and time domains. After verifying the effectiveness of the dispersion analysis, more structures were designed, as outlined in Section 4.

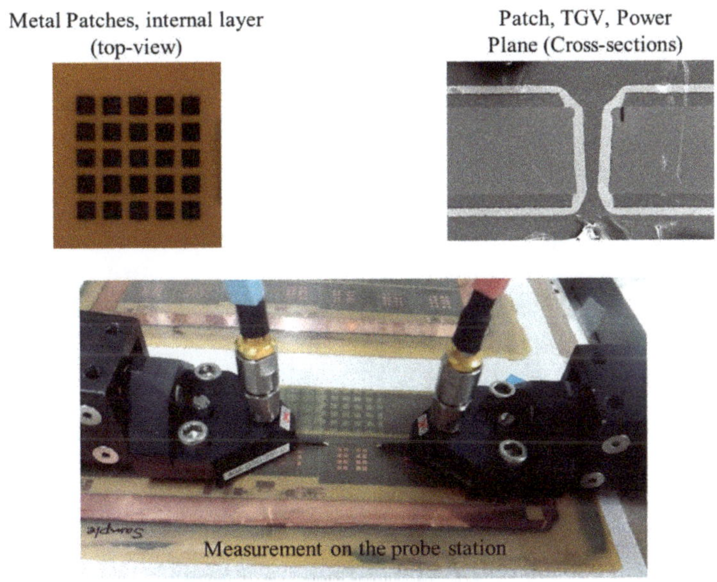

Figure 7. Measurement of fabricated glass interposer test vehicles.

3.2. Verification by Measurement and EM Simulation

Measured power/ground noise couplings (S_{21}) are plotted and compared in Figure 8. Two microprobes (Picoprobe GS type with 250 μm pitch, GGB industries Inc., Naples, FL, USA), two coaxial cables (W.L. Gore & Associates, Inc., Newark, DE, USA), and a calibration kit (CS-14, GGB industries Inc., Naples, FL, USA) were used to measure power/ground noise couplings. A vector network analyzer (VNA) (N5230A, Keysight, Santa Rosa, CA, USA) was used to measure the couplings in the PDN and validate the noise suppression band in the frequency domain up to 20 GHz. As a reference, a PDN without noise suppression structures was also measured. The distance between the two measurement ports (port 1 and port 2 shown in Figure 6c,d) was approximately 15.5 mm ($\approx 5 \times W_u$). In such cases, it is difficult to suppress the generated noise in the PDN due to low-loss substrate and polymer build-up layer. By adopting the proposed structures (Types A and B), −40 dB noise suppression bands were generated. In these frequency bands, the power/ground noise will be significantly attenuated and isolated.

Due to the double patches and paired through vias, the Type A structure had much larger lumped capacitance (C) than the Type B structure. The total lumped inductance (L) of the Type A structure was 130.80 pH, whereas it was 89.65 pH for the Type B structure. As can be seen from (8), the Type A structure had a lower f_L compared to Type B due to larger capacitance. With larger C, the noise suppression band can be expanded by lowering f_L. However, the Type A structure had larger L compared to the Type B structure, which also lowered f_H. To achieve wider noise suppression band, higher f_H is desired. In the following section, the impacts of various design parameters and material properties on the noise suppression band is analyzed. A design direction to achieve wider noise suppression band is also given.

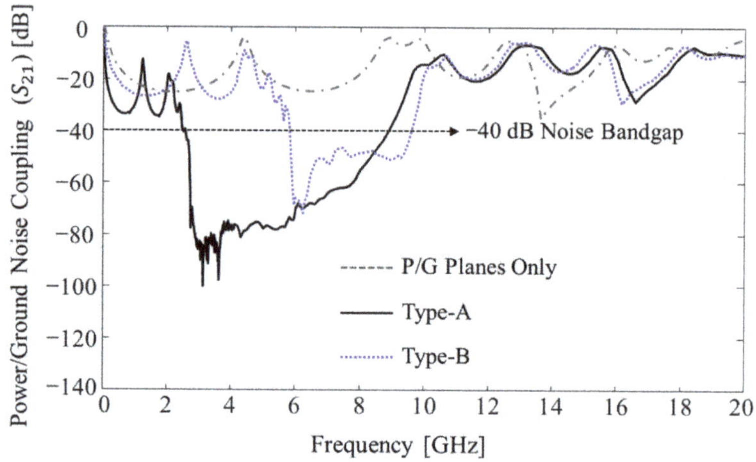

Figure 8. Comparison of the measured power/ground noise couplings (S_{21}) in the PDN.

In Figure 9, the measurement results are compared with simulated results. Using the 3-D EM simulator Ansys high-frequency structure simulator (HFSS) (version 2020 R2), noise coupling (S_{21}) of each structure was estimated and compared. The measurement and simulation results showed good agreement up to 20 GHz in the frequency domain for both structures. To verify the proposed dispersion analysis, estimated stopband edges (f_L and f_U) were compared with the simulated and measured edges. In Table 3, edges obtained by different methods are summarized and compared. The estimated edges showed good correlation with the simulated and measured edges. The accuracy of the proposed dispersion analysis for stopband edge estimation was verified. When fabricating the glass interposer test vehicles, the diameter of the TGV showed process variations associated with substrate drilling and copper plating. If the process becomes more mature, more accurate results are expected.

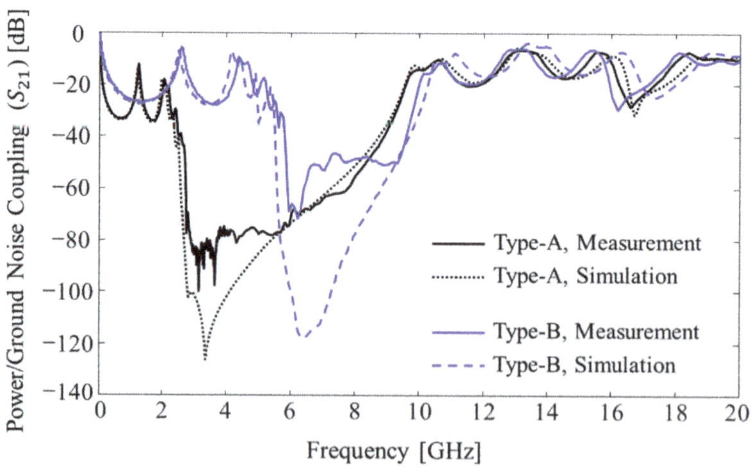

Figure 9. Measured power/ground noise couplings are compared with simulated results.

Table 3. Summary and comparison of measured, simulated, and estimated stopband edges.

Structures	Edges	Measurement	Simulation (Error)	Estimation (Error)
Type A	f_L	2.51 GHz	2.49 GHz (1.00%)	2.51 GHz (0.04%)
	f_U	8.91 GHz	8.75 GHz (1.83%)	8.59 GHz (3.61%)
Type B	f_L	5.82 GHz	5.50 GHz (5.49%)	5.87 GHz (0.86%)
	f_U	9.66 GHz	9.75 GHz (0.93%)	10.4 GHz (7.66%)

Compared to 3-D EM simulation, the proposed method required less time and computational resources to estimate the stopband edges. When designing the structure, the proposed dispersion analysis could effectively estimate the stopband considering the design rules and target noise band. After preliminary analysis, the structure could be designed in the 3-D EM simulator for further analysis and validation before tape-out.

4. Design and Analysis of Noise Suppression Structures with Various Low-Loss Materials

In this section, design directions are discussed based on the proposed dispersion analysis. The impacts of various design parameters and material properties are considered. Some candidates are chosen, and noise stopbands are estimated. Additional measurements and 3-D EM simulations are conducted to verify the impacts of the proposed structures on power/ground noise suppression and decoupling.

4.1. Noise Suppression Band Formulation with Various Materials

In general, broadband noise suppression is desired to cover various applications. To achieve broadband, the lower stopband edge (f_L) should be designed toward lower frequency. At the same time, the upper stopband edge (f_U) should be formed at higher frequency. Design parameters and material have significant impacts, but not all of them can be realized and adopted. It can be limited by process design rules or have a conflict with usages of advanced packages and interposers. In Table 4, the stopband expansion method by changing design parameters and impacts is summarized.

Table 4. Stopband expansion method by changing design parameters and impacts.

Band Expansion	Design Parameters	Impacts	Note
$f_L \downarrow$	High-*K* materials in BU	C ↑	
	Thin BU materials	C ↑	
	Larger pactes (W_{pa})	C ↑	Limited
	Increase package/interposer and PDN thickness ($t_{PDN}\uparrow$)		Not desired
	Add defects in P/G planes	$L\uparrow$	Limited
	Through via diameter ↓	$L\uparrow$	Limited & $1/\sqrt{L}$
$f_U \uparrow$	Through via diameter ↑	$L\downarrow$	$1/L$
	Via arrays or parallel vias	$L\downarrow$	$1/L$
	$Z_{0,PDN}\uparrow(t_{PDN}\uparrow)$	RSH of (10) ↑	Not desired

By analyzing the proposed dispersion equations, the impacts of design parameters can be easily determined. As can be seen from (8), increasing lumped capacitance (C), inductance (L), and PDN thickness will lower f_L. Adopting build-up materials with high permittivity can increase C. At the same time, selecting thinner build-up materials will have the same impact. However, using larger structures can increase the overall x–y dimensions of the packages/interposers. The x–y dimensions directly affect the overall fabrication yield, so this design direction is not desired. Moreover, a design direction that increases the PDN thickness should be avoided. Adding defects in power/ground planes increases L [31]. In terms of power integrity, this can be a good solution. However, adding defects can cause return current discontinuity issues. If such a design is adopted, routings, fan-out, and signal integrity analysis should be carefully conducted as well.

The diameter of the through-substrate via affects both stopband edges, so it is highlighted in Table 4. When the diameter is altered, RHS of (10) is affected more dramatically, which is inversely proportional to L, whereas (8) is inversely proportional to $\sqrt{L}$. Moreover, the diameter of the through substrate is heavily determined by the process design rule. It is difficult to freely modify the diameter when designing packages/interposers. It is more realistic to adopt parallel through via scheme than changing the diameter to achieve lower L. A design that can achieve higher characteristic impedance ($Z_{0,\mathrm{PDN}}$) shifts RSH of (10) and can formulate f_U at higher frequency. However, compared to other parameters, changing $Z_{0,\mathrm{PDN}}$ is not easy as it will have multiple impacts. The easiest method is increasing the PDN thickness, but this direction is not desired in advanced packages/interposers.

In Figure 10, dispersion diagrams with design parameters and material properties are plotted for comparison. For fare comparison, the normalized value in radian ($\beta_x W_u$) is plotted instead of the propagation constant (β_x). Fundamental bandgaps are formed in the bands where $\beta_x W_u$ does not exist. High-K material alumina can be embedded in polymer build-up layers of glass packages and interposers to increase C [40]. Moreover, ultrathin and high-K materials, such as FaradFlex substrate, can be embedded in the build-up layers of packages [18]. These materials can also be used for miniaturization of structures instead of increasing C. By adopting parallel trough via arrays, the suspression band is expanded toward higher frequency.

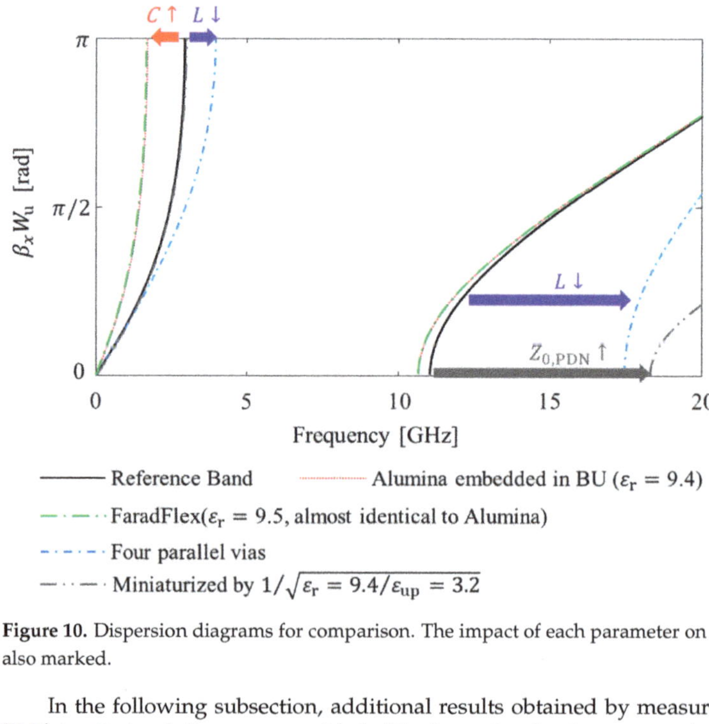

Figure 10. Dispersion diagrams for comparison. The impact of each parameter on stopband edge is also marked.

In the following subsection, additional results obtained by measurement and 3-D EM/circuit simulations are provided. The impacts of noise suppression/isolation are graphically delivered in the time domain.

4.2. Impacts of Power/Ground Noise Decoupling Using the Proposed Structures

In Figure 11, measured power/ground noise coupling results are plotted and compared in the time domain. A pulse-pattern generator (PPG) (Anritsu MP-1763C, Atsugi, Japan) and a digital sampling oscilloscope (Tektronix TDS800B, Beaverton, OR, USA) were used to conduct measurements in the time domain. The Type B structure explained in Section 3 was used for the experiment. Twelve gigabits per second (GB/s) clock signal (0 to 1 V, 30 ps rise-and-fall time and all ports terminated with 50 ohm) was injected to the interposer PDN (port 1 in Figure 6d) as a noise source. Frequency band of the injected noise existed in the suppression band of the Type B structure. From the noise source to the measurement location in the PDN (port 2 in Figure 6d), five unit structures existed, and the distance was approximately 15.5 mm. Without the noise suppression structure, 142 mV peak-to-peak voltage (V_{pp}) was measured, which corresponded to 14.2% of the input voltage. When the proposed noise suppression structure was fabricated in the PDN, 51 mV V_{pp} was observed. Significant noise suppression/isolation was achieved by adopting the proposed structure.

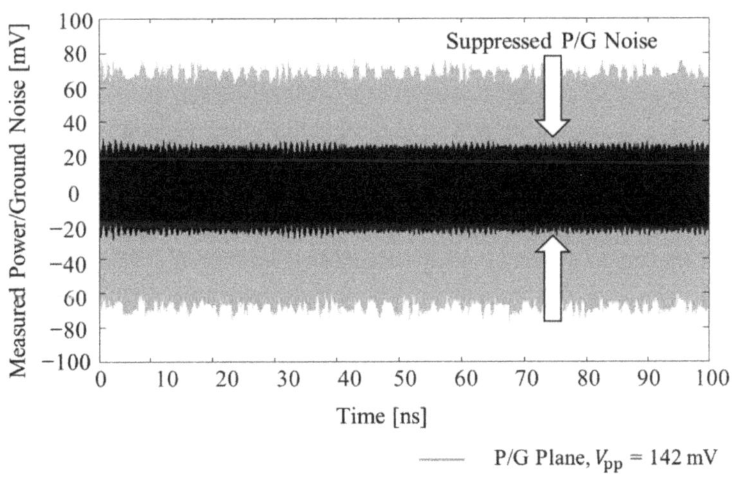

Figure 11. Comparison of the measured power/ground noise coupling results in the time domain without and with the noise suppression structure (Type B). By adopting the proposed structure, power/ground noise was significantly suppressed in the interposer with low-loss substrate.

Additional 3-D EM/circuit simulations were conducted to verify the effectiveness of the proposed structure. In this study, eye diagrams of the through-substrate (glass or FaradFlex) via channel under the influence of power/ground noise coupling were simulated and compared. The through-substrate via channel (victim) was designed to penetrate the PDN, and the noise source was located far away from the victim. In this scenario, the noise induced in the PDN propagates without attenuation, couples to the victim channel, and degrades SI of the victim channel (PDN-induced crosstalk). This scenario is likely to happen as there are thousands of signal through-substrate vias escaping packages or interposers, such as SerDes. Pseudo-random binary sequence (PRBS) of $2^8 - 1$, 0 V to 1.2 V, with rise-and-fall time of 30 ps and data rate (DR) of 2 GB/s was injected to the victim channel. The total length of the victim channel was designed to be approximately 14 mm, including the through via, microvia, interposer channel, and PCB channel located under the package. At the receiving location of the victim channel where eye diagrams were monitored, a capacitive termination was applied. This simulation scenario was graphically depicted in Figure 12.

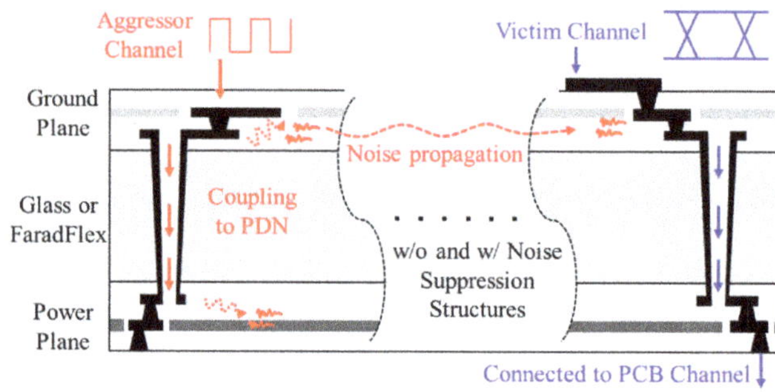

Figure 12. Graphical depiction of the simulation scenario. Aggressor channel escaping the package or interposer induces noise in the PDN. The noise propagates and couples to the victim channel. Eye diagrams of the victim channel are compared without and with noise suppression structures.

In Figure 13, eye diagrams of the victim channel are plotted and compared without and with the proposed structure. In Figure 13, the Type B structure is used as a representative. In Table 5, additional results are also summarized for other structures, and eye-opening voltage, jitter, and maximum V_{pp} noise at logics zero/one are compared. For all three cases, eye diagrams of the through-substrate via channel were improved by adopting the proposed structures. Low-loss substrates for packages and interposers provide various advantages. However, power/ground noise must be isolated and suppressed. In this study, noise suppression was conducted by designing various structures in the PDN. The structures were analyzed and determined by the proposed dispersion analysis. Compared to full 3-D EM simulations, the proposed dispersion analysis is fast and requires smaller computational resources. Therefore, the proposed dispersion analysis is useful at the preliminary PDN design stage.

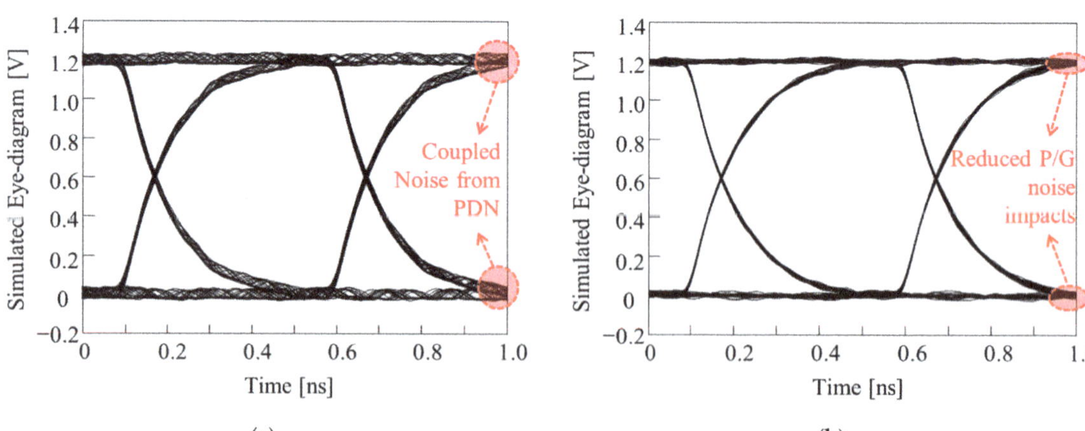

Figure 13. Verifications of the impacts of the proposed structure by eye diagram simulations. By adopting the proposed structure, power/ground noise coupling from the PDN to through-substrate via channel was suppressed. (**a**) P/G planes only. (**b**) embedded Type B structure.

Table 5. Summary of the eye diagram improvement by adopting the proposed structures.

Structures	Eye-Opening Voltage	Jitter (% of UI)	P/G Noise at 0 or 1
Type A	784 mV →838 mV	36.5 ps (7.1) → 22.2 ps (4.4)	60 mV →19 mV
Type B (Figure 13)	1.08 V →1.15 V	18 ps (3.6) → 7 ps (1.4)	62 mV →36 mV
FaradFlex based (Figure 10)	1.05 V →1.11 V	21 ps (4.2) → 13 ps (2.6)	70 mV →39 mV

5. Conclusions

In this study, dispersion analysis was proposed to efficiently design power/ground noise suppression structures for packages and interposers with low-loss substrates. The mechanism of noise suppression/isolation was thoroughly explained based on the proposed dispersion analysis. By conducting the proposed dispersion analysis, the impacts on physical design parameters and material properties on the suppression band could be easily explained. To validate the proposed dispersion analysis, noise suppression structures were designed and fabricated in the glass interposer PDN and measured. It was verified that f_L and f_U estimated based on the proposed analysis method showed good agreement with those acquired from experiments and simulations. Compared to EM simulation, the proposed method required smaller computational resources but showed good accuracy. Various structures were designed and analyzed based on the proposed dispersion analysis. The effectiveness of the proposed structures was further validated by additional experiments and simulations in the time domain. The proposed structures suppressed power/ground noise propagation and coupling.

Low-loss substrates for packages and interposers provide various advantages, especially for high-speed signaling. However, power/ground noise must be isolated and suppressed. To solve issues, this article proposed an efficient dispersion analysis method, fabricated the noise suppression structures, and applied the structures. The proposed structures have minimal impacts on the channel routing, fan-out, and return current path. However, the structures proposed in this article require additional metal layers. These designs increase the fabrication cost. Even though they provide promising solutions toward power/ground noise issues with minimal impacts on channel properties and designs, more cost-effective designs are desired in the near future. Because the proposed dispersion analysis can be expanded to various designs, such as defects in the plane, development of a new structure based on the proposed method without increasing the number of metal layer remains the subject of work for the near future.

Funding: This work was supported in part by JSPS KAKENHI Grant Number 20K14719 and in part by JSPS KAKENHI Grant Number 22K14241. In addition, this work was supported in part by the MKE through the International Collaborative Research and Development Program "Glass interposer based RF FEM for Next Generation Mobile Smart Phone" under Grant N0000899.

Conflicts of Interest: The author declares no conflict of interest.

References

1. Pak, J.; Ryu, C.; Kim, J. Electrical Characterization of Trough Silicon Via (TSV) depending on Structural and Material Parameters based on 3D Full Wave Simulation. In Proceedings of the 2007 IEEE International Conference on Electronic Material and Packaging, Daejeon, Korea, 19–22 November 2007; pp. 1–6.
2. Tummala, R.; Sundaram, S.; Chatterjee, R.; Raj, P.M.; Kumbhat, N.; Sukumaran, V.; Sridharan, V.; Choudury, A.; Chen, Q.; Bandyopadhyay, T. Trend from ICs to 3D ICs to 3D systems. In Proceedings of the IEEE Custom Integrated Circuits Conference, San Jose, CA, USA, 13–16 September 2009; pp. 439–444.
3. Swaminathan, M.; Han, K. *Design and Modeling for 3D ICs and Interposers*; World Scientific Publishing Co. Pte. Ltd.: Singapore, 2014; ISBN 978-981-4508-59-9.
4. Technology Roadmap of DRAM for Three Major Manufacturers: Samsung, SK-Hynix and Micron. Available online: https://docplayer.net/29965569-Technology-roadmap-of-dram-for-three-major-manufacturers-samsung-sk-hynix-and-micron-may-2013.html (accessed on 30 May 2013).
5. Lim, S. *Design for High Performance, and Reliable 3D Integrated Circuits*; Springer: Berlin/Heidelberg, Germany, 2013. [CrossRef]

6. *Standard JESD235A*; High Bandwidth Memory (HBM) DRAM. JEDEC: Arlington, VA, USA, 2015.
7. Cho, K.; Kim, Y.; Lee, H.; Kim, H.; Choi, S.; Song, J.; Kim, S.; Park, J.; Lee, S.; Kim, J. Signal Integrity Design and Analysis of Silicon Interposer for GPU-Memory Channels in High-Bandwidth Memory Interface. *IEEE Trans. Compon. Packag. Manuf. Technol.* **2018**, *8*, 1658–1671. [CrossRef]
8. Kim, H.; Lee, H.; Cho, J.; Kim, Y.; Kim, J. Electrical Design of silicon, glass and organic interposer channels. In Proceedings of the 2014 Pan Pacific Microelectronics Symposium, Kauai, HI, USA, 3 February 2014.
9. Sukumaran, V.; Bandyopadhyay, T.; Sundaram, V.; Tummala, R. Low-cost thin glass interposers as a superior alternative to silicon and organic interposers for packaging of 3-D ICs. *IEEE Trans. Compon. Packag. Manuf.* **2012**, *2*, 1426–1433. [CrossRef]
10. Hu, D.C.; Hung, Y.; Chen, T.; Tain, R.; Lo, W. Embedded glass interposer for heterogeneous multi-chip integration. In Proceedings of the IEEE Electronic Components and Technology Conference (ECTC), San Diego, CA, USA, 26–29 May 2015; pp. 314–317.
11. Lu, H.; Furuya, R.; Sawyer, B.M.; Nair, C.; Liu, F.; Sundaram, V.; Tummala, R.R. Design, modeling, fabrication and characterization of 2-5 μm redistribution layer traces by advanced semi-additive processes on low-cost panel based glass interposers. *IEEE Trans. Compon. Packag. Manuf.* **2016**, *6*, 959–967. [CrossRef]
12. Park, G.; Cho, K.; Son, K.; Park, H.; Lho, D.; Kim, S.; Shin, T.; Kim, Y.; Kim, J.; Watanabe, A.; et al. Design and Measurement of a 28 GHz Glass Band Pass Filter based on Glass Interposers for 5G Applications. In Proceedings of the 2019 IEEE Electrical Design of Advanced Packaging and Systems Symposium, Kaohsiung, Taiwan, 16–18 December 2019.
13. Laakso, M.J.; Bleiker, S.J.; Liljeholm, J.; Martensson, G.E.; Asiatici, M.; Fischer, A.C.; Stemme, G.; Ebefors, T.; Niklaus, F. Through-Glass Vias for Glass Interposers and MEMS Packaging Applications Fabricated Using Magnetic Assembly of Microscale Metal Wires. *IEEE Access.* **2018**, *6*, 44306–44317. [CrossRef]
14. Tschoban, C.; Rossi, M.; Reyes, J.; Ndip, I.; Lang, K. Development of a Glass Technology Based 79 GHz MIMO Radar Front-End Module for Autonomous Driving. In Proceedings of the 2019 IEEE 21st Electronics Packaging Technology Conference, Singapore, 4–6 December 2019; pp. 376–381.
15. Chang, Y.-H.; Chen, J.-C.; Chung, W.; Li, W.-Y.; Shih, P.-T.B.; Ng'Oma, A.; Yang, C.; Huang, M.-C.; Lin, H.-Y.; Wang, C.-H.; et al. A Novel Fabrication Process and Measurement Results of a 28 GHz Glass Antenna with Single TGV for 5G Communication Applications. In Proceedings of the 2019 14th International Microsystems, Packaging, Assembly and Circuits Technology Conference, Taipei, Taiwan, 23–25 October 2019; pp. 112–115.
16. Hwangbo, S.; Yoon, Y.; Shorey, A. Glass Interposer Integrated Dual-band Millimeter Wave TGV Antenna for Inter-/Intra Chip and Board Communications. In Proceedings of the 2016 IEEE International Symposium on Antennas and Propagation, Fajardo, Puerto Rico, 26 June–1 July2016; pp. 1639–1640.
17. Kim., Y.; Kim, Y.; Cho, K.; Park, G.; Kim, S.; Kang, K.; So, K.; Jeon, J.; Kim, C.; Paik, W. Power and Signal Integrity Analysis of Fan-out Wafer Level Package for Mobile Application Processor. In Proceedings of the DesignCon 2017, Santa Clara, CA, USA, 30 January 2017.
18. FaradFlex for PCBs. Available online: https://www.faradflex.com/pcbs/ (accessed on 1 April 2022).
19. Kitazawa, T.; Hayashi, Y.; Fukawa, Y.; Kim, Y. Analysis of the PDN Induced Crosstalk Impacts on the High-speed Signaling in Ultra-thin and High Permittivity Substrates. In Proceedings of the EMC Europe 2022, Gothenburg, Sweden, 5–8 September 2022.
20. Kim, Y.; Kim, K.; Cho, J.; Kim, J.; Sundaram, V.; Tummala, R. Analysis of Power Distribution Network in Glass, Silicon Interposer and PCB. In Proceedings of the 2014 IEEE International Symposium on Electromagnetic Compatibility, Raleigh, NC, USA, 5 August 2014.
21. Hashimoto, M.; Nair, R. *Power Integrity for Nanoscale Integrated Systems*; McGraw-Hill Education: New York, NY, USA, 2014; ISBN 978-0-07-178776-5.
22. Kim, Y.; Wada, S.; Fujimoto, D.; Hayashi, Y. Analysis of Electromagnetic Information Leakage from Overdesigned Power Delivery Network of Cryptographic Devices. In Proceedings of the 2021 Asia Pacific International Symposium on Electromagnetic Compatibility, Hybrid (Bali & On-line), 28 September 2021.
23. Wada, S.; Hayashi, Y.; Fujimoto, D.; Homma, N.; Kim, Y. Measurement and Analysis of Electromagnetic Information Leakage from Printed Circuit Board Power Delivery Network of Cryptographic Devices. *IEEE Trans. Electromagn. Compat.* **2021**, *63*, 1322–1332. [CrossRef]
24. Rogers, S.D. Electromagnetic-bandgap Layers for Broad-band Suppression of TEM modes in Power Planes. *IEEE Trans. Microw. Theory Tech.* **2005**, *53*, 2495–2505. [CrossRef]
25. Toyota, Y.; Engin, A.; Kim, T.; Swaminathan, M. Stopband Analysis Using Dispersion Diagram for Two-Dimensional Electromagnetic Bandgap Structures in Printed Circuit Boards. *IEEE Microw. Wirel. Compon. Lett.* **2006**, *16*, 645–647. [CrossRef]
26. Kim, L.; Schutt-Aine, J. Analysis and Modeling of Hybrid Planar-Type Electromagnetic-Bandgap Structures and Feasibility Study on Power Distribution Network Applications. *IEEE Trans. Microw. Theory Tech.* **2008**, *56*, 178–186. [CrossRef]
27. Hwang, C.; Shim, Y.; Koo, K.; Kim, M.; Pak, J.; Kim, J. An On-Chip Electromagnetic Bandgap Structure using an On-Chip Inductor and a MOS Capacitor. *Microw. Wirel. Compon. Lett.* **2011**, *21*, 439–441. [CrossRef]
28. Wei, P.; Tsai, M.; Hsu, S.; Shen, C.; Wu, T. An Electromagnetic Bandgap Structure Integrated with RF LNA Using Integrated Fan-Out Wafer-Level Package for Gigahertz Noise Suppression. *IEEE Trans. Microw. Theory Tech.* **2018**, *66*, 5482–5490. [CrossRef]
29. Das, R.; Yoo, H. Application of a Compact Electromagnetic Bandgap Array in a Phone Case for Suppression of Mobile Phone Radiation Exposure. *IEEE Trans. Microw. Theory Tech.* **2018**, *66*, 2363–2372. [CrossRef]

30. Zhu, H.; Wang, J.; Sun, Y.; Wu, X.; Mao, J. A Novel Automatically Designed EBG Structure by Improved GA for Ultrawideband SSN Mitigation of System in Package. *IEEE Trans. Compon. Packag. Manuf. Technol.* **2020**, *10*, 123–133. [CrossRef]
31. Kim, M.; Koo, K.; Hwang, C.; Shim, Y.; Kim, J.; Kim, J. A Compact and Wideband Electromagnetic Bandgap Structure Using a Defected Ground Structure for Power/Ground Noise Suppression in Multilayer Packages and PCBs. *IEEE Trans. Electromagn. Compat.* **2012**, *54*, 689–695.
32. Kim, K.; Hwang, C.; Koo, K.; Cho, J.; Kim, H.; Kim, J.; Lee, J.; Lee, H.; Park, K.; Pak, J. Modeling and Analysis of a Power Distribution Network in 3-D Memory IC Including P/G TSVs, On-chip Decoupling Capacitors, and Silicon Substrate Effects. *IEEE Trans. Compon. Packag. Manuf. Technol.* **2012**, *2*, 2057–2070. [CrossRef]
33. Kim, K.; Yook, J.; Kim, J.; Kim, H.; Lee, J.; Park, K.; Kim, J. Interposer Power Distribution Network Modeling Using a Segmentation Method for 3-D ICs with TSVs. *IEEE Trans. Compon. Packag. Manuf. Technol.* **2013**, *3*, 1891–1906. [CrossRef]
34. Tuncer, E.; Serdyuk, Y.V.; Gubanski, S.M. Dielectric mixtures: Electrical properties and modeling. *IEEE Trans. Dielectric. Electr. Insul.* **2002**, *9*, 809–828. [CrossRef]
35. McCann, S.R.; Sato, Y.; Sundaram, V.; Tummala, R.; Sitaraman, K. Prevention of Cracking from RDL Stress and Dicing Defects in Glass Substrates. *IEEE Trans. Device Mater. Reliab.* **2016**, *16*, 43–49. [CrossRef]
36. Kim, Y.; Cho, J.; Kim, J.J.; Kim, K.; Cho, K.; Kim, S.; Sitaraman, S.; Sundaram, V.; Raj, P.M.; Tummala, R.; et al. Measurement and Analysis of Glass Interposer Power Distribution Network Resonance Effects on a High-Speed Through Glass Via Channel. *IEEE Trans. Electromagn. Compat.* **2016**, *58*, 1747–1759. [CrossRef]
37. Kim, Y.; Park, G.; Cho, K.; Raj, P.M.; Tummala, R.R.; Kim, J. Wideband Power/Ground Noise Suppression in Low-Loss Glass Interposers Using a Double-Sided Electromagnetic Bandgap Structure. *IEEE Trans. Microw. Theory Tech.* **2020**, *68*, 5055–5064. [CrossRef]
38. Pozar, D. *Microwave Engineering*, 3rd ed.; John Wiley and Sons: Hoboken, NJ, USA, 2005.
39. Clavijo, S. Design Methodology for Sievenpiper High-impedance Surfaces. M.S. Thesis, Department of Electrical Engineering, Arizona State University, Tempe, AZ, USA, 2002.
40. 2.5D Interposer for High Bandwidth: Design, Fabrication and Characterization. Available online: http://www.prc.gatech.edu/sites/default/files/documents/iab_spring2016/Presentations/Wednesday%20Presentations/09%20IAB_May_2016-DDTV_2.5DGIP%20Brett%20Sawyer.pdf (accessed on 1 December 2016).

micromachines

MDPI

Article

Average Power Handling Capability of Corrugated Slow-Wave Transmission Lines

Zehao Zheng [1], Min Tang [1,*], Haochi Zhang [2] and Junfa Mao [1]

[1] Key Laboratory of Ministry of Education of China for Research of Design and Electromagnetic Compatibility of High Speed Electronic Systems, Shanghai Jiao Tong University, Shanghai 200240, China; zzh1996@sjtu.edu.cn (Z.Z.); jfmao@sjtu.edu.cn (J.M.)

[2] State Key Laboratory of Millimeter Waves, Southeast University, Nanjing 210096, China; hczhang@etnws.com

* Correspondence: tm222@sjtu.edu.cn

Abstract: In this article, the average power handling capability (APHC) of corrugated slow-wave transmission lines (SWTLs) is investigated. Firstly, the attenuation constants of conductor and dielectric are extracted by the multiline method. Secondly, the thermal resistance of corrugated SWTLs is analyzed based on the constant-angle model. To deal with the non-uniform corrugated structure of SWTLs, the concept of average heat-spreading width (AHSW) is introduced. Finally, the APHC of the corrugated SWTL is calculated using the attenuation constant and the thermal resistance. In addition, the APHC considering the temperature-dependent resistivity of metal conductor is also presented. For validation, the APHCs of SWTLs with different geometric parameters are evaluated. The results agree well with those obtained by the commercial software.

Keywords: average power handling capability (APHC); slow-wave transmission line (SWTL); thermal resistance; average heat-spreading width; temperature-dependent resistivity

Citation: Zheng, Z.; Tang, M.; Zhang, H.; Mao, J. Average Power Handling Capability of Corrugated Slow-Wave Transmission Lines. *Micromachines* **2022**, *13*, 961. https://doi.org/10.3390/mi13060961

Academic Editor: Wensheng Zhao

Received: 26 May 2022
Accepted: 15 June 2022
Published: 17 June 2022

Publisher's Note: MDPI stays neutral with regard to jurisdictional claims in published maps and institutional affiliations.

Copyright: © 2022 by the authors. Licensee MDPI, Basel, Switzerland. This article is an open access article distributed under the terms and conditions of the Creative Commons Attribution (CC BY) license (https://creativecommons.org/licenses/by/4.0/).

1. Introduction

Corrugated slow-wave transmission lines (SWTLs), as a kind of TLs with periodic structures, have attracted much attention in the microwave area. With the advantage of controllable dispersion properties, the corrugated SWTLs can achieve a specific phase delay without changing the longitudinal length, which brings much convenience to the circuit design. In recent years, corrugated SWTLs are popular candidates in the design of microwave circuits, such as miniaturized antennas [1–3], miniaturized power dividers [4], antenna feeding networks [5], and broadband baluns [6].

With the increase of integration level and the trend of miniaturization of microwave components, thermal effect is becoming an essential issue that must be considered during the design of microwave systems. In order to avoid thermal failure, the operating temperature of the microwave system should be controlled. Thus, the power transmitted into the system should be limited below the average power handling capability (APHC). The APHC of a microstrip line is first studied, where the parallel-plate model is proposed to calculate the temperature rise [7]. The parallel-plate model is also utilized in estimating the APHCs of multilayer microstrip lines [8] and various microwave passive components [9–11]. However, for the transmission lines with more complicated structures, such as coupled microstrip lines [12,13] and coplanar waveguides [14,15], the constant-angle model is more often used when estimating the APHCs. Although the SWTLs are utilized in the design of various microwave circuits, to the best of the authors' knowledge, their APHCs have not yet been investigated.

In this article, the APHCs of corrugated SWTLs are estimated using the constant-angle model. The thermal resistances of the conductor and dielectric are first analyzed. The concept of average heat-spreading width (AHSW) is employed in the derivation of thermal resistance of the corrugated structure. The effect of temperature-dependent resistivity of metal conductor is also considered. Using the developed closed-form expressions, the APHCs of corrugated SWTLs with different geometric parameters are evaluated efficiently. The validity of the proposed model is verified by the commercial software.

2. Analysis of the Corrugated Slow-Wave Transmission Lines

2.1. Dispersion Characteristics

The configuration of typical corrugated SWTLs is described in Figure 1. The unit structure may be U-shaped or H-shaped. The top view of the U-shaped unit is shown in Figure 1b, with the groove width a, the groove length h, the period length p, the total width w, and the width of main line w_1 ($w_1 = w - h$). The H-shaped unit is obtained by a mirror transformation of the U-shaped unit, as is shown in Figure 1c, where the dashed line is the axis of symmetry.

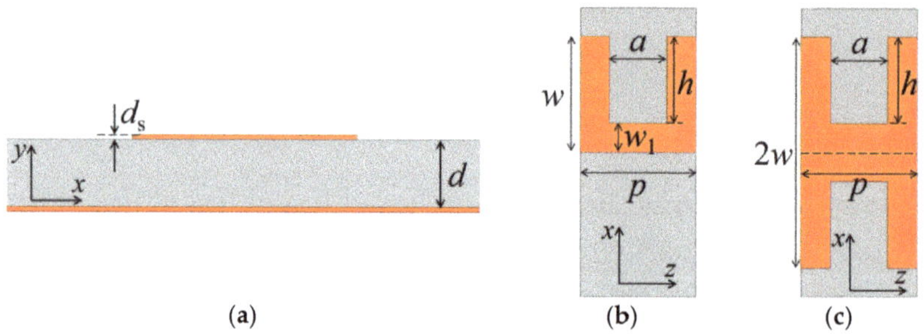

Figure 1. Configuration of the corrugated SWTL: (**a**) cross-sectional view; (**b**) top view of the U-shaped unit; (**c**) top view of the H-shaped unit.

The dispersion curves of the corrugated SWTLs are obtained by the eigenmode simulation, as shown in Figure 2, where k_z is the wavenumber along the z axis. The dimensions of the corrugated SWTLs are $p = 4$ mm, $w_1 = 1$ mm, $d = 0.508$ mm, $d_s = 0.018$ mm and the relative permittivity of the substrate is $\varepsilon_r = 2.2$. For comparison, the dispersion curve of a 50 Ω microstrip line is also provided. It is observed that the wavenumbers of the corrugated SWTLs are larger than that of the microstrip line within the whole frequency range, exhibiting an obvious slow-wave characteristic. For the U-shaped units, the dispersion curve deviates from the microstrip line as the groove length h increases. The increase of asymptotic frequency is observed as the groove width a decreases. In addition, it is found that the dispersion curves of the H-shaped unit and U-shaped unit with the same geometric parameters almost coincide, where the wavenumber of the H-shaped unit is slightly larger.

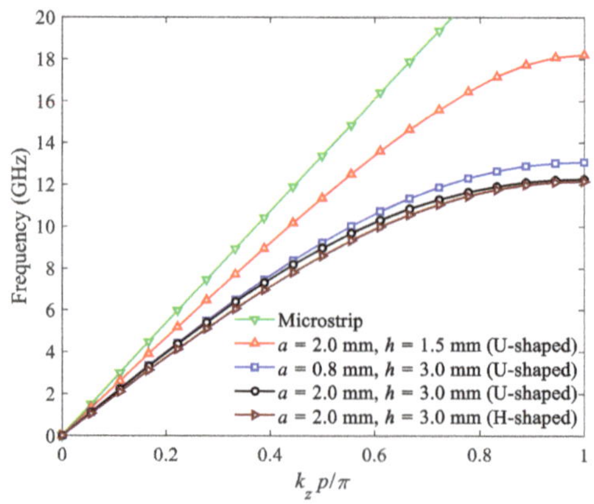

Figure 2. Dispersion curves of the corrugated SWTLs.

2.2. Extraction of Conductor and Dielectric Attenuation

The dispersion curves in Figure 2 show that the wavenumbers of the corrugated SWTLs are larger than that of the microstrip, and the deviation is relieved with the decrease of h. Therefore, to compensate for the impedance and momentum mismatch between the microstrip and SWTL, the transition regions are designed with gradient corrugations, as described in Figure 3.

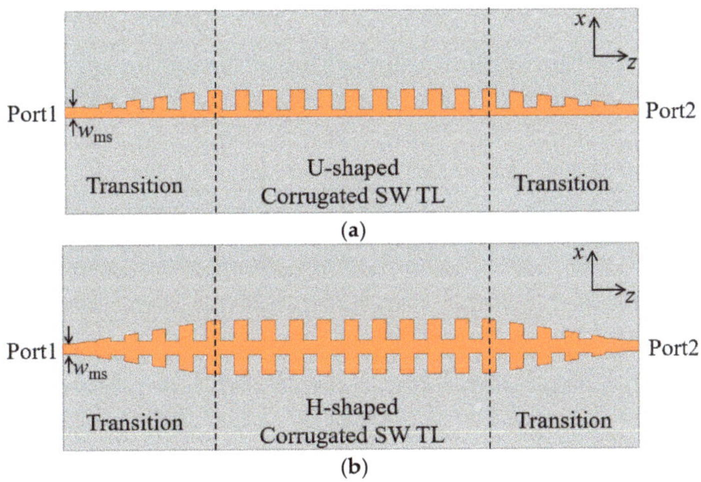

Figure 3. Configuration of the corrugated SWTLs with transitions: (**a**) U-shaped; (**b**) H-shaped.

To accurately characterize the APHC of the corrugated SWTLs, the conductor and dielectric attenuation constants should be extracted first. For the purpose of eliminating the impact of transition regions, the multiline method [16] is utilized. In this method, two corrugated SW structures with different numbers of units are investigated. The total transmission matrices of them can be written as

$$\mathbf{T}_1 = \mathbf{T}_{\text{left}}\mathbf{T}_{N1}\mathbf{T}_{\text{right}} \tag{1}$$

$$\mathbf{T}_2 = \mathbf{T}_{\text{left}}\mathbf{T}_{N2}\mathbf{T}_{\text{right}} \tag{2}$$

where $\mathbf{T}_{N1}$ and $\mathbf{T}_{N2}$ are the transmission matrices of the SWTLs with the unit number of N_1 and N_2, respectively. The transmission matrices of the transition regions on the left and right sides are represented by $\mathbf{T}_{\text{left}}$ and $\mathbf{T}_{\text{right}}$, respectively.

From (1) and (2), we obtain

$$\mathbf{T}_{12}\mathbf{T}_{\text{left}} = \mathbf{T}_{\text{left}}\mathbf{T}_{\text{DUT}} \tag{3}$$

where

$$\mathbf{T}_{12} = \mathbf{T}_1\mathbf{T}_2^{-1} \tag{4}$$

$$\mathbf{T}_{\text{DUT}} = \mathbf{T}_{N1}\mathbf{T}_{N2}^{-1} = \begin{bmatrix} e^{\gamma(N_2-N_1)p} & 0 \\ 0 & e^{-\gamma(N_2-N_1)p} \end{bmatrix} \tag{5}$$

where γ is the complex propagation constant.

Thus, the total attenuation constant is derived by

$$\alpha = \text{Re}(\gamma) = \text{Re}\left\{ \frac{1}{|N_2 - N_1|p} \ln\left[\frac{1}{2}\left(\lambda_1 + \frac{1}{\lambda_2}\right)\right]\right\} \tag{6}$$

where λ_1 and λ_2 ($|\lambda_1| > |\lambda_2|$) are the two eigenvalues of $\mathbf{T}_{12}$ matrix.

Note that with the special settings of material, the conductor attenuation constant α_c and the dielectric attenuation constant α_d can be extracted separately based on the above technique. For example, when the loss of metal is omitted, the dielectric attenuation constant α_d can be extracted alone. Similarly, the conductor attenuation constant α_c can be obtained if the loss of substrate is ignored.

3. Modeling of Thermal Resistance

3.1. Thermal Resistance for U-Shaped Corrugated SWTL

Assume that the ground plane of SWTL is the heat sink with constant temperature and the other boundaries are adiabatic, as shown in Figure 4. According to the constant-angle model [17,18], the heat flow region locates within a heat-spreading angle θ, and $w_h(y)$ is the width of the heat flow region at height y.

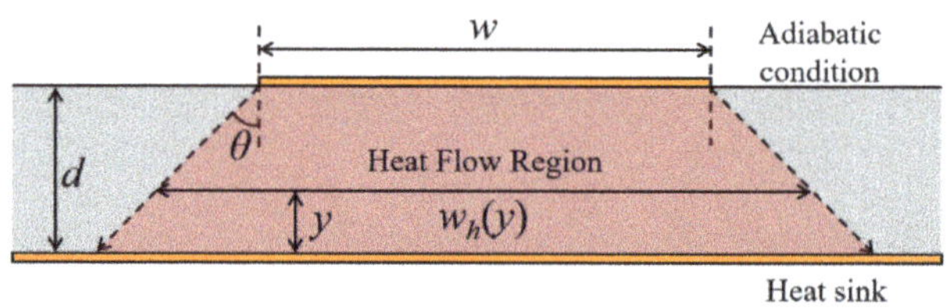

Figure 4. Cross-sectional view of the heat flow region in the constant-angle model.

Thus, the thermal resistances for a transmission line per unit length can be derived as

$$R_{thc} = \frac{1}{K}\int_0^d \frac{\mathrm{d}y}{w_h(y)} \tag{7}$$

$$R_{thd} = \frac{1}{K}\int_0^d \frac{(1 - y/t)\mathrm{d}y}{w_h(y)} \tag{8}$$

where R_{thc} and R_{thd} are the thermal resistances of the conductor and dielectric, respectively, and K is the thermal conductivity of the substrate.

For the U-shaped corrugated SWTLs, the top view of the heat flow region is shown in Figure 5. Applying $\theta = 45°$, the horizontal distance between the boundary of the heat flow region and the metal conductor is $(d - y)$ at the height of y, as shown in Figure 5a. As a result, $w_h(y)$ varies along the propagation direction (z-axis) because of the non-uniform corrugated structure of the conductor, making it difficult to calculate the thermal resistance.

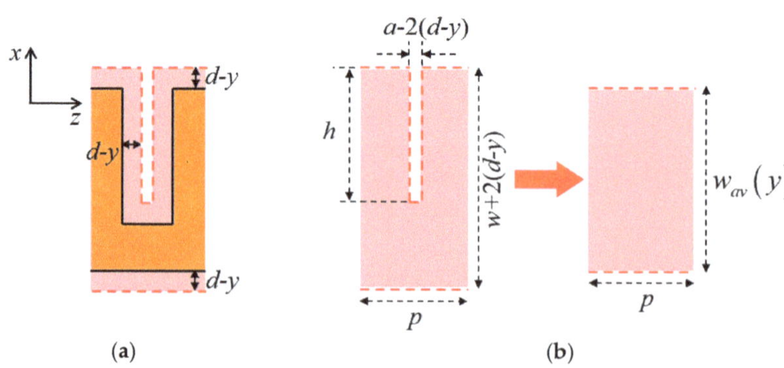

Figure 5. Top view of the heat flow region of U-shaped corrugated SWTL: (**a**) Boundary of the heat flow region; (**b**) schematic of $w_{av}(y)$.

To overcome this problem, the concept of AHSW is introduced in this work, where the heat flow region of a unit U-shaped SWTL is replaced by a uniform rectangular region with the same area and longitudinal length, as shown in Figure 5b. The width $w_{av}(y)$ of the rectangular region is defined as AHSW.

In the situation of $d \leq a/2$, a rectangular gap always exists in the heat flow region. Therefore, the AHSW can be expressed as

$$w_{av}(y) = w + 2(d-y) - [a - 2(d-y)]h/p \tag{9}$$

Otherwise, the rectangular gap in the heat flow region vanishes when $y < (d - a/2)$, and the expression of AHSW is

$$w_{av}(y) = \begin{cases} w + 2(d-y), & 0 < y < d - a/2 \\ w + 2(d-y) - [a - 2(d-y)]h/p, & d - a/2 \leq y < d \end{cases} \tag{10}$$

Then, replacing $w_h(y)$ by $w_{av}(y)$ in (7) and (8), the thermal resistances for the U-shaped corrugated SWTLs are

$$R_{thc} = \begin{cases} \frac{1}{2K(1+h/p)} \ln\left(1 + \frac{2d(1+h/p)}{w-ah/p}\right), & d \leq a/2 \\ \frac{1}{2K}\left(\ln\frac{w+2d}{w+a} + \frac{1}{1+h/p}\ln\frac{w+a}{w-ah/p}\right), & d > a/2 \end{cases} \tag{11}$$

$$R_{thd} = \begin{cases} \frac{1}{2K(1+h/p)}\left[1 - \frac{w-ah/p}{2d(1+h/p)}\ln\left(1 + \frac{2d(1+h/p)}{w-ah/p}\right)\right], & d \leq a/2 \\ \frac{1}{2K}\left[\left(1 - \frac{a}{2d} - \frac{w}{2d}\ln\frac{w+2d}{w+a}\right) + \frac{1}{1+h/p}\left(\frac{a}{2d} - \frac{w-ah/p}{2d(1+h/p)}\ln\frac{w+a}{w-ah/p}\right)\right], & d > a/2 \end{cases} \tag{12}$$

3.2. Thermal Resistance for H-Shaped Corrugated SWTL

Applying the same constant-angle model with $\theta = 45°$, the heat flow region of a H-shaped corrugated SWTL is shown in Figure 6. The derivation of AHSW is similar to that of the U-shaped one, as is shown in Figure 6b.

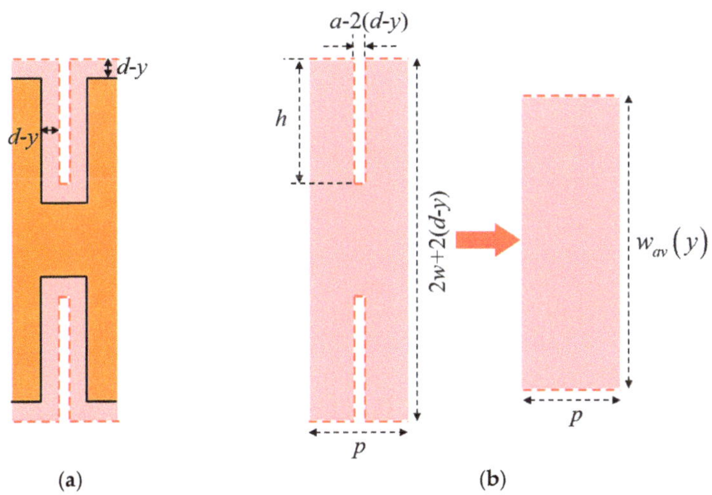

Figure 6. Top view of the heat flow region of H-shaped corrugated SWTL: (**a**) Boundary of the heat flow region; (**b**) schematic of $w_{av}(y)$.

When $d \leq a/2$, the AHSW of the H-shaped corrugated SWTL can be expressed as

$$w_{av}(y) = 2w + 2(d-y) - 2[a - 2(d-y)]h/p \tag{13}$$

Otherwise,

$$w_{av}(y) = \begin{cases} 2w + 2(d-y), & 0 < y < (d - a/2) \\ 2w + 2(d-y) - 2[a - 2(d-y)]h/p, & (d - a/2) \leq y < d \end{cases} \tag{14}$$

Comparing (13)–(14) with (9)–(10), it is observed that the AHSW of the H-shaped corrugated SWTL can be obtained from the U-shaped one by substituting w and h with $2w$ and $2h$. Moreover, the thermal resistances for the H-shaped corrugated SWTLs can be obtained with the same substitution.

4. Derivation of Average Power Handling Capability

For a lossy transmission line, the propagated power along the line satisfies the following expression

$$P(z) = P_0 e^{-2\alpha z} \tag{15}$$

where P_0 is the input power. Using the first-order approximation, the power loss per unit length is calculated as

$$\Delta P = \frac{P(0) - P(l)}{l} = \frac{P_0\left(1 - e^{-2\alpha l}\right)}{l} \approx 2\alpha P_0 \alpha \tag{16}$$

where l is the length of the transmission line. As the power loss is treated as heat source, the temperature rise per unit input power of the SWTL is calculated as

$$\Delta T = \frac{\Delta P_c R_{thc} + \Delta P_d R_{thd}}{P_0} = 2\alpha_c R_{thc} + 2\alpha_d R_{thd} \tag{17}$$

where ΔP_c and ΔP_d are the power loss due to conductor and dielectric attenuation.

Thus, the APHC of the corrugated SWTL can be calculated by

$$P_{av} = \frac{T_{max} - T_{amb}}{\Delta T} = \frac{T_{max} - T_{amb}}{2\alpha_c R_{thc} + 2\alpha_d R_{thd}} \tag{18}$$

where T_{max} is the maximum operating temperature, and T_{amb} is the ambient temperature.

In practice, the resistivity of metal conductor usually changes with temperature, resulting in the variation of conductor attenuation. In order to take this impact into account, the expression of APHC in (18) need to be modified. The temperature-dependent resistivity of most metals is expressed by the following linear approximation [19]

$$\rho(T) = \rho_0[1 + \alpha_T(T - T_{amb})] \tag{19}$$

where ρ_0 is the resistivity at T_{amb}, and α_T is the temperature coefficient of resistivity.

As the conductor attenuation is proportional to the square root of resistivity [20], the impact of temperature on the conductor attenuation is expressed by

$$\alpha_c(T) = \alpha_{c0}\sqrt{1 + \alpha_T(T - T_{amb})} \tag{20}$$

where α_{c0} is the conductor attenuation at T_{amb}.

Thus, considering the effect of temperature-dependent resistivity, the APHC is modified as

$$P_{av} = \frac{T_{max} - T_{amb}}{2\alpha_{c0}\sqrt{1 + \alpha_T(T_{max} - T_{amb})}R_{thc} + 2\alpha_d R_{thd}} \tag{21}$$

5. Numerical Results and Discussion

5.1. Temperature Rise of U-Shaped Corrugated SWTL

In this section, the temperature rise of the U-shaped corrugated SWTL is evaluated using the proposed model. The substrate is Rogers RT5880 (ε_r = 2.2, tanδ = 0.0009, K = 0.2 W/(m°C)) with the thickness of 0.508 mm. The metal conductor is copper ($\sigma = 5.8 \times 10^7$ S/m) with the thickness of 0.018 mm. The effect of temperature-dependent resistivity is not considered initially. The geometrical parameters of the U-shaped corrugated SWTLs are listed in Table 1, where three samples with different values of groove width a and groove length h are investigated.

Table 1. Geometric parameters of the U-shaped corrugated SWTLs (unit: mm).

	a	h	p	w_1
Sample 1	2	3	4	1
Sample 2	2	1.5	4	1
Sample 3	0.8	3	4	1

Firstly, the thermal resistances of the three samples are calculated by (11) and (12), and the results are given in Table 2. It is observed that both R_{thc} and R_{thd} increase with the decrease of h and the increase of a.

Table 2. Thermal resistances of U-shaped corrugated SWTLs (unit: m°C/W).

	Sample 1	**Sample 2**	**Sample 3**
R_{thc}	0.7674	1.0670	0.6027
R_{thd}	0.3495	0.4816	0.2811

Then, the attenuation constants of the samples are extracted. The results of conductor and dielectric attenuation constants as functions of frequency are depicted in Figure 7. It is observed that the groove width a has small impact on the attenuation constants, especially for the dielectric attenuation case. In contrast, with the increase of operating frequency, the impact of groove length h is significant on the attenuation constants of SWTLs.

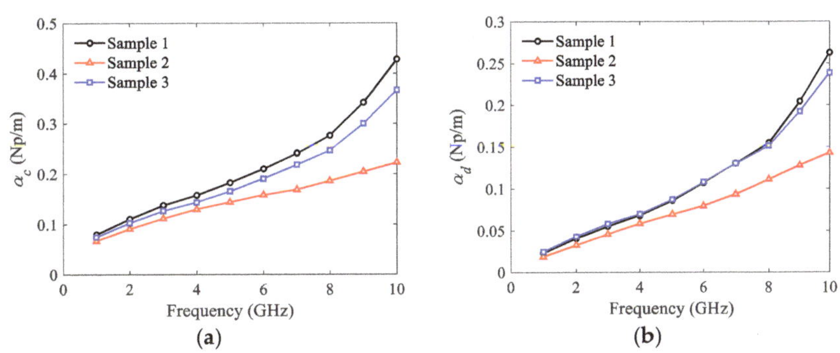

Figure 7. Attenuation constants of U-shaped corrugated SWTLs: (**a**) conductor attenuation; (**b**) dielectric attenuation.

Based on the above results of thermal resistances and attenuation constants, the temperature rises can be calculated directly by (17). To validate the proposed model, the temperature rises are also simulated using a commercial solver ANSYS. The input power is 50 W, and the ground plane is treated as heat sink with a fixed temperature of T_{amb} = 25 °C. The temperature rises per watt are shown in Figure 8. Good agreement is observed between the calculated results and the simulated ones.

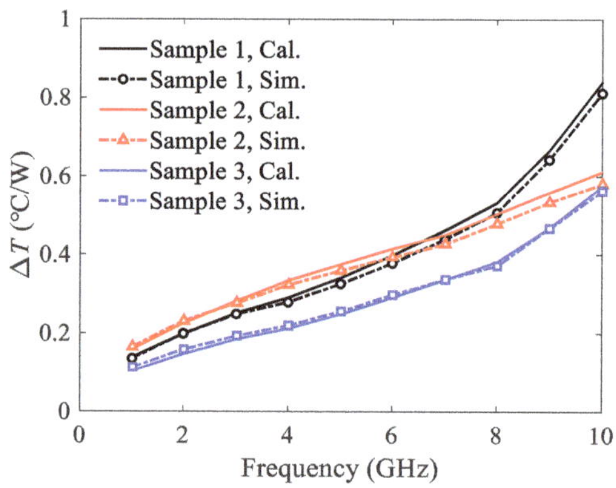

Figure 8. Temperature rises per watt of the U-shaped corrugated SWTLs.

In Figure 8, there is an intersection in the simulated results of sample 1 and sample 2 at around 6.5 GHz. A brief explanation of this phenomenon is given as below. The thermal resistance of sample 1 is smaller, which leads to the less temperature rise per watt in the low frequency range. However, the attenuation constant of sample 1 is much larger than that of sample 2 at high frequencies. Therefore, the resulting temperature rise per watt is higher. In addition, the temperature rise per watt of sample 3 is the smallest due to its lowest thermal resistance.

It should be mentioned that for all the above results, the top surfaces of the structures are assumed to be adiabatic. In practice, however, there is convective heat transfer from the top surface to the ambient. In order to investigate the effect of heat convection on the top surface, the temperature rise of the samples are simulated by ANSYS, where the convection coefficient is set to 10 W/(m^2°C) to represent the natural convection situation. The results are compared with those under the adiabatic assumption, as shown in Figure 9. It is observed that the temperature rise per watt will not be affected apparently when the natural convection condition is imposed. Therefore, the heat convection on the top surface is negligible and the assumption of adiabatic condition is reasonable.

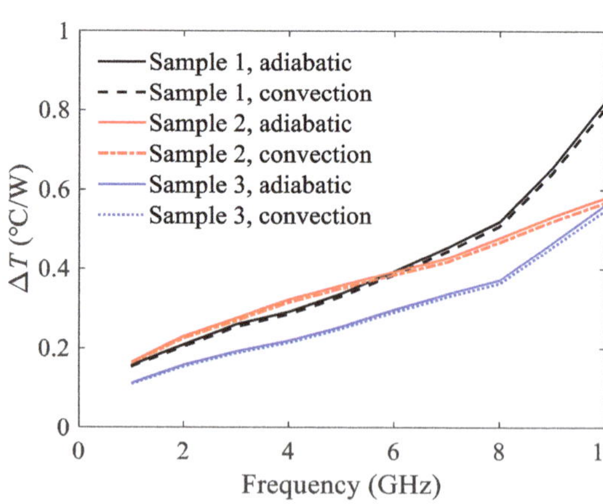

Figure 9. Temperature rises per watt of the U-shaped corrugated SWTLs when the top surface is under adiabatic or natural convection condition.

For more detailed illustration, the thermal profiles at 10 GHz are shown in Figure 10. It is shown that the maximum temperature of sample 1 is the highest. The maximum temperature of sample 2 is lower due to the decrease of attenuation constants, while that of sample 3 is the lowest due to the smallest thermal resistance.

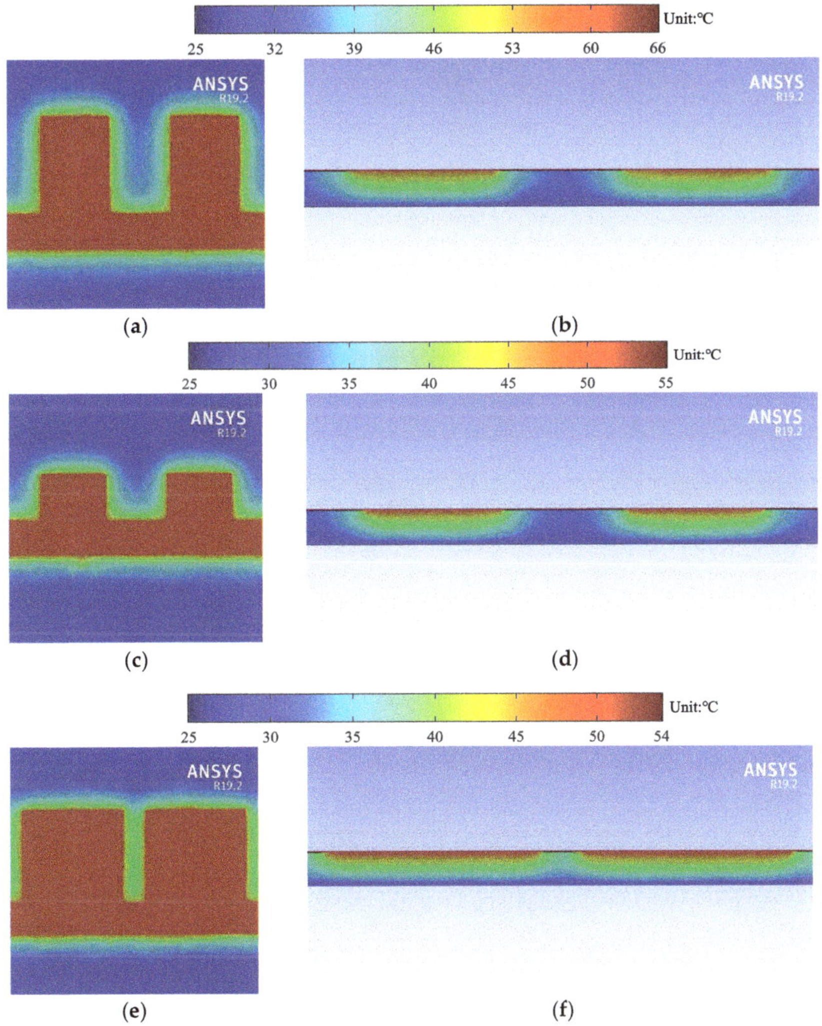

Figure 10. Simulated thermal profile of the U-shaped corrugated SWTLs: (**a**) top view of sample 1; (**b**) cross-sectional view of sample 1; (**c**) top view of sample 2; (**d**) cross-sectional view of sample 2; (**e**) top view of sample 3; (**f**) cross-sectional view of sample 3.

5.2. Temperature Rise of H-Shaped Corrugated SWTL

The temperature rise of the H-shaped corrugated SWTL is studied in this section. The material of conductor and dielectric are the same as the U-shaped ones, and the geometric parameters are listed in Table 3. Note that this H-shaped corrugated SWTL can be formed by the mirror transformation of the U-shaped one (sample 1) in the previous section.

Table 3. Geometric parameters of the H-shaped corrugated SWTL (unit: mm).

a	h	p	w
2	3	4	4

The thermal resistances of the H-shaped corrugated SWTL are calculated using (11) and (12) by replacing w and h with $2w$ and $2h$, as shown in Table 4. For comparison, the thermal resistances of the U-shaped SWTL (sample 1) are also listed in Table 4. The results show that the thermal resistances of the H-shaped SWTL are lower than those of the U-shaped one.

Table 4. Thermal resistances of corrugated SWTL (unit: m°C/W).

	H-Shaped	U-Shaped (Sample 1)
R_{thc}	0.4108	0.7674
R_{thd}	0.1914	0.3495

The conductor and dielectric attenuation constants of the H-shaped and U-shaped corrugated SWTLs are simulated and plotted in Figure 11. It is observed that the conductor attenuation constant of the H-shaped SWTL is very close to that of the U-shaped one. Moreover, the dielectric attenuation constant of the H-shaped SWTL is slightly higher than that of the U-shaped one because of the larger wavenumber depicted in Figure 2.

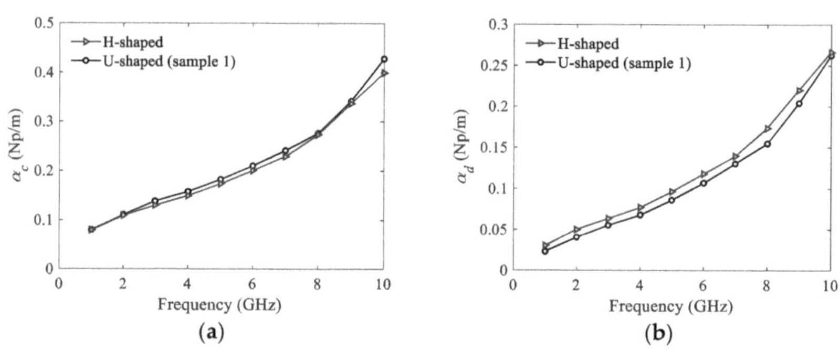

Figure 11. Attenuation constants of the corrugated SWTLs: (**a**) conductor attenuation; (**b**) dielectric attenuation.

Similarly, the temperature rises of the H-shaped and U-shaped corrugated SWTLs are calculated by (17) and simulated by ANSYS, as shown in Figure 12. The thermal profile of the H-shaped corrugated SWTL is shown in Figure 13, where the simulation condition is the same as those for the U-shaped ones. It is observed that the temperature rise of the H-shaped SWTL is obviously lower than the U-shaped one, which is due to the similar attenuation constants but smaller thermal resistances. Furthermore, the calculated results agree well with the simulated ones, proving the validity of the proposed model.

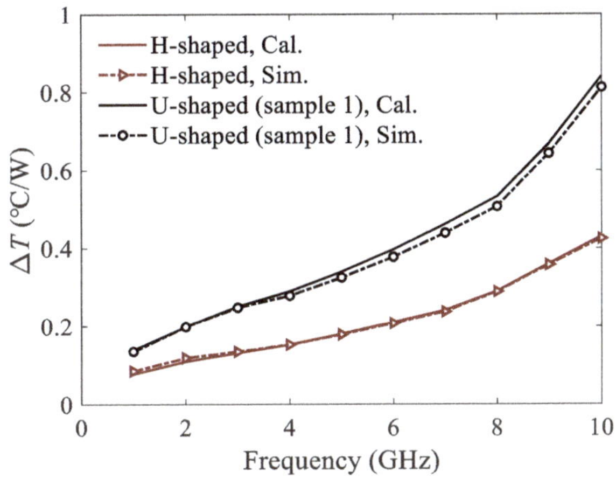

Figure 12. Temperature rises per watt of the corrugated SWTLs.

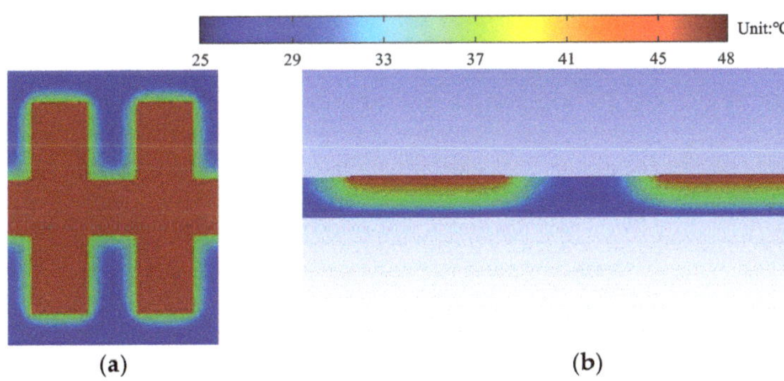

Figure 13. Simulated thermal profile of the H-shaped corrugated SWTL: (**a**) top view; (**b**) cross-sectional view.

5.3. APHCs of the Corrugated SWTLs

When the temperature rise is known, the APHCs of the corrugated SWTLs are readily obtained by (18). To consider the impact of temperature-dependent resistivity, the APHCs should be calculated using (21), where the temperature coefficient of resistivity for copper is α_T = 0.0039 °C^{-1}. For validation, the APHCs are also obtained by the electro-thermal co-simulation of ANSYS. The results at the operating frequency of 10 GHz are compared in Table 5. It is observed that the APHC of the H-shaped corrugated SWTL is the highest, due to the smallest thermal resistances. Among the U-shaped ones, sample 1 has the lowest APHC because of the largest attenuation constants. Sample 2 has a higher APHC due to the smaller attenuation constant, and sample 3 holds the highest APHC because of the smallest thermal resistance.

Furthermore, the calculated APHCs with the consideration of temperature-dependent resistivity are in accordance with the results of ANSYS, where the maximum relative error is less than 5%. In contrast, the APHCs will be overestimated when the temperature-dependent resistivity is ignored, and the maximum relative error is up to 16.0%. Therefore, the consideration of temperature-dependent resistivity is rather necessary

Table 5. APHCs of the corrugated SWTLs.

	ANSYS	Proposed Model		Proposed Model (TDR [1])	
	APHC (W)	APHC (W)	RE [2]	APHC (W)	RE
U-shaped (sample 1)	78.5	89.2	13.6%	80.6	2.6%
U-shaped (sample 2)	105.5	122.4	16.0%	110.6	4.9%
U-shaped (sample 3)	115.3	130.1	12.8%	117.8	2.1%
H-shaped	153.9	174.2	13.2%	157.8	2.5%

[1] TDR: temperature-dependent resistivity. [2] RE: relative error.

6. Conclusions

The closed-form expressions to estimate the APHCs of corrugated SWTLs are provided in this article. The thermal resistances of both U-shaped and H-shaped SWTLs are calculated based on the constant-angle model, where the AHSW is introduced to deal with the non-uniform corrugated structure. The temperature-dependent resistivity of the metal conductor is also considered in the evaluation of APHC. Good agreement of the APHC results is achieved from the proposed model and the commercial software.

Author Contributions: Conceptualization, Z.Z. and M.T.; methodology, Z.Z. and H.Z.; validation, Z.Z.; writing, Z.Z. and M.T.; supervision, M.T. and J.M.; funding acquisition, M.T. and J.M. All authors have read and agreed to the published version of the manuscript.

Funding: This research was funded by the Guangdong Provincial Key-Field Research Program (2018B010115001), National Natural Science Foundation of China (62131014, 61831016), and Science and Technology Commission of Shanghai Municipality (20501130400).

Conflicts of Interest: The authors declare no conflict of interest.

References

1. Lu, J.; Zhang, H.C.; He, P.H.; Zhang, L.P.; Cui, T.J. Design of Miniaturized Antenna Using Corrugated Microstrip. *IEEE Trans. Antennas Propag.* **2020**, *68*, 1918–1924. [CrossRef]
2. Wang, M.; Tang, M.; Zhang, L.P.; Zhang, H.C.; Xu, J.; Cui, T.J.; Mao, J. Miniaturization of Frequency-Reconfigurable Antenna Using Periodic Slow-Wave Structure. *IEEE Trans. Antennas Propag.* **2021**, *69*, 7889–7894. [CrossRef]
3. Zhao, B.; Tang, M.; Shao, Z.; Zhang, Y.; Mao, J. Design of Broadband Compact Grid Array Antennas Using Gradient Slow-Wave Structures. *IEEE Antennas Wirel. Propag. Lett.* **2022**, *21*, 620–624. [CrossRef]
4. Rawat, K.; Ghannouchi, F.M. A Design Methodology for Miniaturized Power Dividers Using Periodically Loaded Slow Wave Structure with Dual-Band Applications. *IEEE Trans. Microw. Theory Tech.* **2009**, *57*, 3380–3388. [CrossRef]
5. Li, M.; Zhang, Z.; Tang, M.-C.; Yi, D.; Ziolkowski, R.W. Compact Series-Fed Microstrip Patch Arrays Excited with Dolph–Chebyshev Distributions Realized with Slow Wave Transmission Line Feed Networks. *IEEE Trans. Antennas Propag.* **2020**, *68*, 7905–7915. [CrossRef]
6. Zhang, L.P.; Zhang, H.C.; Tang, W.; Shao, C.; He, P.H.; Tang, M.; Mao, J.-F.; Cui, T.J. A Broadband 90° Balun with Low-Phase-Imbalance Performance Based on Periodic Slow Wave Structure. *IEEE Trans. Antennas Propag.* **2021**, *69*, 4681–4687. [CrossRef]
7. Bahl, I.J.; Gupta, K.C. Average Power-Handling Capability of Microstrip Lines. *IEE J. Microw. Opt. Acoust.* **1979**, *3*, 1–4. [CrossRef]
8. Bahl, I.J. Average Power Handling Capability of Multilayer Microstrip Lines. *Int. J. RF Microw. Comput.-Aided Eng.* **2001**, *11*, 385–395. [CrossRef]
9. Sanchez-Soriano, M.A.; Quere, Y.; Le Saux, V.; Quendo, C.; Cadiou, S. Average Power Handling Capability of Microstrip Passive Circuits Considering Metal Housing and Environment Conditions. *IEEE Trans. Compon. Packag. Manuf. Technol.* **2014**, *4*, 1624–1633. [CrossRef]
10. Sanchez-Soriano, M.A.; Quere, Y.; Le Saux, V.; Marini, S.; Reglero, M.S.; Boria, V.E.; Quendo, C. Peak and Average Power Handling Capability of Microstrip Filters. *IEEE Trans. Microw. Theory Tech.* **2019**, *67*, 3436–3448. [CrossRef]
11. Wu, L.-S.; Zhou, X.-L.; Yin, W.-Y.; Tang, M.; Zhou, L. Characterization of Average Power Handling Capability of Bandpass Filters Using Planar Half-Wavelength Microstrip Resonators. *IEEE Microw. Wirel. Compon. Lett.* **2009**, *19*, 686–688.
12. Yin, W.-Y.; Dong, X.; Gan, Y.B. Frequency-Dependent Maximum Average Power-Handling Capabilities of Single and Edge-Coupled Microstrip Lines on Low-Temperature Co-Fired Ceramic (LTCC) Substrates. *Int. J. RF Microw. Comput.-Aided Eng.* **2006**, *16*, 103–117. [CrossRef]

13. Yin, W.-Y.; Kang, K.; Mao, J.-F. Electromagnetic-Thermal Characterization of on On-Chip Coupled (A)Symmetrical Interconnects. *IEEE Trans. Adv. Packag.* **2007**, *30*, 851–863. [CrossRef]
14. Yin, W.-Y.; Zhang, Y.; Dong, X.; Gan, Y.B. Average Power-Handling Capability of the Signal Line in Coplanar Waveguides on Polyimide and GaAs Substrates Including the Irregular Line Edge Shape Effects. *Int. J. RF Microw. Comput.-Aided Eng.* **2005**, *15*, 156–163. [CrossRef]
15. Tai, T.-C.; Wu, H.-W.; Wang, Y.-H.; Chiu, C.-T. Microwave Characteristics of Thin-Film Passivation on Ceramic Substrate by Using DC Reactive Magnetron Sputtering. *IEEE Trans. Compon. Packag. Manuf. Technol.* **2018**, *8*, 1800–1806. [CrossRef]
16. Marks, R.B. A Multiline Method of Network Analyzer Calibration. *IEEE Trans. Microw. Theory Tech.* **1991**, *39*, 1205–1215. [CrossRef]
17. Holway, L.H.; Adlerstein, M.G. Approximate Formulas for the Thermal Resistance of IMPATT Diodes Compared with Computer Calculations. *IEEE Trans. Electron Devices* **1977**, *24*, 156–159. [CrossRef]
18. Masana, F.N. A Closed Form Solution of Junction to Substrate Thermal Resistance in Semiconductor Chips. *IEEE Trans. Compon. Packag. Manuf. Technol. Part A* **1996**, *19*, 539–545. [CrossRef]
19. Holman, J.P. *Heat Transfer*, 10th ed.; McGraw-Hill: New York, NY, USA, 2010.
20. Pozar, D.M. *Microwave Engineering*, 4th ed.; Wiley: Hoboken, NJ, USA, 2012.

micromachines

MDPI

Article

Reliability Evaluation Based on Mathematical Degradation Model for Vacuum Packaged MEMS Sensor

Guizhen Du [1,2], Xianshan Dong [2,*], Xinglong Huang [2,3], Wei Su [2] and Peng Zhang [1,*]

1 Institute of Advanced Wear & Corrosion Resistance and Functional Materials, Jinan University, Guangzhou 510632, China
2 Science and Technology on Reliability Physics and Application Technology of Electronic Component Laboratory, China Electronic Product Reliability and Environmental Testing Research Institute, Guangzhou 511370, China
3 College of Mechanical and Electrical Engineering, Central South University, Changsha 410083, China
* Correspondence: dongxs@pku.edu.cn (X.D.); tzhangpeng@jnu.edu.cn (P.Z.)

Abstract: Vacuum packaging is used extensively in MEMS sensors for improving performance. However, the vacuum in the MEMS chamber gradually degenerates over time, which adversely affects the long-term performance of the MEMS sensor. A mathematical model for vacuum degradation is presented in this article for evaluating the degradation of vacuum packaged MEMS sensors, and a temperature-accelerated test of MEMS gyroscope with different vacuums is performed. A mathematical degradation model is developed to fit the parameters of the degradation of Q-factor over time at three different temperatures. The results indicate that the outgassing rate at 85 °C is the highest, which is 0.0531 cm^2/s; the outgassing rate at 105 °C is the lowest, which is 0.0109 cm^2/s; and the outgassing rate at 125 °C is in the middle, which is 0.0373 cm^2/s. Due to the different mechanisms by which gas was released, the rate of degradation did not follow this rule. It will also be possible to predict the long-term reliability of vacuum packaged MEMS sensors at room temperature based on this model.

Keywords: MEMS sensor; reliability evaluation; vacuum degradation; mathematical model

Citation: Du, G.; Dong, X.; Huang, X.; Su, W.; Zhang, P. Reliability Evaluation Based on Mathematical Degradation Model for Vacuum Packaged MEMS Sensor. *Micromachines* **2022**, *13*, 1713. https://doi.org/10.3390/mi13101713

Academic Editor: Wensheng Zhao

Received: 8 September 2022
Accepted: 3 October 2022
Published: 11 October 2022

Publisher's Note: MDPI stays neutral with regard to jurisdictional claims in published maps and institutional affiliations.

Copyright: © 2022 by the authors. Licensee MDPI, Basel, Switzerland. This article is an open access article distributed under the terms and conditions of the Creative Commons Attribution (CC BY) license (https://creativecommons.org/licenses/by/4.0/).

1. Introduction

The micro-electro-mechanical system (MEMS) is a micro-intelligent system utilizing microelectronics, micromechanics, and other technologies to integrate sensors, actuators, and signal transmission units [1]. Micro-electro-mechanical sensors possess the characteristics of small volume, light weight, and low cost, and are on the verge of achieving passivity, miniaturization, and anti-interference capabilities [2]. As technology advances, MEMS sensors have become widely used in a variety of fields, including consumer electronics, aerospace, military equipment, biomedicine, and others [3–6]. It is common for MEMS sensors to have movable structures, such as gyroscopes, accelerometers, and filters that are based on resonant structures. Vacuum packaging can reduce the air damping of their movable structures, improving their performance. In spite of this, the vacuum in the MEMS chamber will gradually degenerate over time, affecting directly the performance of vacuum packaged MEMS sensors, while seriously affecting their long-term reliability. Vacuum degradation and failure are common problems associated with vacuum-packaged MEMS sensors. As a consequence, the study of the vacuum degradation and long-term reliability of vacuum packaged MEMS sensor is of great importance for the development and application of vacuum packaged MEMS sensors.

In the present day, some scholars have conducted in-depth research into the vacuum degradation of MEMS devices. Through vacuum reflow technology, Liang et al. [7] were able to ensure good sealing of the packaging cavity and placed getters within the cavity to maintain a stable vacuum inside the cavity. As a result of testing the resonant gyroscope for nine months, Wu et al. [8] prepared multilayer getters, which were activated at 300 °C, and the getters showed effective performance. As reported by Tepolt et al. [9], the initial outgassing of the material in the packaging cavity, as well as the vacuum pressure required to stabilize the devices, was measured. In order to optimize the processing technology and minimize the exhaust amount of the MEMS devices cavity without serious degradation, the data was compared with the specific getter capacity data. A mathematical model has been developed by Lu et al. [10] that provides a theoretical basis for directly comparing the behavior of different gases entering the sealing cavity. According to the principles of thermodynamics and hydrodynamics, He et al. [11] determined the relationship between Q-factor, air pressure, and gas number of MEMS sensor cavities. For MEMS resonators, Candler proposed a single-chip packaging method [12]. Using thick polysilicon packaging, the MEMS resonator was packaged under a pressure of less than 1 Pa, and the reliability test was conducted under high temperatures. As a result, the majority of research focuses on improving the packaging process and developing packaging materials for MEMS devices. Some researchers devote a great deal of time to assessing the long-term reliability of vacuum-packaged sensors. Currently, there are few mature theoretical models of the effect of deflation of the internal materials of MEMS devices on the long-term reliability of these devices.

As part of this paper, an evaluation of the reliability of vacuum-packaged MEMS sensors is presented. In order to analyze the degradation failure of vacuum packaged MEMS sensors, a degradation failure model was developed. Additionally, the parameters of the model were obtained by fitting data obtained from MEMS sensors with varying vacuum degrees at multiple temperature points. According to the results, the model is well suited to the experimental data. Further, the model was used to predict the vacuum degradation of a vacuum-packaged MEMS sensor at room temperature based on the model obtained.

2. Materials and Methods

2.1. Model of Vacuum Degradation

In vacuum packaged MEMS sensors, vacuum degree is degraded mainly due to the release of gas inside the cavity [13], including gas molecules adsorbed on the inner wall of the cavity, gas molecules in the material inside the cavity, etc. High temperatures are the main source of sensitive stress [14]. It is generally true that vacuum-packaged MEMS sensors have resonant structures, which can be used as a measure of the level of vacuum inside the cavity of the package [15]. Hence, this paper first establishes a mathematical model for predicting the degradation of the Q-factor of vacuum packaged MEMS sensors when exposed to high temperatures. Here is a description of how the mathematical model is constructed.

Material outgassing from vacuum-packaged MEMS sensors is considered an unsteady process, and the diffusion coefficient in different directions is regarded as a constant. Based on Fick's second law:

$$\frac{\partial_n}{\partial_t} = D\nabla^2 n \tag{1}$$

In this formula, n represents the number of gas molecules per unit cross-sectional area, t represents the diffusion time, and D represents the diffusion coefficient. The Gilliland formula can be used to express the diffusion coefficient D as follows:

$$D = \frac{435.7\sqrt{2}\sqrt{T^3}}{2p\sqrt{\mu}\sqrt[3]{V}} \tag{2}$$

where T denotes the thermodynamic temperature, p is the total pressure, μ is the molecular weight, and V is the liquid gram molar volume of the gas at its normal boiling point.

After vacuum packaging, the MEMS sensor cannot maintain an absolute vacuum state in its cavity, so there will be a certain amount of gas in the initial state, which is directly related to the initial Q-factor of the samples. Considering this, assuming there are N_0 gas molecules in the sealed cavity at the beginning, and N_{tot} is the total number of gas molecules released, the boundary condition for the number of molecules inside the cavity $N(t)$ over time is:

$$N(t) = \begin{cases} N_0, t = 0 \\ N_0 + N_{tot}, t = \infty \end{cases} \quad (3)$$

Based on the above conditions, the Fourier series method may be used to calculate the diffusion equation as follows:

$$n(z,t) = \frac{N_{tot}}{V} + \sum_{i=1}^{\infty} \frac{2N_{tot}}{V} \times e^{-\lambda_i D t} \times \cos\left(\frac{i\pi z}{l_z}\right) \quad (4)$$

As a result, the number of free gas molecules can be expressed as follows:

$$\begin{aligned} \frac{1}{Q(t)} &= A \cdot N(t) = A \cdot [N_0 + N_{tot}(t)] = A \cdot \left[\frac{A}{Q_0} + N_{tot} \cdot \left(1 - e^{-D_T \cdot t}\right)\right] \\ &= \left(\frac{1}{Q_0} + A \cdot N_{tot}\right) - A \cdot N_{tot} \cdot e^{-D_T \cdot t} \end{aligned} \quad (5)$$

MEMS sensors that are vacuum packaged can be described by the following vacuum degradation model:

$$p(t) \propto \frac{1}{Q(t)} \propto N(t) = a - b \times e^{-c \times t} \quad (6)$$

In the formula, $a = 1/Q_0 + A \times N_{tot}$, $b = A \times N_{tot}$, $c = D_T$, and the physical meanings of model parameters a, b, and c are as follows:

Parameter a represents the final total number of gases inside the sample cavity, which determines the final Q-factor of the sample. Depending on parameter b, the degree of degradation of the sample is determined by the number of gases released inside the sample cavity. Parameter c characterizes the degradation speed of the sample, which is related to temperature.

With time, the vacuum degree of the vacuum-packaged MEMS sensor decays exponentially. By fitting the experimental data to the above-mentioned model, parameters a, b, and c can be obtained. The above analysis also reveals that $1/(a-b)$ is the initial value of the Q-factor of the sample.

2.2. Experiment of Reliability Evaluation

2.2.1. Sample

As it is difficult to detect and activate the MEMS microstructure of the MEMS products electrically, we customized a batch of samples. It is typical to find MEMS gyroscopes in commercial applications. The structure of the device must be in a resonant state, and it is usually packaged in a vacuum. In order to investigate vacuum degradation of vacuum packaged MEMS devices, this paper selects the sensitive structure of a MEMS gyroscope as the subject of experimental research.

As shown in Figure 1b, MEMS gyroscopes have comb structures as their internal structure. During the operation of the gyroscope, the position of the combs will change, as will the capacitance between them. It is possible to determine the value based on the change in capacitance. As a result, the vacuum encapsulated gyroscope will have difficulty moving the comb, which will adversely affect the gyroscope's performance.

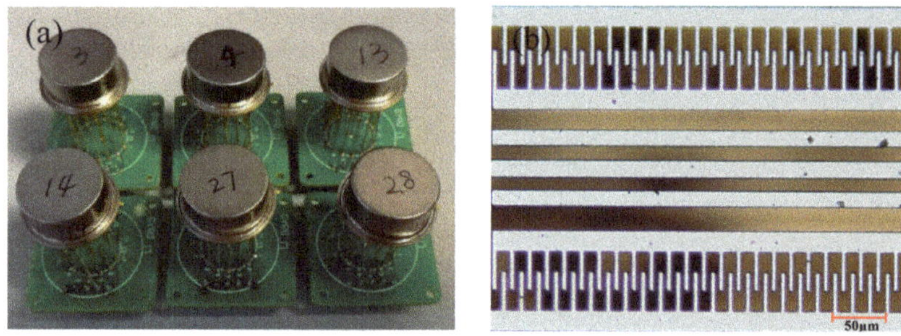

Figure 1. Schematic illustration of (**a**) test sample and (**b**) internal structure diagram.

As shown in Figure 1a, three types of device-level packaged MEMS gyroscopes were customized with different vacuum degrees.

Table 1 provides information regarding the vacuum packaging of the samples. There were a total of 18 samples, which were divided into three groups. Samples that were highly vacuum packaged were H1 and H6, medium vacuum packaged samples were M1 and M6, and low vacuum packaged samples were L1 and L6.

Table 1. Design values of package air pressure and Q-factor for three kinds of samples.

Sample Serial Number	Package Air Pressure	Q-Factor
H1~H6	0.5×10^{-3} Torr	20,000
M1~M6	1.5 Torr	8000
L1~L6	20 Torr	1700

2.2.2. Test Method and System

Considering that the Q-factor can be used to characterize the vacuum of a MEMS cavity, the Q-factor of our sample was evaluated. As shown in Figure 2, we developed a Q-factor test method based on transient excitation [16] for the above-mentioned test samples with different levels of vacuum.

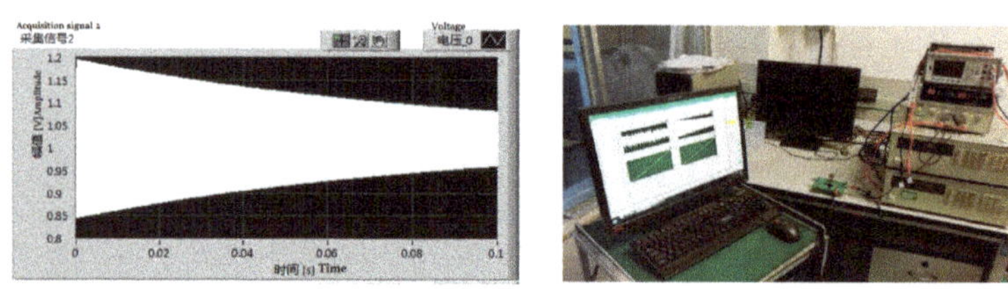

Figure 2. Q-factor test frequency response diagram and test system.

Q value measurement is based on the oscillating principle [17]. Labview software, an NI PXI 4461 acquisition card, an analog circuit, and a MEMS sensor are included in the test system. A part of the test process is as follows: by applying excitation signals to the resonant structure of the MEMS sensor using the acquisition card, Labview software causes the sensor to enter the steady state. After stopping the excitation, the sensor is in a state of damped vibration. As a final step, the acquisition card collects the transient response voltage signal and passes it to Labview for analysis and processing. Using Labview software, it is possible to determine the resonant frequency and Q value rapidly by using wave peak detection technology, the cycle average method, and exponential fitting algorithm.

2.2.3. Experiment

The environment has a significant impact on the reliability of vacuum-packaged MEMS sensors. It has been reported that the Q-factor of vacuum packaged MEMS gyroscopes will degrade with time at room temperature [13]. Based on theory and analysis, it has been tentatively determined that temperature affects the outgassing of internal materials.

At the beginning of the research, a vacuum packaged MEMS gyroscope was baked at three different high temperatures and the Q-factor was monitored over time. According to Figure 3, the Q-factor of the MEMS gyroscope dropped sharply at high temperatures. Accordingly, the analysis of the theoretical and experimental results indicates that high temperature is the main stress contributing to the degree of vacuum degradation of MEMS sensors packaged in a vacuum.

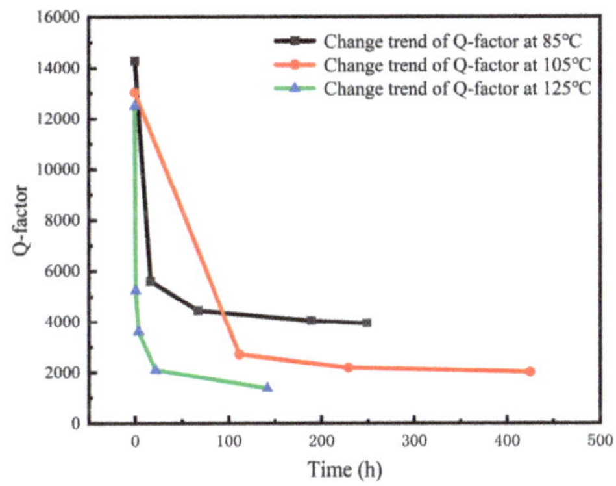

Figure 3. Degradation of Q-factor of vacuum packaged gyroscope with time at high temperature.

At different temperatures, we conducted accelerated degradation experiments on the samples using the high-temperature accelerated test method based on MIL-STD-883. As shown in Table 2, samples with high vacuum, medium vacuum, and low vacuum were tested at three temperatures of 85 °C, 105 °C, and 125 °C.

Table 2. Test conditions and grouping.

Environmental Stress	Samples	Test Parameter
85 °C	H1, H2, M1, M2, L1, L2	
105 °C	H3, H4, M3, M4, L3, L4	Q-factor
125 °C	H5, H6, M5, M6, L5, L6	

In order to conduct a reliability test, the following steps must be followed:

(1) Prior to the test, the Q-factor of the samples is measured at room temperature;

(2) Incubator temperature is raised to the specified temperature, and samples are placed in the incubator after the temperature reaches the specified temperature;

(3) The samples are baked at a high temperature, and then taken out of the incubator after t_1 min;

(4) The Q-factor of the samples is determined after they have stood at room temperature for two hours;

(5) After the measurements have been completed, the samples are placed back into the incubator;

(6) Repeat steps (3) and (4) until the test is complete.

3. Results

3.1. Test Results

In Tables 3–5, reliability test results of samples at 85 °C, 105 °C, and 125 °C under high-temperature stress are presented.

Table 3. Changes of Q-factor of different samples under high-temperature baking at 85 °C.

Time	H1	H2	M1	M2	L1	L2
0 h	14,281	19,390	7572	7822	2118	2018
16.50 h	5603	7578	4407	4563	1816	1766
66.75 h	4457	5914	3735	3856	1706	1673
189.25 h	4042	5163	3496	3579	1659	1624
248.75 h	3944	5034	3478	3550	1654	1621

Table 4. Changes of Q-factor of different samples under high-temperature baking at 105 °C.

Time	H3	H4	M3	M4	L3	L4
0 h	13,048	16,388	7710	7744	2072	2061
111.5 h	2721	3200	2572	2535	1439	1407
229 h	2201	2591	2202	2157	1320	1292
425 h	2030	2403	2076	2029	1265	1242

Table 5. Changes of Q-factor of different samples under high-temperature baking at 125 °C.

Time	H5	H6	M5	M6	L5	L6
0 h	12,504	19,571	8582	8405	1967	1933
1 h	5245	7459	5211	5134	1745	1721
4 h	3604	4737	3809	3740	1556	1535
22 h	2100	2432	2196	2198	1210	1192
142 h	1399	1467	1366	1437	895	891

As a result of the reliability test, we used the mathematical model presented in this paper for fitting. As can be seen from Figure 4, the typical fitting curve reaches a fitting degree of 1.00, indicating that the mathematical model established in this paper is more capable of reflecting the changes in vacuum degree over time under a high-temperature environment.

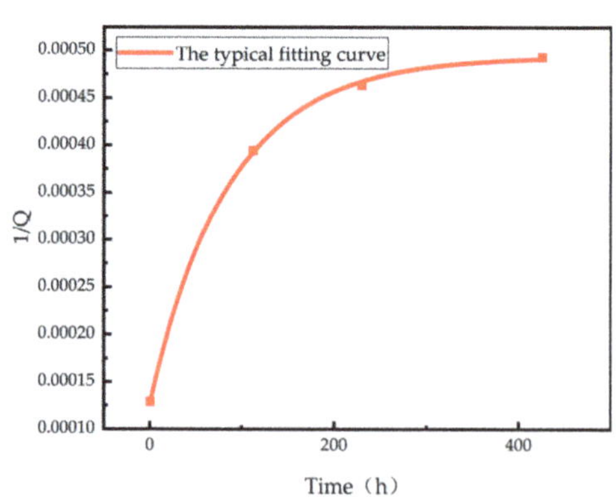

Figure 4. Q-factor degradation fitting curve of M4 sample at 105 °C.

Having analyzed the data obtained from the experiment, the fitting parameters for the test samples with different vacuum degrees at three temperature points are determined, as shown in Tables 6–8.

Table 6. Parameter values of Q-factor degradation fitting curve baking at 85 °C.

Sample	Q_0	*a*	*b*	*c*	R^2
H1	14,281	2.44×10^{-4}	1.73×10^{-4}	0.0552	0.97
H2	19,390	1.91×10^{-4}	1.37×10^{-4}	0.0452	0.95
M1	7572	2.82×10^{-4}	1.49×10^{-4}	0.0576	0.98
M2	7822	2.76×10^{-4}	1.47×10^{-4}	0.0547	0.97
L1	2118	6.00×10^{-4}	1.27×10^{-4}	0.0544	0.98
L2	2018	6.12×10^{-4}	1.15×10^{-4}	0.0517	0.97

Table 7. Parameter values of Q-factor degradation fitting curve baking at 105 °C.

Sample	Q_0	*a*	*b*	*c*	R^2
H3	13,048	4.96×10^{-4}	4.19×10^{-4}	0.0105	1.00
H4	16,388	4.18×10^{-4}	3.57×10^{-4}	0.0108	1.00
M3	7710	4.82×10^{-4}	3.52×10^{-4}	0.0118	1.00
M4	7744	4.93×10^{-4}	3.64×10^{-4}	0.0115	1.00
L3	2072	7.92×10^{-4}	3.09×10^{-4}	0.0102	1.00
L4	2061	8.06×10^{-4}	3.21×10^{-4}	0.0107	1.00

Table 8. Parameter values of Q-factor degradation fitting curve baking at 125 °C.

Sample	Q_0	*a*	*b*	*c*	R^2
H5	12,504	7.10×10^{-4}	5.73×10^{-4}	0.0452	0.95
H6	19,571	6.82×10^{-4}	5.88×10^{-4}	0.0379	0.97
M5	8582	7.34×10^{-4}	5.79×10^{-4}	0.0354	0.98
M6	8405	6.95×10^{-4}	5.38×10^{-4}	0.0393	0.97
L5	1967	1.12×10^{-3}	5.80×10^{-4}	0.0325	0.98
L6	1933	1.13×10^{-3}	5.76×10^{-4}	0.0335	0.98

The curve fitting degree for 18 samples exceeds 1.00 or is close to it, which confirms the accuracy of our mathematical model. As a result, it can also be concluded that the established model can accurately predict the degree of vacuum degradation of MEMS sensors that are vacuum packaged.

Based on the physical meaning of the fitting parameters, this paper analyses and discusses the obtained parameter values.

Parameter *a* specifies the total number of gases inside the sample cavity and determines the final Q-factor of the sample. According to the parameter values above for the fitting curve parameters, there are 18 samples at three different temperature points, and there is no significant difference between the *a* values of the high vacuum and medium vacuum samples at the same temperature point. This is due to the fact that the amount of gas released from high and medium vacuum samples is significantly greater than the amount of gas released from the initial sample. Compared to the high and medium vacuum samples at the same temperature, the low vacuum sample has a higher *a* value. Since the initial gas number of the low vacuum samples is higher, it is comparable with the total amount of gas released.

In addition, parameter *b* determines the degree of sample degradation by characterizing the total number of gases released inside the sample cavity. At the same temperature, the outgassing amount represented by the *b* value is essentially the same, which means that it has nothing to do with the vacuum degree of the sample, but only with its temperature. The higher the temperature, the greater the amount of outgassing.

Again, *c* characterizes the degradation speed of the samples, that is, the outgassing rate inside the sample, which is related to temperature. In high, medium, and low vacuums, the outgassing rates of samples remain essentially the same, indicating this parameter is not related to the sample's vacuum degree, but only to its temperature.

3.2. Data Analysis

Parameters b and c of the six samples were relatively consistent at every temperature point, and the average value was taken as the value of parameter b and parameter c and summarized in Table 9.

Table 9. Test sample parameters b and c at three temperature points.

Test Temperature	b	c
85 °C	1.41×10^{-4}	0.0531
105 °C	3.54×10^{-4}	0.0109
125 °C	5.72×10^{-4}	0.0373

Parameter b characterizes the total amount of gas released inside the sample, and the total amount of released gas inside the sealed cavity is related to the temperature.

According to the relationship between the total amount of gas out of non-metallic materials and temperature [18]:

$$b = b_0 \times e^{-B \times T} \quad (7)$$

The above equation can be transformed into a linear equation:

$$\text{In}b = \text{In}b_0 - B \times T \quad (8)$$

The data of parameter b and temperature T at three temperature points are shown in Table 10.

Table 10. Data table of parameter b and temperature T at three temperature points.

T	Inb
358 K	−8.8644
378 K	−7.9472
398 K	−7.4663

Taking T as the abscissa and Inb as the ordinate, the linear fitting curve is shown in Figure 5, and the fitting equation is $y = 0.0350 \times x - 21.3138$.

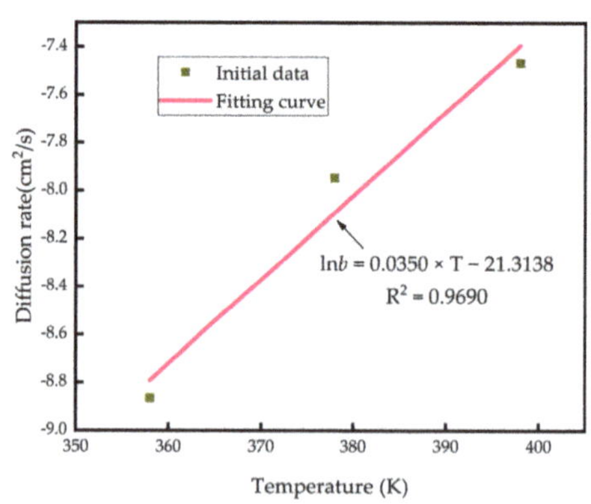

Figure 5. Linear fitting curve of T and ln$b$.

The following parameters can be calculated from the equation of total gas output with temperature:

$$\ln b_0 = -21.3138; \ B = 0.0350 \quad (9)$$

Total outgassing volume and temperature are related as follows:

$$b = e^{-21.3138+0.0350\times T} \tag{10}$$

In a sealed cavity, the rate of gas release is determined by parameter c, and the rate of gas release is related to the temperature inside the sealed cavity. According to theory, the higher the temperature, the greater the rate of outgassing. Figure 6 illustrates the results of the tests conducted in this paper. The outgassing rate at 85 °C is the highest, which is 0.0531; the outgassing rate at 105 °C is the lowest, which is 0.0109; and the outgassing rate at 125 °C is in the middle, which is 0.0373.

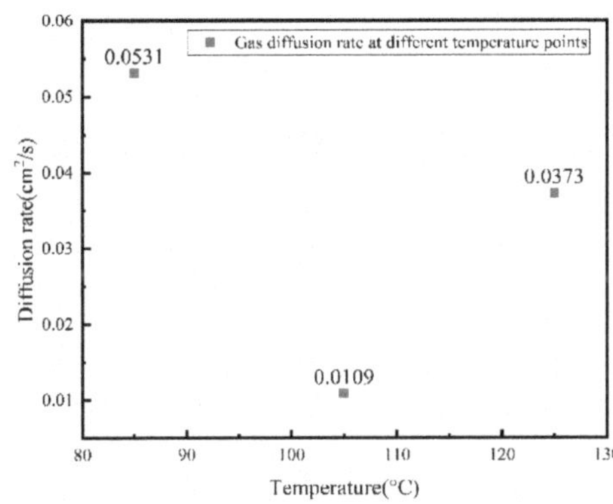

Figure 6. Outgassing rate at different temperature points.

As a result of the experiments conducted, this paper concludes that the outgassing rate of the gas inside the sample is equal to the desorption rate of the gas minus the adsorption rate of the gas. It is important to note that the maximum outgassing rate occurs at 85 °C as a result of the desorption of physically adsorbed gas atoms or molecules from the surface of the material, whereas a higher temperature is required for gas in the deeper part of the material. The collision frequency between gas molecules or atoms increases at 105 °C, which increases the adsorption rate. As a result, the outgassing rate decreases at 105 °C. Gas atoms and molecules are desorbed at 125 °C by physical adsorption and chemical adsorption, resulting in an increase in the rate at which they are outgassed.

3.3. Reliability Prediction

Thus, we can obtain a mathematical model for describing the vacuum degradation of MEMS sensors as follows:

$$p(t) \propto \frac{1}{Q(t)} \propto N(t) = a + \left(e^{-21.3138+0.0350\times T}\right) \times \left(\mathrm{e}^{-c\times t}\right) \tag{11}$$

The majority of vacuum packaged MEMS gyroscopes and other MEMS sensors are used at room temperature, and the obtained mathematical model is used to predict the long-term reliability of MEMS gyroscopes under 25 °C vacuum degradation conditions.

The following parameters can be determined at a room temperature of 25 °C using the formula deduced above:

$$b = e^{-21.3138+0.0350\times 298} = 0.00001876 \tag{12}$$

Using the previous analysis, it can be determined that $1/(a-b)$ can obtain the initial value of the Q-factor of samples, and we also know the value of b. Therefore, parameter a, which represents the final gas volume of the cavity, determines the final value of the MEMS sensor. Since the value of c is uncertain, $t = \infty$ is taken.

Table 11 shows the changes in Q-factor during storage (T = 25 °C) with time for different initial Q-factor.

Table 11. Changes of different initial Q-factor with time during storage (T = 25 °C).

Time	Q_0 = 1000	Q_0 = 10,000	Q_0 = 100,000	Q_0 = 1,000,000
$t = 0$	1000	10,000	100,000	1,000,000
∞	982	8420	34,771	50,607

4. Discussion

According to the established mathematical degradation model of vacuum-packaged MEMS sensors, the long-term reliability of vacuum-packaged MEMS sensors at room temperatures (25 °C) was predicted. The data are shown in Table 11, the vacuum degree of samples with different initial Q-factor degrades continuously over time; whereas samples with small initial Q-factor do not exhibit obvious degradation trends, samples with larger initial Q-factors demonstrate more obvious degradation trends. Furthermore, the degradation degree of the Q-factor of all samples tends to be flat with the increasing storage time.

At room temperature, the gas molecules adsorbed on the surface material of the sensor cavity are continuously released, which leads to the decrease of the vacuum degree of the sensor, thus reducing the reliability of the sensor. In the process of high-temperature accelerated aging, physical and chemical desorption of gas will occur, especially the gas released by some polymers in the cavity, which has a great impact on the reduction of vacuum degree.

5. Conclusions

In this paper, a theoretical model for the prediction of vacuum degradation of vacuum packaged MEMS sensors is established. Furthermore, combination with the experiments, the MEMS gyroscopes with high vacuum degree, medium vacuum degree, and low vacuum degree are used as examples to study the vacuum degradation law at different high temperatures and get the parameters of the model. The reliability life prediction of the vacuum packaged MEMS sensor under a room temperature environment (25 °C) is also carried out. The results show that our theoretical model can achieve a good prediction of vacuum degradation of vacuum packaged MEMS sensors. This is of great significance for improving the reliability of vacuum-packaged MEMS sensors and promoting the commercial application of vacuum-packaged MEMS sensors. In the next step, we will conduct further research on the variation law of the outgassing rate in the cavity with temperature, such as using a lower temperature test to obtain the relationship between the outgassing rate and temperature and predict the variation of the vacuum degree with time.

Author Contributions: Conceptualization, G.D. and X.D.; methodology, G.D. and X.D.; software, X.H.; validation, G.D. and X.H.; formal analysis, G.D. and X.D.; investigation, X.D. and W.S.; resources, X.D. and W.S.; data curation, G.D.; writing—original draft preparation, G.D. and X.D.; writing—review and editing, G.D. and X.D.; visualization, G.D. and P.Z.; supervision, P.Z. and X.D.; project administration, P.Z. and X.D.; funding acquisition, X.D. All authors have read and agreed to the published version of the manuscript.

Funding: This research was funded by Guangzhou Science and Technology Plan Project (Grant Number 202102080190), Industrial Technology Foundation Public Service Platform Project (Grant Number 2021-H021-1-1), and Ministry of Science and Technology Key R&D Program (Grant Number 2020YFB2008900).

Data Availability Statement: Not applicable.

Conflicts of Interest: The authors declare no conflict of interest.

References

1. Liu, Y. Development Status and Key Technology Analysis of MEMS 2019. *New Mater. Ind.* **2019**, *3*, 51–55.
2. Pan, J.H. The Development Trend of MEMS Sensor and Its Application 2021. *Electron. World* **2021**, *5*, 37–38.
3. Lammel, G. The Future of MEMS Sensors in Our Connected World. In Proceedings of the 2015 28th IEEE International Conference on Micro Electro Mechanical Systems (MEMS), Estoril, Portugal, 18–22 January 2015; pp. 61–64.
4. Javed, Y.; Mansoor, M.; Shah, I.A. A Review of Principles of MEMS Pressure Sensing with Its Aerospace Applications. *SR* **2019**, *39*, 652–664. [CrossRef]
5. Ruffin, P.B.; Burgett, S.J. *Recent Progress in MEMS Technology Development for Military Applications*; Varadan, V.K., Ed.; Proceedings of the SPIE: Newport Beach, CA, USA, 2001; pp. 1–12.
6. Ou, J.; Liu, K.; Jiang, J.; Wilson, D.A.; Liu, L.; Wang, F.; Wang, S.; Tu, Y.; Peng, F. Micro-/Nanomotors toward Biomedical Applications: The Recent Progress in Biocompatibility. *Small* **2020**, *16*, 1906184. [CrossRef] [PubMed]
7. Liang, C.; Zappella, P. Advanced Packaging Solution to Hermetically Packaging Microelectronic Devices. *IEEE Trans. Compon. Packag. Manufact. Technol.* **2021**, *11*, 1055–1062. [CrossRef]
8. Wu, M.; Moulin, J.; Bosseboeuf, A. Multilayer Thin Film Getter for Sustainable Vacuum in MEMS Packaging. *ECS Trans.* **2018**, *86*, 95–104. [CrossRef]
9. Tepolt, G.B.; Mescher, M.J.; LeBlanc, J.J.; Lutwak, R.; Varghese, M. *Hermetic Vacuum Sealing of MEMS Devices Containing Organic Components*; Kullberg, R.C., Ramesham, R., Eds.; Proceedings of the SPIE: San Francisco, CA, USA, 2010; p. 759207.
10. Fang, L.; Menk, L.A. On Gas Ingression of Hermetic Packages. *IEEE Trans. Compon. Packag. Manufact. Technol.* **2019**, *9*, 1038–1044. [CrossRef]
11. He, C.H.; Wang, Y.P.; Huang, Q.W.; Zhao, Q.C.; Yang, Z.C.; Zhang, D.C.; Yan, G.Z. Research on the Packaging Reliability and Degradation Models of Quality Factors for a High Vacuum Sealed MEMS Gyroscope. In Proceedings of the 2017 19th International Conference on Solid-State Sensors, Actuators and Microsystems (TRANSDUCERS), Kaohsiung, Taiwan, 18–22 June 2017; pp. 1077–1080.
12. Candler, R.N.; Hopcroft, M.A.; Kim, B.; Park, W.-T.; Melamud, R.; Agarwal, M.; Yama, G.; Partridge, A.; Lutz, M.; Kenny, T.W. Long-Term and Accelerated Life Testing of a Novel Single-Wafer Vacuum Encapsulation for MEMS Resonators. *J. Microelectromech. Syst.* **2006**, *15*, 1446–1456. [CrossRef]
13. Gu, Z.Y. Research on Degradation and Failure Characteristics of MEMS Gyroscope Under Environmental Stress. Master's Thesis, South China University of Technology, Guangzhou, China, 2017.
14. Jin, J.; Cai, Y.Z. Analysis of Residual Gas Pressure and Composition in Vacuum Packaged Microcavity. *Vac. Cryog.* **2020**, *26*, 5.
15. Li, Y.S. Research on Packaging of Micromachined Gyroscope. Master's Thesis, North Central University, Taiyuan, China, 2011.
16. Guo, B.Z. Simulation of AC Characteristics of Semiconductor Devices by Transient Excitation Analysis. *Semicond. Inf.* **1998**, *35*, 39–42.
17. Wang, Y.Z.; Teng, L.; Sun, X.Z.; Wang, G.; Guo, Z.X.; Yu, C.J. Calibration of quality factor of vacuum packaged silicon micro gyroscope 2014. *Opt. Precis. Eng.* **2014**, *22*, 6. [CrossRef]
18. Wang, Z.Y.; Chen, X. *Vacuum Technology*, 2nd ed.; Beihang University Press: Beijing, China, 2007; p. 129.

micromachines

MDPI

Review

Recent Progress in Physics-Based Modeling of Electromigration in Integrated Circuit Interconnects

Wen-Sheng Zhao [1,*], Rui Zhang [2] and Da-Wei Wang [1]

[1] Zhejiang Provincial Key Lab of Large-Scale Integrated Circuit Design, School of Electronics and Information, Hangzhou Dianzi University, Hangzhou 310018, China; davidw.zoeq@hdu.edu.cn

[2] Cadence Design Systems, Inc., San Jose, CA 95134, USA; tangrui.zhang@gmail.com

* Correspondence: wshzhao@hdu.edu.cn

Abstract: The advance of semiconductor technology not only enables integrated circuits with higher density and better performance but also increases their vulnerability to various aging mechanisms which occur from front-end to back-end. Analysis on the impact of aging mechanisms on circuits' reliability is crucial for the design of reliable and sustainable electronic systems at advanced technology nodes. As one of the most crucial back-end aging mechanisms, electromigration deserves research efforts. This paper introduces recent studies on physics-based modeling of electromigration aging of on-chip interconnects. At first, the background of electromigration is introduced. The conventional method and physics-based modeling for electromigration are described. Then studies on how electromigration affects powers grids and signal interconnects are discussed in detail. Some of them focus on the comprehensiveness of modeling methodology, while others aim at the strategies for improving computation accuracy and speed and the strategies for accelerating/decelerating aging. Considering the importance of electromigration for circuit reliability, this paper is dedicated to providing a review on physics-based modeling methodologies on electromigration and their applications for integrated circuits interconnects.

Keywords: integrated circuit interconnects; aging; reliability; electromigration; physics-based modeling

Citation: Zhao, W.-S.; Zhang, R.; Wang, D.-W. Recent Progress in Physics-Based Modeling of Electromigration in Integrated Circuit Interconnects. *Micromachines* **2022**, *13*, 883. https://doi.org/10.3390/mi13060883

Academic Editor: Aiqun Liu

Received: 16 May 2022
Accepted: 29 May 2022
Published: 31 May 2022

Publisher's Note: MDPI stays neutral with regard to jurisdictional claims in published maps and institutional affiliations.

Copyright: © 2022 by the authors. Licensee MDPI, Basel, Switzerland. This article is an open access article distributed under the terms and conditions of the Creative Commons Attribution (CC BY) license (https://creativecommons.org/licenses/by/4.0/).

1. Introduction

Although technology scaling enables integrated circuits (ICs) with higher density and better performance, it is still faced with serious vulnerability to various aging mechanisms appearing from front-end to back-end [1–10]. These aging mechanisms include Bias Temperature Instability (BTI), Hot Carrier Injection (HCI), Random Telegraph Noise (RTN), Gate-Oxide Breakdown (GOBD) at the front-end, Middle-of-line (MOL) time-dependent dielectric breakdown (TDDB), Back-end-of-line (BEOL) TDDB, and Electromigration (EM). BTI, HCI, RTN, and GOBD cause device parameter deviations. MOL and BEOL TDDBs cause *short circuit* between interconnects, while EM increases interconnects' resistance and it eventually results into *open circuit*. These phenomena lead to malfunction in circuits. From BTI to BEOL TDDB, there are numerous studies for their impact on device, circuit, and system performance and reliability [11–30], but they are not the topic discussed here. In this paper, our attention is focused on recent progress in physics-based modeling of EM in on-chip interconnects.

EM is the migration of interconnects' metal atoms after they obtain momentum from moving electrons. Since the interconnects are wrapped by barrier layer, the metal atoms' movement causes depletion regions and it, eventually, causes void nucleation and growth. EM induced degradation and failure is one of the most critical reliability issues for deeply scaled ICs. Such a degradation is expected to get even worse with further technology scaling. It is reported by ITRS-2015 that the operating current density has exceeded 1 MA/cm^2 and is rapidly approaching to 10 MA/cm^2. Figure 1a shows the experiment and model of lifetime scaling versus interconnect geometry, where an effective scaling model has been established by assuming the void is located at the cathode end of the interconnect which contains a single via with drift velocity dominated by interfacial diffusion [31]. The interconnects' EM lifetime is predicted to decrease by half for each new generation. Figure 1b shows evolution of J_{max} (the maximum equivalent dc current density) and J_{EM} (the current density for target EM lifetime) [31]. Both J_{max} and J_{EM} are limited by interconnect geometry. J_{max} increases with the increase of operating frequency and the reduction of interconnect cross-section. Although there are several process options, such as usage of Cu alloys seed layer and the short length effect to overcome EM severity, EM is still an inescapable topic since the operating current density is entering J_{EM} region, as shown in Figure 1b. It is necessary to apply novel models and methodologies to study the strategies which can help mitigate EM degradation by incorporating the innovations in material and process.

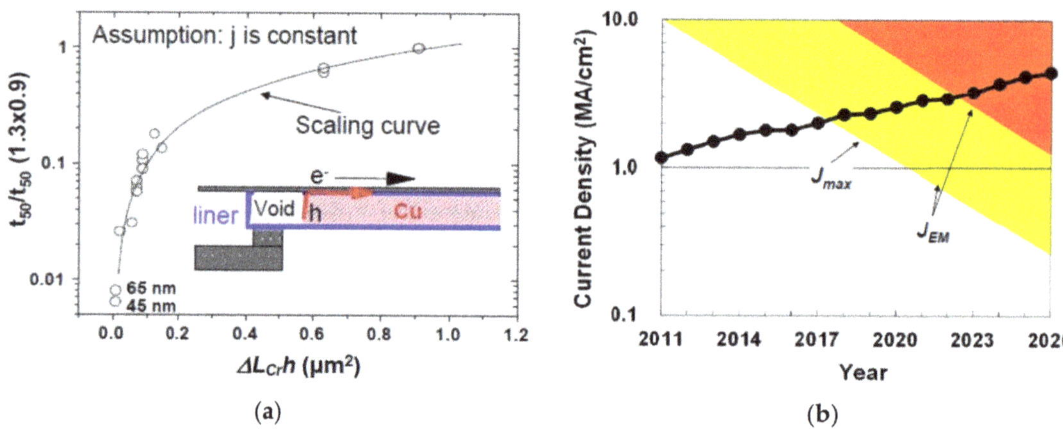

Figure 1. (**a**) Experiment and model of lifetime scaling versus interconnect geometry; (**b**) evolution of J_{max} (the maximum equivalent dc current density) and J_{EM} (the current density for target EM life-time). They are predicted by ITRS-2015 [31].

Depending on their functionality, on-chip interconnects can propagate signal within and between cells and deliver power to sub-circuits. Most of the previous EM studies have focused on the reliability of power grids and clock/signal nets. In traditional approach, the Blech limit and Black's equation are applied to investigate interconnects' EM reliability. The Blech limit or Blech product is [32]

$$(j \times L) \leq (j \times L)_{crit} = \frac{\Omega \sigma_{crit}}{eZ\rho} \tag{1}$$

where j is the current density, L is the interconnect branch length, Ω is the atomic volume, σ_{crit} is the critical stress for void nucleation, e is the electron charge, eZ is the effective charge of migrating atoms, and ρ is the metal resistivity.

The interconnect branches with $(j \times L) \leq (j \times L)_{crit}$ are filtered out as EM immortal. The mean time to failure of remaining EM mortal branches is characterized with Black's equation [33]

$$MTTF = Aj^{-n}exp\{E_a/k_BT\} \tag{2}$$

where A and n are assumed to constants, E_a is the activation energy, k_B is the Boltzmann's constant, and T is the absolute temperature.

A and n are measured through accelerated test with a higher current density injected under higher temperature. Then the $MTTF$ under normal use condition is extrapolated as a function of $MTTF$ under accelerated test,

$$MTTF_{use} = MTTF_{stress}\left(\frac{j_{stress}}{j_{use}}\right)^{n} exp\left\{\frac{E_a}{k_B}\left(\frac{1}{T_{use}} - \frac{1}{T_{stress}}\right)\right\} \tag{3}$$

where j_{stress} and T_{stress} are current density and absolute temperature under accelerated test, while j_{use} and T_{use} are current density and absolute temperature under normal-use condition, respectively.

Based on $MTTF_{use}$, the probability of failure (at a specific stress time) of each branch is obtained from cumulative distribution function of its lifetime distribution (lognormal or weibull). Then the overall probability of failure of studied branches is computed by weakest-link statistics. Although this traditional approach is convenient to use, it does not provide accurate estimation on EM failure due to the following reasons: first, n is not really a constant under different current density especially when j_{stress} is much larger than j_{use}; second, the Belch limit is accurate only for single branch interconnect. It is inapplicable for a general interconnect tree with multiple segments/branches. As shown in Figure 2, since the segments are connected without barriers, it is necessary to consider the atom transportation between them [34]. Third, weakest-link statistics considers a single branch failure leads to failure of studied interconnect trees which is not true in state-of-the-art ICs, especially for power grids with a mesh structure [35]. Therefore, some physics-based EM compact models have been proposed to incorporate the atom movement between segments [36–44]. It is noted that there are already several excellent reviews on relevant topics in [34,45–48], but they are mainly on power grids. In this review article, we will summarize the studies on physics-based EM compact modeling and its applications on various on-chip interconnects and their most recent updates. The remainder of this paper is organized as follows. Section II introduces the physics-based three-phase compact EM model. The model should be imposed on interconnects with corresponding current density and boundary conditions. The applications of this compact EM model in current and future interconnects are discussed as well. Section III summaries recent progress in EM studies mainly on power grids. These studies cover from Black's equation-based simulations to physics-based simulations. Section IV shows the application of physics-based modeling on interconnects in cache memory. The impact of physical dimensions and cache configurations on EM reliability is discussed. Section V concludes this paper.

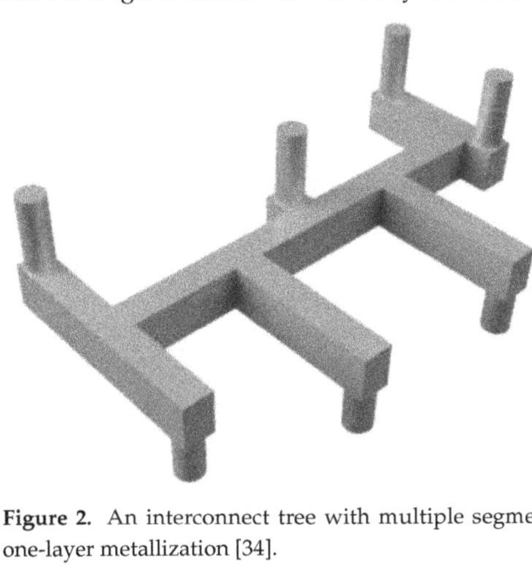

Figure 2. An interconnect tree with multiple segments confined by diffusion barriers/liners in one-layer metallization [34].

2. Physics-Based Three-Phase Compact Model

EM is the phenomenon of metal atoms migration due to the momentum obtained from electrons when there is current flowing over interconnects. Since the momentum is also transferred between metal atoms, hydrostatic stress would appear in the interconnects confined by barrier material and capping layer. Although the stress gradient may prevent atoms to move, the maximum stress continues to increase until it reaches the saturation value or a critical stress value for void generation. A void is considered to be nucleated at the position where the maximum hydrostatic stress reaches critical stress value. Then the void is incubated and it grows under further stimulus by electrons' movement. A void will never appear in an interconnect if the saturation stress value is smaller than the critical value. Such an interconnect is EM immortal. Figure 3a shows a Cu dual damascene conductor structure [49]. The trench is lined with a Cobalt-based liner and the Cu is capped with either a dielectric or metallic layer. Figure 3b shows the voids possibly formed in an interconnect. The void formed under a via can cause *early failure* when the size is large enough to cover the via bottom and it leads to an open circuit status, while the void formed above a via causes *conventional failure* when the gradually growing void introduces an obvious resistance shift which causes circuit's functional failure with a significant performance deviation. Both *early* and *conventional failures* need to be checked during EM analysis. The EM process during void generation and growth can be described with a three-phase model which consists of nucleation phase, incubation phase, and growth phase. To calculate the time-dependent EM process, Korhonen's model has been proposed to capture the time-varying stress. The Korhonen's model together with appropriate initial condition and boundary conditions are able to accurately emulate hydrostatic stress evolution in interconnects. Thereafter, the time-dependent stress and atomic flux are combined to obtain resistance shift due to EM. The resistance shift can be inserted into circuit simulation to show its impact on performance. This section introduces Korhonen's model and the three-phase compact EM model.

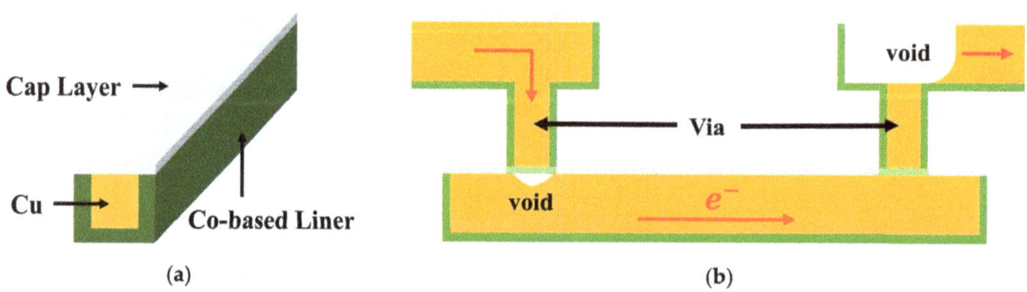

Figure 3. (**a**) Cu dual damascene conductor structure, where the trench is lined with a Cobalt-based liner and the Cu is capped with either a dielectric or metallic layer. (**b**) Void formation in a Cu wire connected with a via-below (left) and a via-above (right) [49].

2.1. Korhonen's Model

In general, multi-physics 3-D simulation is able to present accurate estimation on hydrostatic stress and resistance shift in interconnects. However, since it requires vast computation resource and the simulation speed gets very slow when the studied interconnect structure (such as power grids for a microprocessor) is relatively large, 3-D numerical simulation is not always preferred for EM analysis. Fortunately, Korhonen's 1-D model offers a reasonable trade-off between accuracy and efficiency. Let us take a branch with a length of L as an example. The evolution of 1-D hydrostatic stress follows [50]

$$\frac{\partial \sigma}{\partial t} = \frac{\partial}{\partial x}\left[\frac{D_a B}{k_B T}\left(\Omega \frac{\partial \sigma}{\partial x} - eZ\rho_{Cu} j\right)\right] \tag{4}$$

$$J_a = \frac{D_a C_a \Omega}{k_B T}\left[\frac{\partial \sigma}{\partial x} - \frac{eZ\rho_{Cu} j}{\Omega}\right] \tag{5}$$

where the effective atomic diffusivity $D_a = D_0 exp(-E_a/k_B T)$. D_0 is a constant, E_a is the effective activation energy, k_B is Boltzmann's constant, T is absolute temperature, B is the effective bulk elasticity modulus, Ω is the atom lattice volume, eZ is the effective charge of migrating atoms, ρ_{Cu} is copper resistivity, j is the current density over the studied interconnect, C_a is the atom concentration, J_a is the atomic flux, and t is the stress time.

Initial conditions and boundary conditions are necessary for a solution of hydrostatic stress. They are decided by the material difference in the coefficients of thermal expansion (*CTE*) and the difference between stress free annealing temperature and the circuit temperature. The position dependent (x ranges from *0* to L, $x = 0$ is at the branch's left end, and $x = L$ is the branch's right end when it is assumed to be placed horizontally) initial value (at $t = 0$) is expressed as [51]

$$\sigma(x, t = 0) = B\left(a_M - a_{Conf}\right)(T_{ZS} - T(x, t = 0)) \tag{6}$$

where T_{ZS} is the stress free annealing temperature, x denotes the node position, $T(x, t = 0)$ is the specific node temperature at $t = 0$, and a_M and a_{Conf} are the *CTE* of the metal and confinement materials, respectively.

The studied branch may be connected to other branches at its left and right ends. The boundary conditions at the two ends depend on whether they are connected to other branches and whether the voids have appeared at their locations. At the left and right ends, the boundary conditions before and after void nucleation are given as [41,44]

$$\begin{cases} \sum\limits_{k=0}^{K_l} J_{a,\,k} w_k h_k = 0, \ for\ left\ node\ if\ it'\ in\ nucleation\ phase \\ \frac{\partial \sigma(x=0,t)}{\partial x} = \frac{\sigma(x=0,t)}{\delta}, \ for\ left\ node\ if\ it's\ in\ incubation\ and\ growth\ phases \\ \sum\limits_{k=0}^{K_r} J_{a,\,k} w_k h_k = 0, \ for\ right\ node\ if\ it's\ in\ nucleation\ phase \\ \frac{\partial \sigma(x=L,t)}{\partial x} = -\frac{\sigma(x=L,t)}{\delta}, \ for\ right\ node\ if\ it's\ in\ incubation\ and\ growth\ phases \end{cases} \tag{7}$$

where K_l and K_r are the total number of branches connected to the left and right nodes, respectively, w_k and h_k are the width and thickness of their k_{th} connected branch, and δ is the thickness of the void surface. It is noted that the maximum hydrostatic stress (during void nucleation) in one branch appears at its boundary nodes and the void may not appear at the boundary nodes if the hydrostatic stress at their position never exceeds critical stress (σ_{th}), then the boundary condition should be always applied as the one in nucleation phase.

2.2. Three-Phase Compact Model

The three-phase compact model includes nucleation, incubation, and growth phases, as shown in Figure 4 [45]. When the interconnect branches are flown over by currents, the hydrostatic stress on them evolves with the initial value described in Section 2.1. In the nucleation phase, the resistance of a studied branch remains unchanged before the maximum stress exceeds a threshold (σ_{th}), while after it exceeds the threshold, the model enters the incubation phase where the void size is increasing but it is still smaller than a threshold size. The interconnect resistance is stays unchanged in nucleation and incubation phases. Later, the model enters growth phase if the time-dependent void size grows large enough to cover the interconnects' cross section. The resistance jump is caused by Joule heating. In the growth phase, the branch's time-dependent resistance shift can be expressed as [49]

$$\Delta R = L_v(t)\left(\frac{\rho_{liner}}{(W + 2H + 2t_{liner})t_{liner}} - \frac{\rho_{cu}}{WH}\right) \tag{8}$$

where W and H are the width and height of the studied branch, respectively, ρ_{liner} and t_{liner} are the resistivity and thickness of the liner. $L_v(t)$ is the time-dependent void length, and its growth can be calculated with $\Delta L_v(t) = J_a(t) \cdot \Omega \cdot \Delta t$.

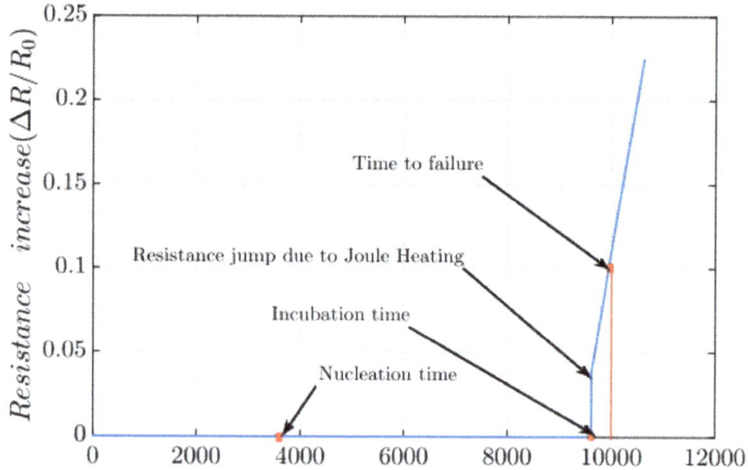

Figure 4. Resistance change over time under the three-phase EM model [45].

The three-phase model can be validated by experimental data. Figure 5 shows resistance trace of interconnects in experimental measurements [52]. Obviously, most of the traces follow resistance shift behavior described by the three-phase EM model.

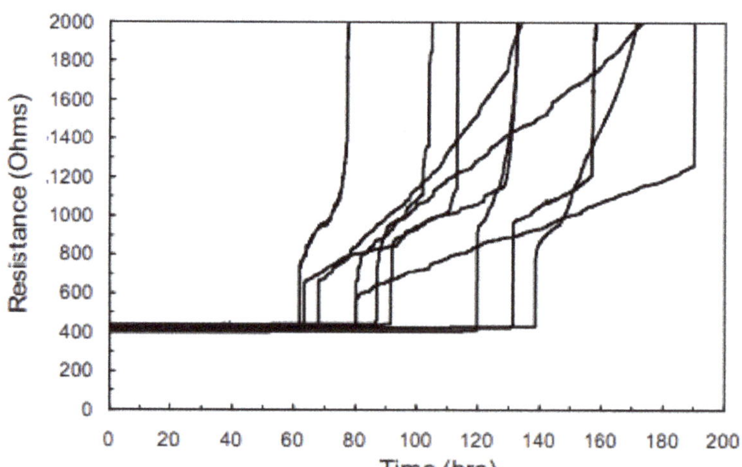

Figure 5. Resistance trace of interconnects under EM testing [52].

2.3. Model Applications

The Korhonen's model describes time-dependent evolution of hydrostatic stress in specific branch. Since the interconnects are confined by diffusion barriers/liners in one-layer metallization, the whole interconnect network can be divided into many individual interconnect trees which are similar as the one shown in Figure 2. For interconnect tree, their hydrostatic stress evolution can be evaluated independently. It should be noted that the current density over the branches may vary under the appearance of resistance shift of some branches. For each branch in a specific tree, the Korhonen's model can be applied with appropriate initial condition and boundary conditions to accurately compute the hydrostatic stress evolution. Based on this, the time-dependent resistance shift of the EM mortal branch is obtained conveniently. The model has been applied to check the reliability of state-of-the-art Cu interconnects. More details are given in the following sections. With the continuous scaling, the Cu interconnects in sub-10 nm technology node suffer from high resistance due to serious surface scattering of the electrons flowing over them. It is found in [53,54] that reducing the linewidth to 10 nm results in a drop of j_{max} to below 1 MA/cm^2 and scaling linewidth from 25 nm to 10.5 nm leads to a 90% drop of j_{fail} i.e. the current density that induces failure at 10 y. Under such a scenario, interconnects based on Ru and Co are potential replacements with better reliability than Cu because of their lower resistivity and higher EM activation energies [55–60]. The barrierless Ru interconnect together with an integration scheme have been identified to be more EM reliable than Cu [61]. And it is found that full Ru vias have no risk of voiding after long thermal storage at high temperature [62]. With respect to Co interconnects, the first estimation of an effective D_0Z^* (~1.72×10^{-10}) for Co is performed [63]. It is two orders of magnitude lower than Cu. While Co vias may be EM immortal, the lack of a barrier may induce diffusion along the Co/dielectric interfaces and Co/Cu intermixing [64]. The physics-based compact EM model introduced in this section is also applicable to interconnects based on new materials such as Co and Ru only if the diffusion coefficients and microstructures are extracted from experiment result [65].

3. Modeling of EM Impact on Power Grids

In the most previous studies of EM modeling, the Black's equation and Blech limit are applied to analyze the reliability of signal interconnects and power grids [66–73]. In [66–68], by incorporating the Joule heating effect, Gracieli Posser et al. have developed approaches for modeling and efficient characterization of cell-internal EM and have simulated EM effects on different metal layers at different wire lengths. On the one hand, they found the cell-internal EM reliability can be optimized with layout modification and constraints on output pin position. On the other hand, it is concluded that larger metal layers have smaller EM effects and, consequently, a higher EM lifetime for the wires. Palkesh Jain et al. proposed a SoC-level logic-IP-internal EM verification methodology which provides on-the-fly retargeting capability for reliability constraints [69,70]. The proposed approach is demonstrated on a 28-nm design. Meanwhile, they presented a fast and stochastic analysis methodology to overcome the lifetime under-estimation by conventional methodologies based on weakest-link assumption for EM assessment of click grids and power grids [71,72]. Vidya et al. also applied Black's equation and Blech limit to estimate the self-heating impact on EM reliability of FinFET and GAAFET designs [73]. In order to overcome the reliability under-estimation due to the traditional series model for EM checking and the pessimistic assumptions about the chip workload and the corresponding supply currents, Mohammad Fawaz and Sandeep Chatterjee et al. proposed a framework for EM checking that allows users to specify conditions-of-use type constraints which help capture realistic chip workload and which includes the use of a novel mesh model for EM prediction in the grid, instead of the traditional series model [74–76]. They developed a framework to estimate the change in statistics of interconnects as their effective-EM current varies and developed a novel vector less mesh model technique to estimate the average minimum time-to-failure of a power grid under uncertain workload. Their results indicate that the series model causes pessimistic estimation of power grid MTF and reliability by a factor of 3–4 [75,76].

Although the novel methodologies/frameworks based on Black's equation can estimate EM reliability, since the atom flow within segment trees is ignored there, they are not able to provide suitable and accurate results as physics-based methods. Sandeep Chatterjee et al. found that the power grid's lifetime estimated by their physics-based approach is on average 2.75× longer than those based on Black's model [44]. Xin Huang et al. verified that the lifetime of IBMPG2 predicted by the traditional approach with the series and mesh model is 7.82 y and 10.67 y while the lifetime predicted by a physics-based EM model is 15.66 y [37]. It means the frameworks based on conventional estimation method is too conservative, therefore, the physics-based EM model is preferred for designs with space limited. In this section, our attention is focused on the studies which applied physics-based model to investigate the EM impact on power grids. This section consists of three subsections. The first subsection covers modeling and simulation methodologies proposed to study power grids' reliability. The second subsection focuses on the strategies adopted to improve EM evaluation speed while ensuring good accuracy and the optimization methodologies for better EM reliability. The third subsection is mainly on techniques for EM acceleration and deceleration.

3.1. Modeling and Simulation Methodologies

There are many studies on modeling and simulation methodologies for EM reliability of power grids with applying physics-based models. Some of them are based on numerical simulations while the others are based on analytical solutions. At an early time, Vivek Mishra et al. modeled the impact of EM in Cu interconnects on power grid integrity with using probability analysis [36,38–40,43]. Figure 6 shows the CDF plots for IR drop of a studied power grid for different circuit lifetime [40]. For t_{life} = 5 y, the studied power grid remains functional because all samples' IR drop are under a 10% threshold, however, it has a worst resistance degradation of 48% which is much more than a typical 10%~20% resistance increase criteria used by circuit designers. It verifies that power grids have inherent resilience to EM failures. They also studied circuit delay variability due to interconnects' resistance shift under AC EM [38]. It shows that even non-catastrophic EM on critical paths can cause serious performance degradation which ultimately result into circuit malfunction. As shown in Figure 7, the impact of EM on absolute delay shifts increases under technology scaling [38]. It is mainly because the higher current density in smaller interconnects exacerbate EM degradation. Under a specific technology node, EM effect becomes more obvious because EM void size and number increase with stress time. Meanwhile, they developed methods to evaluate transient stress evolution in interconnects, and presented simple and practical criteria for EM mortality checking. It is demonstrated that the number of EM mortal interconnects highly depend on lifetime target and reliability expectations [39]. It is also observed in [43] that power grids' EM degradation is impacted by configuration of via arrays which connect interconnects at different layers.

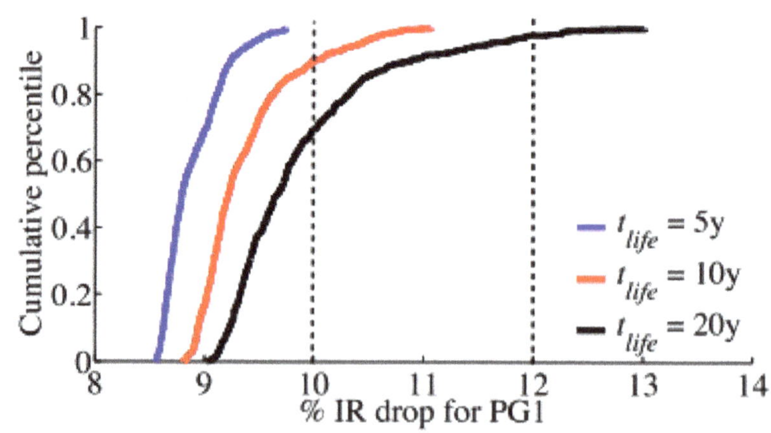

Figure 6. CDF plots for IR drop of a studied power grid for different circuit lifetimes, t_{life} [40].

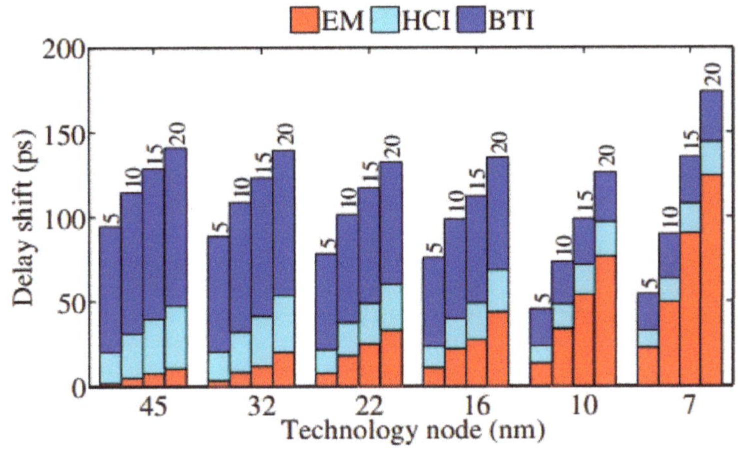

Figure 7. Absolute delay shifts due to various aging mechanisms for advanced technology nodes at circuit operation time of 5, 10, 15, and 20 y [38].

In order to accurately model EM degradation in power grids, Xin Huang et al. proposed a new physics-based assessment method [37,42]. This method incorporates power grids' redundancy by assuming the circuits get failed only when the IR drop reaches a threshold value. The hydrostatic stress in each interconnect tree is evaluated with considering atom transportation between connected branches. The experimental result not only verifies that the result obtained from Black's equation is too pessimistic, it but also shows that IBM P/Gs' EM failure is more likely to happen at the places with large initial stress value and is more likely to happen at longer time when the void volume saturation phenomenon is taken account. The time-dependent IR drop can be captured by placing the EM induced resistance shift into the P/G circuit model. Figure 8 shows the time-dependent voltage drop of the first failed node and maximum voltage drop in IBMPGNEW1 [42]. The proposed method is also applied to study the impact of cross-layout temperature and thermal stress distributions on full-chip EM assessment.

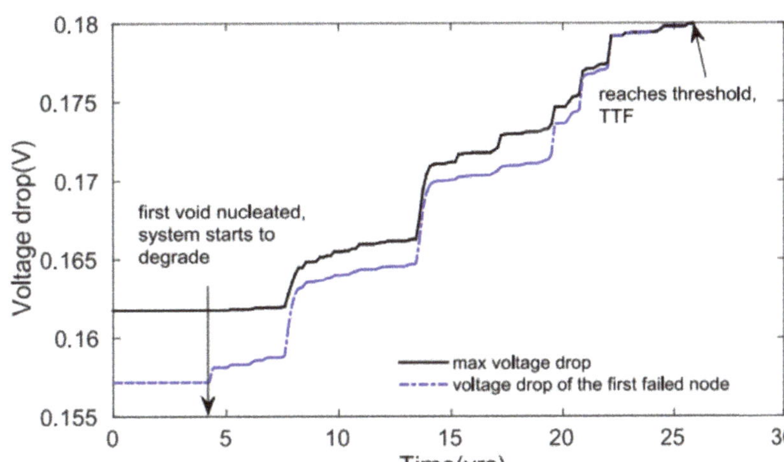

Figure 8. Time-dependent voltage drop of the first failed node and maximum voltage drop in IBMPGNEW1 [42].

Since most of the previous studies are based on the uniform temperature assumption, Xin Huang et al. implemented a flow of EM assessment for multi-layer P/G in a 32 nm test chip [77,78]. The cross-layout temperature variation due to devices' power consumption and interconnects' Joule heating are characterized and incorporated into EM assessment. It is found that uniform temperature assumption causes inaccurate prediction on TTF and the thermal stress variation results into better evaluation on EM reliability. It means the on-chip temperature variation is needed to get more reasonable EM assessment results. With applying the same physics-based EM model, Kai He et al. proposed a lightweight on-chip aging sensor [79]. The interconnects in this sensor are designed to have detectable EM failure at specific time. A number of parallel interconnects are used in the sensor to mitigate inherent variations. The EM-based aging sensor can provide more accurate prediction of chip usage time and offers simpler circuit implementation and smaller area footprints than the ring-oscillator based sensor. Chase Cook et al. has applied finite difference method to solve 1-D EM problem in multi-branch interconnects [80]. The new method can easily accommodate non-uniformly distributed residual stress and time-dependent temperature and current during circuit operation. The numerical results match well with that of COMSOL which is based on finite element method. Taeyoung Kim et al. presented an approach for system-level EM reliability management for multi/many core microprocessors [81]. They proposed a task migration method to balance EM resource consumption by all the cores. It treats TTF as a resource to consume during task execution and uses task migration to balance TTF consumption across the cores. It equalizes the probability of failure of each core to maximize the lifetime of multi-core system. The simulation results show a balanced TTF consumption by the cores and the system's EM reliability has been maximized.

Since the current in P/G is unidirectional, the interconnects' EM immortality can be determined by checking steady-state stress distribution. If the maximum stress is larger than critical value, the void is nucleated in a mortal interconnect, otherwise, the interconnect is immortal. However, since there is not closed form for steady state stress in multi-branch interconnect trees, numerical solutions generally need long time simulation. New techniques for convenient immortality checking are necessary for EM study. In [82], Zeyu Sun et al. proposed a new parameter called Critical EM voltage ($V_{\mathrm{Crit,EM}}$) to evaluate EM immortality at steady-state stress in multi-branch interconnect tree. The $V_{\mathrm{Crit,EM}}$ is an extension of Belch limit concept. The difference is that Blech limit is for single branch while $V_{\mathrm{Crit,EM}}$ is applicable to multi-branch tree. Since this voltage-based EM (VBEM) method overcomes the problem of current-density-based criteria, it can handle the impact of interconnect tree structure on EM-induced stress with easy implementation. With this method, the EM voltage at the ground node or cathode node of a tree is determined with the total area of branches in the tree, the total area of the branches connected to each node in the tree, and the nodal voltage at each node. The EM immortality of studied tree depends on whether the EM voltage is smaller than the $V_{\mathrm{Crit,EM}}$. This criterion is applicable for immortality checking in void nucleation phase. The VBEM analysis not only agrees with results from finite difference method at steady state but also matches well with COMSOL and XSim results, as shown in Figure 9. Since the VBEM analysis assumes that current density is evenly distributed in one branch, the impact on current crowding is investigated. Comparison with COMSOL result shows that current crowding effect is unobvious if the length of a branch is much greater than its width. Void saturation volume is another important issue for EM immortality checking. Since previous saturation volume model only works for single branch, Zeyu Sun et al. derived a void saturation volume model for multi-branch tree, as shown in Figure 10a [83]. The model follows atom conservation at steady state of void growth phases. The overall void saturation volume in a tree at steady state is the sum of saturation volume contributed by each branch. The tree is considered as EM immortal if the overall void volume is smaller than the critical size. This criterion is applicable for immortality checking in void incubation phase. Transient analysis is necessary on the tree only when it is EM mortal. The authors proposed a new EM immortality checking flow which considers the checking criteria in both void nucleation phase and void incubation phase. The algorithm is given in Figure 10b. The new flow reduced conservativeness of existing EM assessment methods, and it helps the designer quickly identify the new type of immortal branches which have void nucleated but with a size smaller than the critical value.

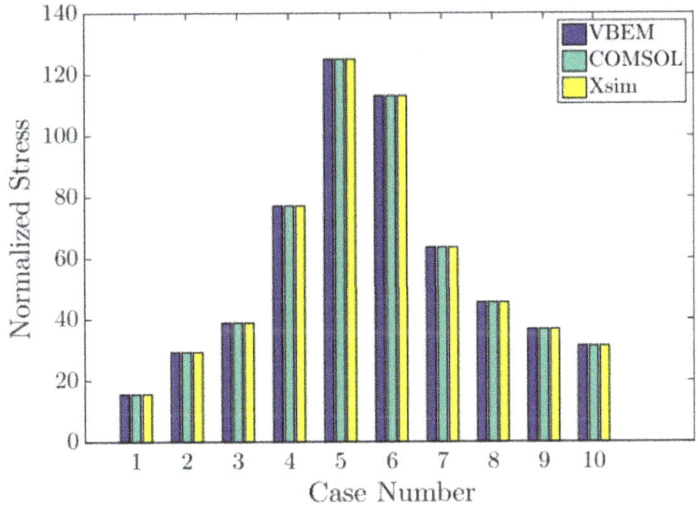

Figure 9. Steady-state EM stress comparisons for each straight-line 3-terminal interconnect case [82].

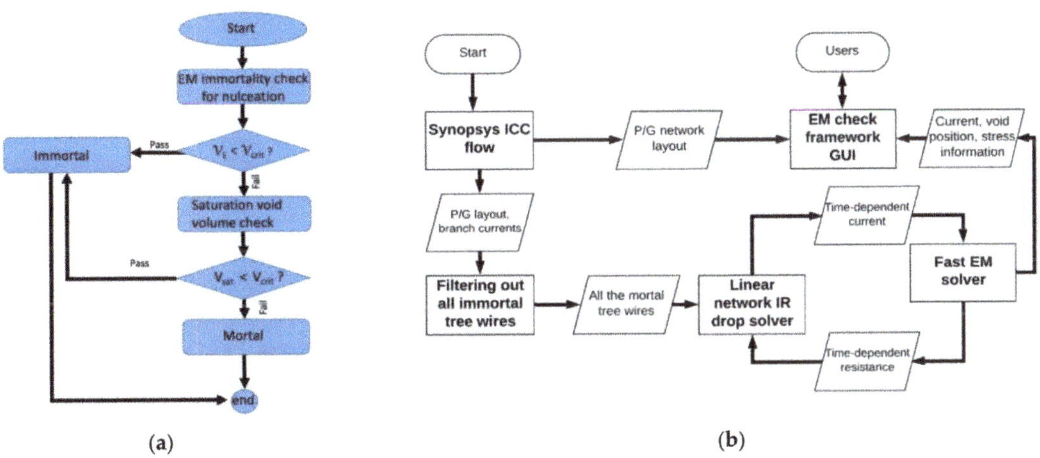

Figure 10. (**a**) EM immortality check algorithm flow, and (**b**) simulation framework for *EMSpice* simulator [83,84].

Based on their EM immortality check algorithm shown in Figure 10a and the existing transient EM analysis theories, Zeyu Sun et al. developed a new full-chip EM simulator called *EMSpice* to evaluate EM reliability of P/Gs [84]. Figure 10b shows the simulation flow of *EMSpice*. It starts from P/G layout information from Synopsys IC Compiler. In the first step, it disregards immortal trees by considering the immortality criteria in nucleation and nucleation phases. Then, the mortal trees are applied with a FDTD solver to extract time-dependent hydrostatic stress in both nucleation and post-voiding phases. Since the EM-induced resistance shift cause current variation in P/G, the EM analysis is interacting with IR drop analysis of a whole P/G at each time step to ensure the comprehensiveness of *EMSpice*. *EMSpice* simulator can reduce the over-conservation in EM assessment. It predicts the failed tree number 76.7% less than the Black's method and 66.7% less than another full-chip EM analysis method.

EM postvoiding analysis attracts researchers' attention because it is hard to handle the interactions between current density, hydrostatic stress, and temperature when the target is to investigate detailed void growth. Hengyang Zhao et al. proposed a multi-physics finite-element-method-based (FEM-based) analysis method for void growth simulation in confined copper interconnect [85,86]. The method considers time-varying interactions between hydrostatic stress in the confined interconnects structure illustrated in Figure 11a, the current density and Joule heating induced temperature rise. The interactions are realized by solving a set of coupled partial differential equations, including the Korhonen's equation, the phase field equation, the Laplace equation, and the heat diffusion equation. The FEM-based EM solver is capable to predict unique transient resistance change for copper interconnects. Figure 11b shows the time-dependent total resistance, hotspot temperature, resistance at the hotspot, and void size. It verifies the statement in Section 2.2 that the resistance jump between nucleation phase and growth phase is due to Joule heating. Meanwhile, the lifetime distribution from this EM solver can provide a higher fitting accuracy on the current density exponent parameter (n) described in Black's equation in the previous study, as shown in Figure 11c.

Sandeep Chatterjee et al. also proposed a methodology to check P/G EM by using physics-based model [41,44]. They firstly listed the detailed steps to extend the physical models for EM in branches to compute the hydrostatic stress evolution in multi-branch interconnect trees. The boundary condition for branches under different scenarios are provided as well. Then filtering and predictor-based schemes are designed to speed up the overall EM assessment. It enables the statistical computation on IBM benchmarks finish in 2.3 hrs. Therefore, the proposed method has potential to be applied on large-scale circuits. The simulation results verified the inaccuracy of Black's model and explained the importance of *early failures* for EM assessment. As shown in Figure 12, exclusion of early failures leads to optimistic evaluation on voltage drop increase and on MTF of P/Gs [44]. In [87], a finite difference method-based EM analysis methodology is applied to 3-D IC test structure. Its comparison with finite element analysis and experiment measurement demonstrates that EM in 3D IC structures can be suitably evaluated with finite difference simulation. In [88], the authors presented a systematic approach to resize the grid metal lines to achieve a design target lifetime at the minimal extra cost of metal area. With the help of this approach, on a grid with 1.2 million nodes, the authors can increase its MTF from 10.5 y to 12.2 y under a cost of 0.02% extra metal area by scaling only 14 interconnect trees. Current density variation is an important factor for EM evaluation. To ensure the comprehensiveness of their EM modeling, Adam Issa et al. has investigated EM checking by using stochastic effective current model [89]. It is observed that current variations bring us worse EM reliability. Based on his EM modeling experience, Farid N. Najm derived the equivalent circuits for EM under different model phases [90]. It is shown that the dynamic behavior of stress and flux in metal line is identical to dynamic behavior of voltage and current in RC circuit, thus EM assessment can be executed by simulating its equivalent RC circuit. It has potential to drastically improve EM assessment capability on large circuits. In [91], an industry-level physics-based tool for EM assessment in commercial-grade PDNs was introduced. As shown in Figure 13, after analysis the tool can highlight voided metal line segments with a voiding probability. The tool's accuracy has been validated with the experimental data [92]. There is good fit between lifetime statistics derived from measurement ($MTTF_{EXP}$ = 62,305 s, Δ_{EXP} = 14,012 s) and simulation ($MTTF_{SIM}$ = 60,344 s, Δ_{SIM} = 12,613 s).

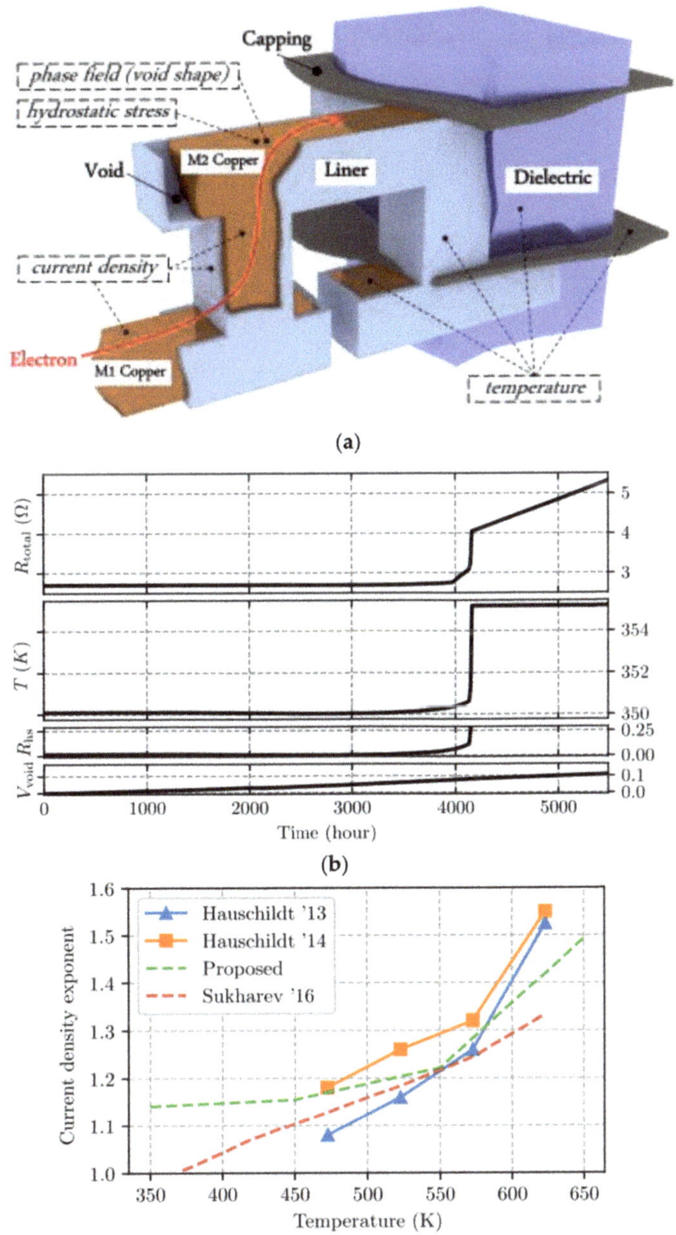

Figure 11. (**a**) 3D illustration of up-stream interconnect structure and simulated physical systems; (**b**) simulation of Joule heating effect. R_{total}: total wire resistance. T: temperature at the hotspot. R_{hs}, in ohms: copper resistance at the hotspot. Void, in cubic micrometer: simulated void size; (**c**) extracted current density exponent compared to experiment data and previous postvoiding EM analysis work [85,86].

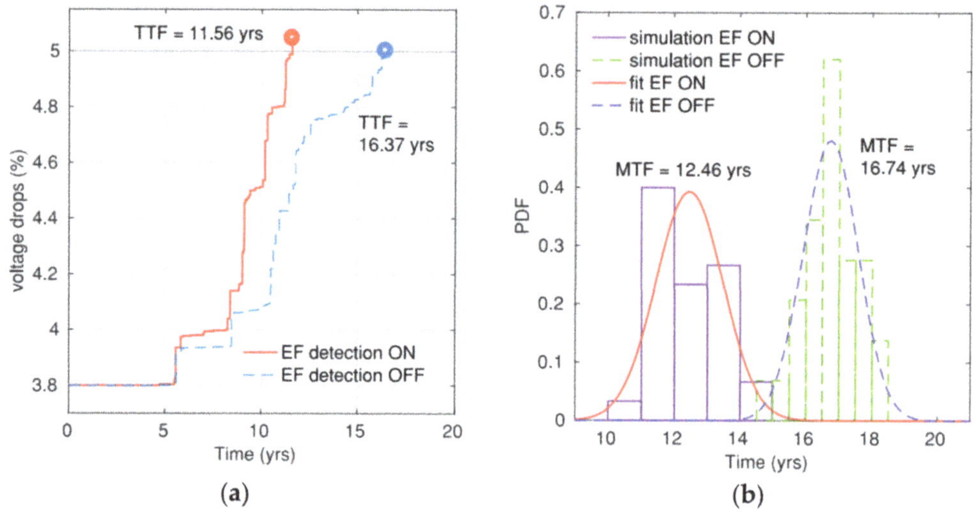

Figure 12. Impact of early failures on (**a**) voltage drop of a sample grid and (**b**) estimated mesh MTF for ibmpg2 [44].

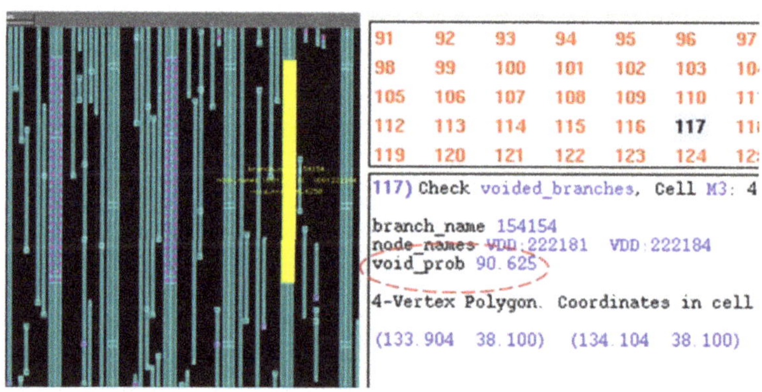

Figure 13. Highlighted wire with high voiding probability on the layout of a metal layer [91].

Meanwhile, Houman Zahedmanesh et al. investigated the EM limits of Cu nano-interconnects by using a novel hybrid physics-based model [93]. The modeling framework incorporates variations of materials, dimensions, interfaces, and operation conditions. It considers void dynamics and resistance shift by using a local cellular automation module with a resistive network. The simulation result only shows that the nucleation phase gets more significant under a narrower linewidth, it but also predicts complex R-shift signatures which match well with experimental data. Sarath Mohanachandran Nair et al. proposed a variation-aware physics-based EM modelling which is experimentally calibrated [94]. The model can be used to explore the impact of material and dimension on design space and to study failure time variation at various operating conditions. Then the model was extended to handle both *early* and *conventional failures* [95]. In [96,97], the system-level simulation on EM under 3 nm technology node was performed. It is shown that Ru rails reduced IR-drop penalty by a factor of ~0.6 than the Cu rails. Although EM voids appear in multiple PDN segments, the EM induced IR-drop always stay below 3.3% without any failed operation of standard cells.

Except for the studies on EM reliability of full power grids, there are other novel works on EM by using physics-based model. In [98–100], the authors explored an approach to enhance the TSV grid reliability. The main idea is to allow the nonfaulty TSVs to be temporarily deactivated so that it can take advantage of EM recovery property. To achieve this goal, a reconfigurable routing network for a (4:2) TSV group was adopted, as depicted in Figure 14. Depending on a recovery schedule, all TSVs cab operate under active mode, recovery mode. The recovery-aware proactive repair approach helps improve EM lifetime of the entire TSV grid by up to 12 times relative to conventional reactive method without an extra area cost.

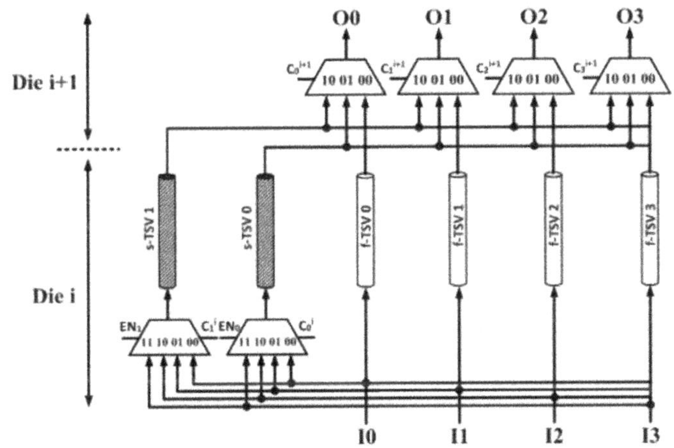

Figure 14. The reconfigurable routing network for a (4:2) TSV group consisting of four f-TSVs and two s-TSVs [100].

3.2. Fast EM Assessment and Optimization

Although FDM and FEM methods can be applied for transient EM analysis with good accuracy, they are not always preferred when the studied interconnect structures are too large or the computation resource is limited. To overcome this problem, a number of analytical and semi-analytical solutions and speed up techniques have been proposed. At the very beginning, Valeriy Sukharev and Xin Huang et al. derived the analytical equation for transient hydrostatic stress in a confined metal wire [101–103]. The analytical equation captures the impact of time-dependent current density and the stress recovery effect, as shown in Figure 15. The calculation results not only show that in the case of high frequency currents with periods much smaller than the characteristic time of stress evolution, the pulse duty factor decides stress buildup, it but also demonstrates that temperature oscillation can cause notable resistance increase in a short metal line with preexisted voids. On the other hand, the stress recovery effect is applicable to improve on-chip interconnect lifetime by properly managing driving powers at run time.

Later, Hai-Bao Chen et al. developed a first principle based analytical solution for hydrostatic stress evolution in 3 specific interconnect trees [104,105]. It solves stress evolution in multi-branch tree by de-coupling individual branches with suitable boundary conditions which account for interactions between adjacent branches. The solution is based on Laplace transformation technique. Since time-varying temperature and current density and branch length have non-negligible impact on stress evolution, they incorporated these factors into their analytical solution under the same method [106,107]. However, since these analytical solutions only work for specific tree structures, Xiaoyi Wang et al. applied eigenfunction technique for stress evolution in multi-branch trees, and they have extended it to EM analysis in full-chip P/Gs [108–110]. This method handles different current densities and nonuniform thermal distribution. Since this method does not require discretization, except for its excellent scalability for large-scaled interconnect trees, it brings 10–100 times of computation speed improvement over FDM method. In order to improve EM analysis speed, Liang Chen et al. proposed accelerated separation of variables (ASOV) method which offers improvements over the existing plain SOV-based method [111]. It exhibits 3–5 times of speedup on a number of multi-branch interconnects benchmarks. Furthermore, the SOV-based approach was adopted to obtain a semi-analytical stress transient analysis method which considers the impact of temperature gradient in the studied trees [112]. This method is about an order of magnitude faster than COMSOL with 10× less memory footprint and negligible error loss. It shows that the impact of Joule heating on EM process is significant. Mohammadamir Kavousi et al. studied EM immortality check with considering Joule heating for multi-branch trees [113]. They improved EM stress analysis speed with the benefit of Krylov subspace-based reduction technique which reduces the size of system matrices [114]. Their analysis on interconnect with up to 1000 branches for both void nucleation and growth phases has been accelerated by 28 times. Last, but not least, Mohammad Abdullah Al Shohel et al. found a linear-time approach for immortality check on general tree/mesh interconnects and developed an analytical model of transient stress based on boundary reflections [115,116]. With respect to speedup techniques different from analytical solution, Sandeep Chatterjee et al. presented a fast and scalable methodology for P/G EM verification [117]. The PDE system was converted to a succession of homogeneous linear time invariant (LTI) system. Then the LTI system was solved with an optimized backward differentiation formulas (BDF) solver. Under further help from preconditioned conjugate gradient and parallel programming, this method gives around 23 times of speedup. In [118], a Krylov subspace-based method was proposed for fast stress evolution under finite difference method. After discretization, the original system matrices are reduced so that they can be simulated more efficiently in time domain. This optimized method brings 1–2 orders of magnitude speedup over ordinary finite difference time domain methods.

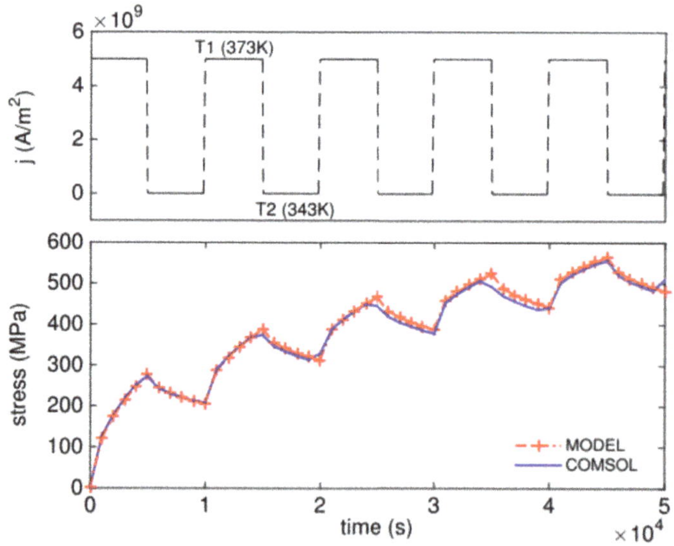

Figure 15. Stress evolution caused by periodic unipolar pulse current densities at cathode end of the metal line under varying temperature [103].

Recently, machine learning techniques have been applied to speed up studies on P/G performance and reliability. In [119], a generative adversarial networks-based (GAN-based) tool (called EM-GAN) was built to do fast analysis on transient stress in multi-branch trees. This work was inspired by the image synthesis feature of GAN. As shown in Figure 16a, the GAN's inputs include P/G topology, current density distribution at a given aging time. Its output is EM stress distribution. This tool can achieve high prediction accuracy with an average error of 6.6%, and it exhibits 8.3 times speedup over analytic EM solver. In order to achieve even higher accuracy, another graph convolution network-based (GCN-based) tool (called EMGraph) was developed for transient EM stress estimation. Its basic framework is shown in Figure 16b. It shows less than 1.5% average error compared with training data and orders of magnitude faster than COMSOL. Moreover, EMGraph surpasses EM-GAN with 4 times higher prediction accuracy and 14 times faster speed.

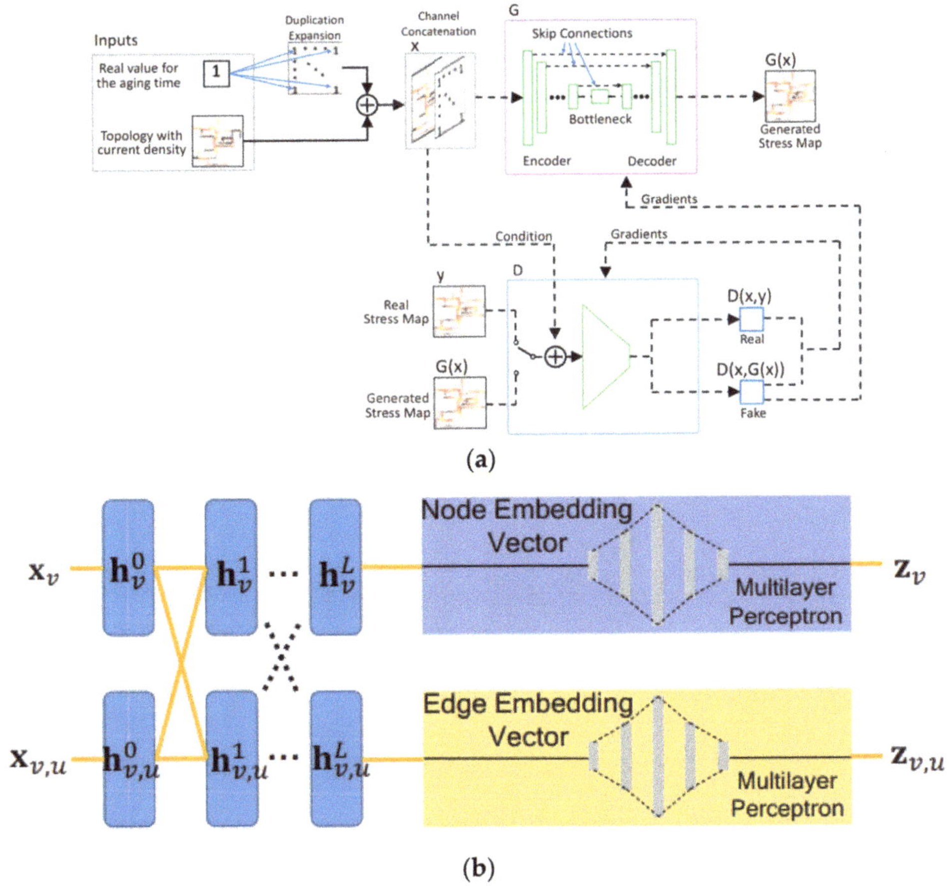

Figure 16. (**a**) EM-GAN framework for stress estimation and (**b**) framework of EMGraph with multilayer perceptron network [119,120].

In [121], a conditional generative adversarial networks-based (CGAN-based) framework (called *GridNet*) was developed, as shown in Figure 17, to accelerate the incremental full-chip EM-induced IR drop analysis and the optimization for IR drop violation fixing. GridNet provides accurate prediction on IR drop as compared with the ground truth obtained from EMSpice. Since GridNet also provides sensitivity information of node voltage with respect to branch resistance, it expediates localized IR drop violation fixing for P/G design.

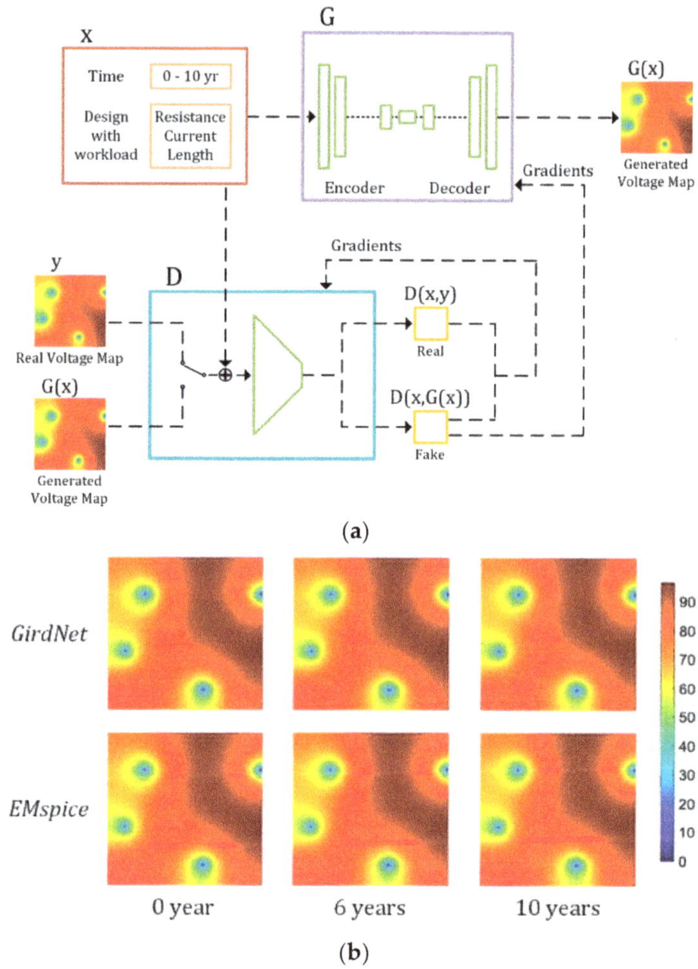

Figure 17. (**a**) CGAN architecture for EM-induced voltage prediction and (**b**) comparison of inference results from GridNet to EMSpice [121].

Successful P/G design targets at a good enough EM lifetime with a reasonable cost on area. P/G design optimization is an important process to get a balance between EM reliability and area cost. Han Zhou et al. proposed P/G optimization techniques based on fast EM nucleation phase immortality check method for multi-branch interconnect trees [122,123]. They, firstly, verified that the issue can be formulated as a sequence of linear programming problem. Then they proposed an aging-aware optimization method which improves mortal wires' lifetime by adding reservoir branches and allows some interconnects to age/breakdown then just optimizes EM reliability of remaining branches. This strategy ensures the optimization operate effectively. Numerical results demonstrated that the new method can effectively reduce P/G area while ensuring immortality or target lifetime of all the wires. Later, the EM incubation phase immortality check method is introduced into the optimization framework as well [124,125]. The updated method can fix IR drop violation due to EM in minutes for P/Gs from ARM core designs.

3.3. EM Acceleration

In order to effectively validate the reliability of dual damascene interconnect trees under specific process and structure, it is necessary to detect EM failures in a relatively short testing time. Since the traditional acceleration techniques mainly focus on high temperature and current density, they lead to higher probability of failure due to not only EM but also the other aging mechanisms such as BTI and HCI, which leads to less distinguishability between the mechanisms. In [126–128], several techniques have been proposed to accelerate and decelerate EM failure with the help of reservoir branches and sink branches. Figure 18 shows an active branch with a reservoir branch enabled/disabled and the time-dependent hydrostatic stress at cathode. Obviously, the reservoir branch highly extends void nucleation time at the cathode which indicates much better EM reliability. The impact of reservoir could be disabled by applying a current to it in a reversed direction. And the current density decides how much the main branch's lifetime is suppressed.

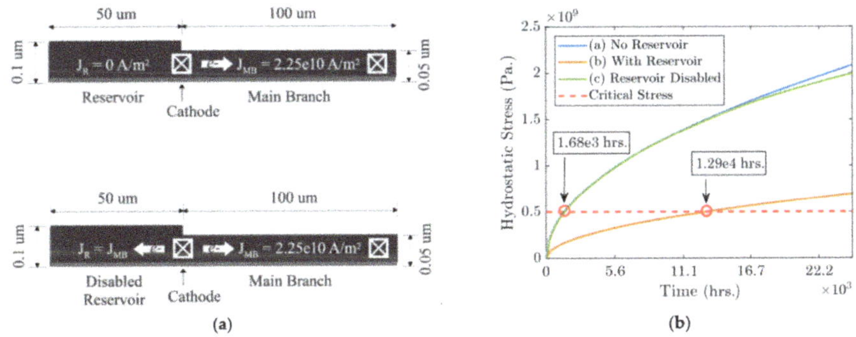

Figure 18. (**a**) An active interconnect branch with a reservoir branch enabled and disabled and (**b**) Hydrostatic stress at cathode under various scenarios [128].

In [127,128], a sink branch was proposed to accelerate EM lifetime of the main branch. As shown in Figure 19, the void nucleation can be accelerated by an active sink branch because the atomic flux gets improved when the currents over main branch and sink are in the same direction [128]. Then the authors proposed a hybrid structure with both sink and reservoir applied to accelerate EM failure. The proposed structures can achieve desired very short TTF under acceleration mode while the main branch itself has 10+ y lifetime under normal usage mode. These novel structures together with temperature control can further accelerate EM testing about 10^5 time under the 150 °C temperature limit.

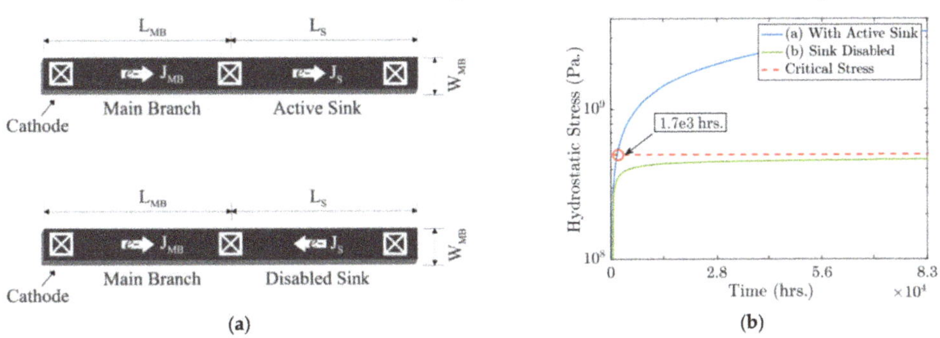

Figure 19. (**a**) An active interconnect branch with a sink branch enabled and disabled and (**b**) hydrostatic stress at cathode with the sink branch enabled and disabled [128].

4. Modeling of EM Impact on Interconnects in Cache Memory

EM not only affects P/G but also worsen the reliability of bitlines which is frequently stressed by currents during read/write operations of cache memory. During read/write operations, the unbalanced currents flowing over interconnects cause voids which ultimately lead to operation failure by introducing large delay. In [129–131], the authors designed a methodology for SRAM EM reliability assessment with considering process variations. The equivalent current distributions over bitlines are calculated with an AFD-based current conversion scheme. As shown in Figure 20, the equivalent current is much different from average value of pulsed DC which indicates different conclusion on EM reliability. The current distribution is combined with process variations including threshold voltage variation, gate length variation, and bitline edge roughness, to evaluate bitlines' reliability by using statistical modelling methodology. The authors adjusted bitline width to find an optimal value to minimize the total probability of failure for an SRAM array and to maximize its yield. The experiment results indicate that a 22 nm technology-based 256 rows × 128 columns SRAM array suffer from serious EM issue if the bitline width is chosen as ½ metal pitch, as shown in Figure 21. And a tradeoff between functional failure and EM failure can be reached for a 46 nm width bitline when the edge roughness is incorporated.

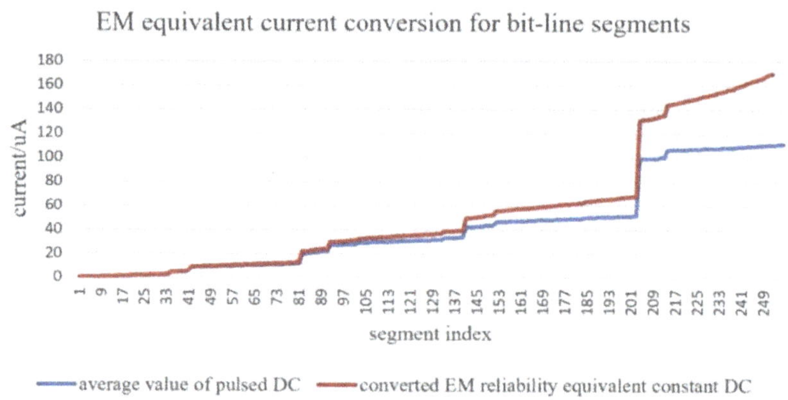

Figure 20. Typical current distribution comparison between time-average value and converted value for all segments within an SRAM bitline [131].

The current distributions over bitlines are decided by activity of cells attached to them. The impact of cell activity on EM reliability of an SRAM cell array was studied in [132]. However, since none of the previous prior works have taken SRAM workload in realistic usage scenarios to evaluate interconnects' reliability, a simulator called CacheEM was proposed for reliability analysis on cache memory aging due EM by considering the realistic application scenarios of cache in an ARM microprocessor [49]. CacheEM includes five parts: microprocessor emulation, memory cell array activity extraction, computation of current in long interconnects, evaluation on time-dependent hydrostatic stress and the resistance shift of interconnects, and characterization of the interconnect EM lifetime distribution of a cache memory.

Figure 22a shows a simplified schematic of an SRAM cell array which is handled in CacheEM with currents corresponding to specific operations (read 0 and1 and write 0 and1) marked. Figure 22b,c show interconnect-array in SRAM cell array and the representative interconnects corresponding to a column of cells [49]. L_{Start} and L_{End} are the segments from the array to the pre-charge and write drivers, respectively. In order to perform time-dependent EM assessment on these interconnects, the current density distributions over them need to be determined in the first step. To obtain the current density distributions, it is required to know the number of load and store to each cell, which means the cells' activity in cache memory needs to be extracted. On the other hand, the activity of cells in the cache memory highly depends on the application of the microprocessor where the cache is configured. It needs a simulation flow for CacheEM from a top–down perspective as described in Figure 23. In the first step, several testbenches on the microprocessor which is configured with specific cache memory settings are executed. The trace of the target cache memory is recorded. In the second step, the recorded memory trace is fed into a cache simulator to extract the cells' activity. Then, the variations in effective atomic diffusivity and threshold stress are incorporated into CacheEM with a Monte Carlo simulation to obtain the cache EM lifetime distribution. For each sample, the current flowing in each branch is derived individually with considering the cells' activity and the currents during each operation (read 0/1 and write 0/1).

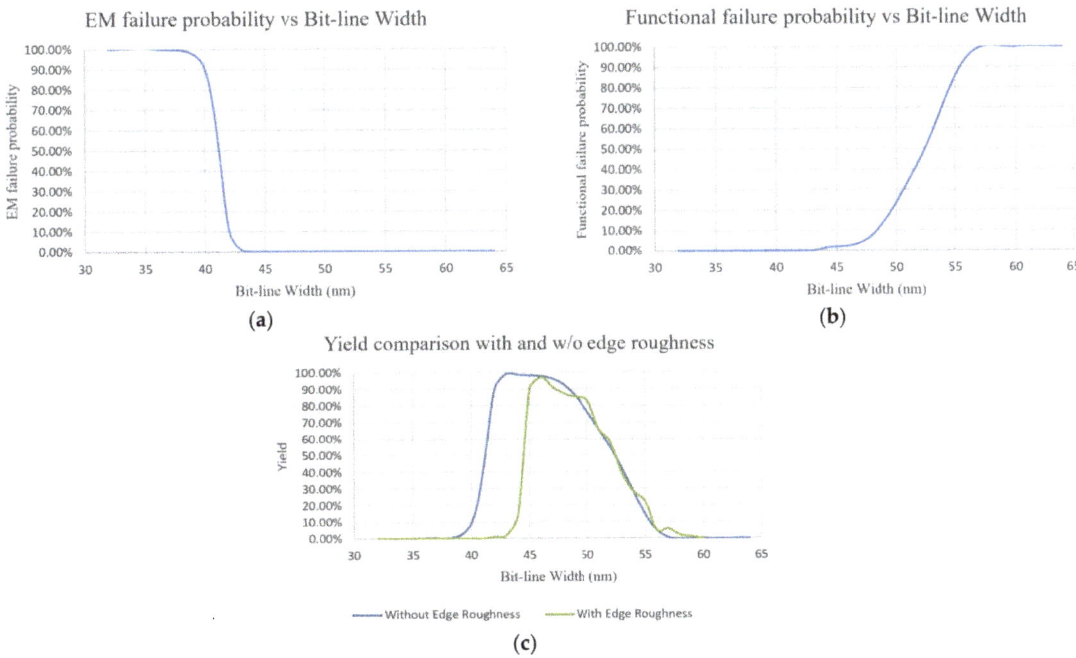

Figure 21. (**a**) EM probability of failure, (**b**) functional probability of failure, and (**c**) yield comparison for 256 rows × 128 columns 22-nm SRAM array w/and w/o considering edge roughness [131]

Figure 24 shows the representative operation distributions on the L1 I-Cache cells in an ARMv8 core while it is running the *sjeng*, *specrand*, and *patricia* benchmarks [49]. The cells' activity and currents during read/write operations under user-defined parameters, including gate length, bit-line unit capacitance, row number of the cells, supply voltage, and temperature are combined together to calculate the effective current density distributions over each BL, BLB, VDD, and GND interconnects. Then they are provided to FDM-based EM solver to evaluate time-dependent stress evolution. The EM lifetime distribution is measured under predefined threshold value. Figure 25 shows the hydrostatic stress distribution of VDDs, GNDs, BLs, and BLBs in randomly selected caches [49]. The current density in the interconnect which suffers the most serious EM stress is plotted on the basis of the left *y*-axis in each subplot. In BL and BLB, the voids are most likely formed in the middle of the lines. The exact void position depends on the ratio between read and write operations. The voids most likely appear at the ends of VDDs and GNDs. However, the voids most likely appear near the via for GNDs, but far from the vias for the VDDs due to the difference of current direction. Stress is highest when current flows into the vias, and lowest when current flows away from the vias. Under a two-port implementation and bi-directional flow, the stress is highest around the middle of the VDDs and at the ends of the GNDs. After the maximum stress reaches σ_{th}, the voids are generated, and then after the void size becomes larger than the threshold size, the interconnect resistance increases. An interconnect's lifetime is obtained when the resistance shift reaches a pre-defined threshold value.

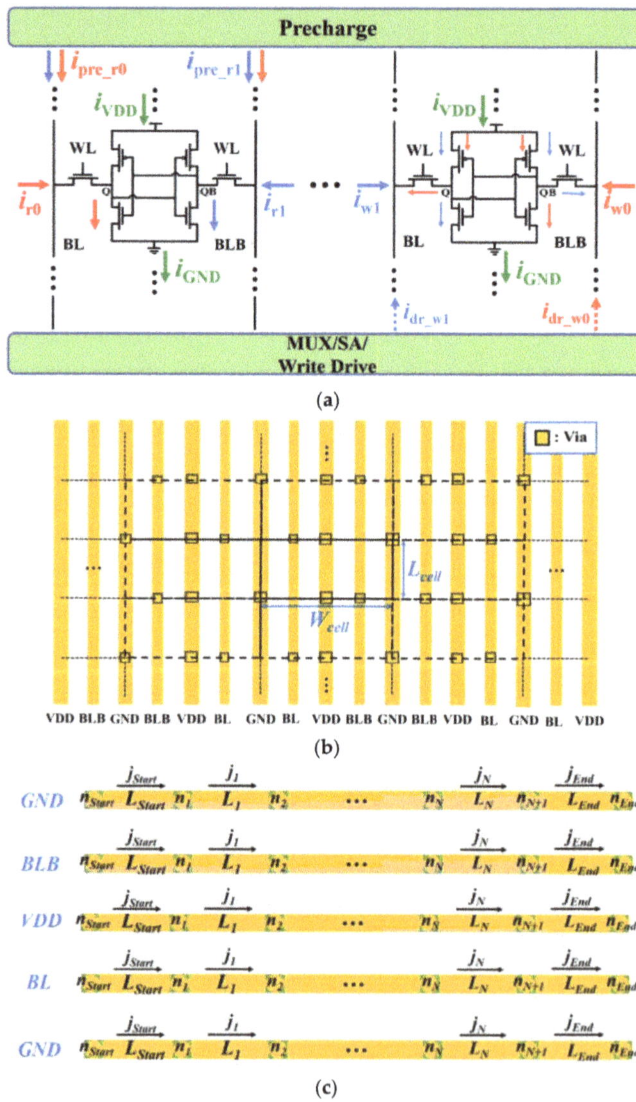

Figure 22. (**a**) Simplified schematic of SRAM cell array with currents relevant with various operation marked. The left and right parts show the currents relevant with read 0 and 1 and write 0 and 1, respectively. The currents for read 0 and write 0 in red, while the currents for read 1 and write 1 are in blue, (**b**) interconnect array in SRAM cell array, with periodic cells shown functional probability of failure, and (**c**) representative interconnects corresponding to a column of cells [49].

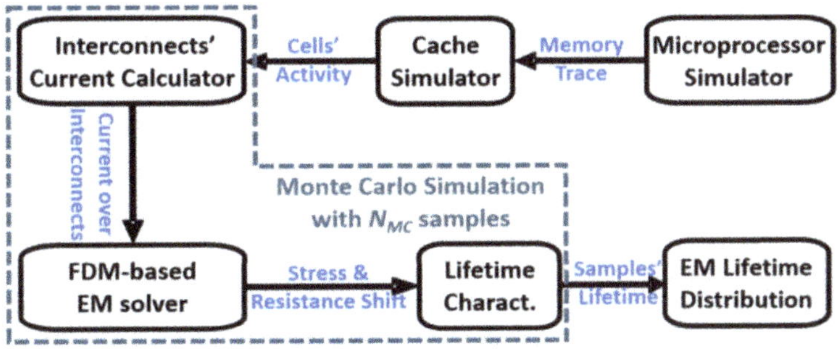

Figure 23. Simulation flow of CacheEM [49].

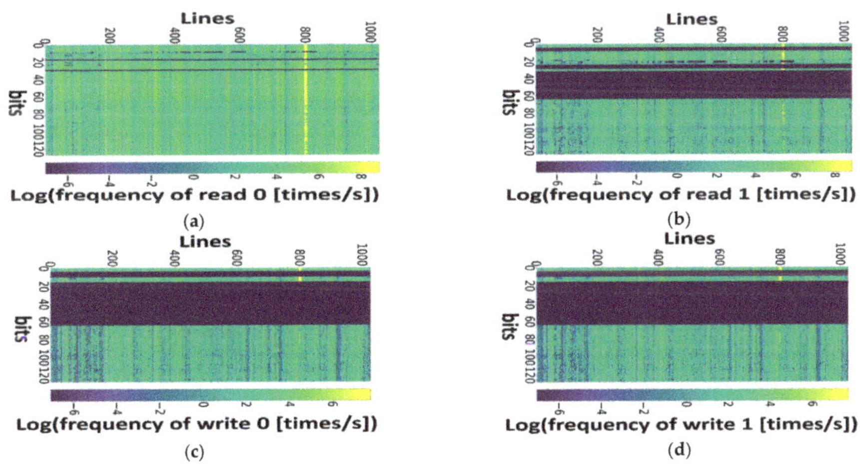

Figure 24. (**a**) Read 0, (**b**) read 1, (**c**) write 0, and (**d**) write 1 distributions of L1 I-Cache cells in ARMv8 core which has run sjeng, specrand, and *patricia* benchmarks [49].

Figure 26a shows the lifetime distribution of BL, BLB, and GND interconnects which suffer from the most stress in 100 samples of an I-Cache. With the help of the two-ports connection, GND has a much better EM reliability than BL and BLB. BL is the most vulnerable interconnect in the I-Cache. Figure 26b shows the time-dependent resistance shift of the corresponding BLs in 100 cache samples [49]. BL is not always more vulnerable than BLB. Their relative vulnerability might change with the cache configuration. CacheEM was applied to investigate the impact of configuration parameters on performance and EM degradation of an I-Cache. The configuration parameters include cache line size, replacement policy, set associativity, latency, and cache size. Cache line size has options of 8B, 16B, and 32B. The I-Cache hit rate is 94.43%, 93.65%, and 93.75%, when the line size is 8B, 16B, and 32B, respectively. 32B brings better lifetime than the other two options because the interconnect length is much shorter while the current density is smaller. Replacement policy, associativity, and latency do not have obvious impact on I-Cache performance and reliability. Cache size has a non-negligible impact on the hit rate and the lifetime distribution of the studied I-Cache. The hit rate is 89.16% under 8 KB, 93.65% under 16 KB, 97.39% under 32 KB, and 99.19% 64 KB, respectively. 8 KB cache has better lifetime than the other three options because the interconnect length is shorter than the other cases and the current density is smaller. For the remaining three cases, EM reliability improves with the increase of the cache size.

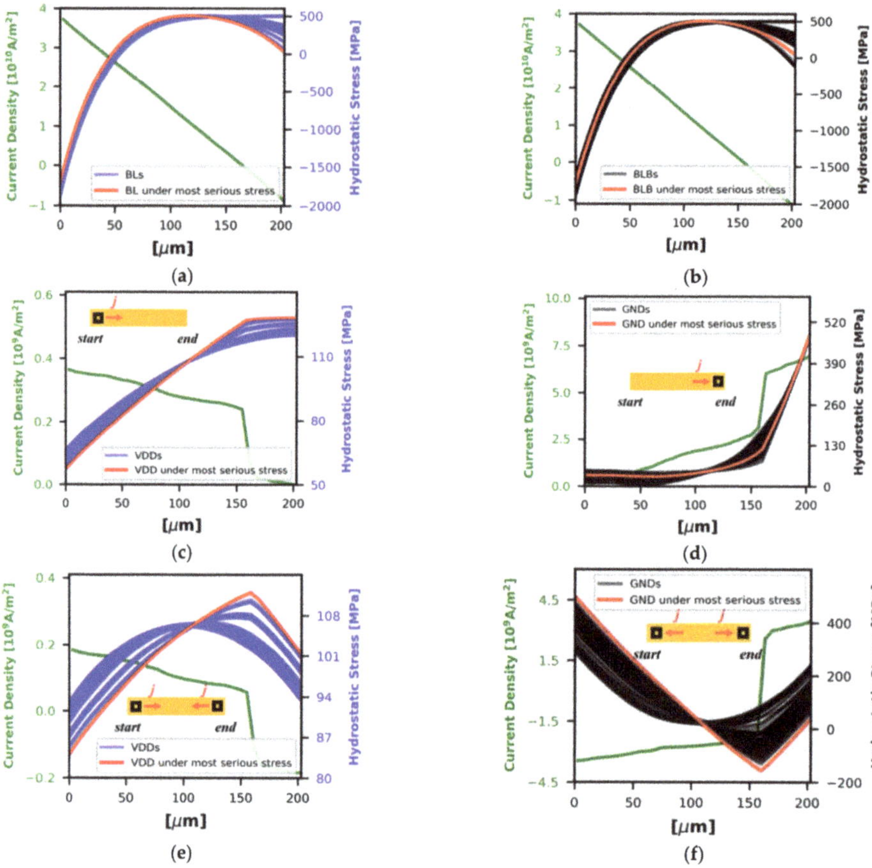

Figure 25. The hydrostatic stress distribution before a void appears in (**a**) BLs, (**b**) BLBs, (**c**) VDDs in a one port implementation, (**d**) GNDs in a one port implementation, (**e**) VDDs in a two-port implementation, and (**f**) GNDs in a two-port implementation [49].

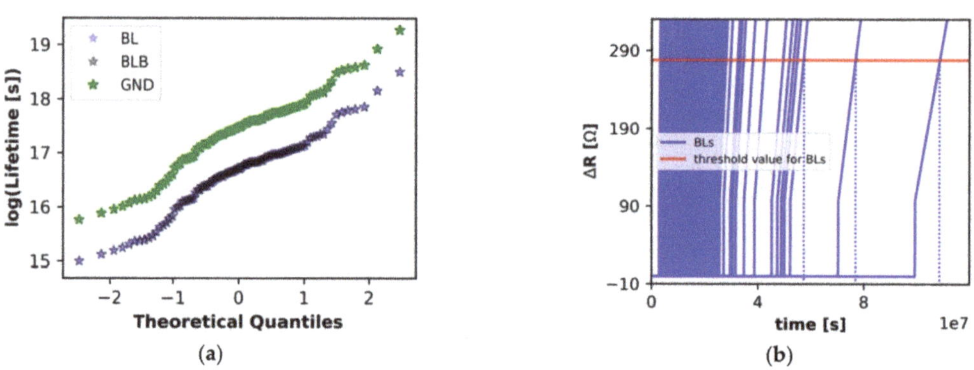

Figure 26. (**a**) Lifetime distribution of the BL, BLB, and GND interconnects which suffer from the most serious stress in samples of an I-Cache, and (**b**) the time-dependent resistance shift of the corresponding BLs [49].

5. Conclusions

Recent progress in physics-based modeling of EM aging of on-chip interconnects has been reviewed in this paper. Due to the over-conservative EM assessment on multi-branch interconnect tree by conventional Black's equation, physics-based modeling methodologies which consider the atom flow between adjacent branches are required to provide more accurate EM assessment. These modeling methodologies can be achieved with analytical equations or numerical computations. Voltage-based immortality check can simplify EM assessment by eliminating the immortal trees which do not need time-dependent evaluation on stress and resistance. The speedup techniques, such as separation of variable method and Krylov subspace-based reduction technique, are applicable for fast assessment on large-scaled circuits. Machine learning algorithms like GAN also help improve assessment efficiency. With respect to interconnects in cache memory, their EM reliability depends on physical dimensions, cache configurations, and workloads under specific benchmarks.

Since the integration density in SoC is always getting higher to achieve better system performance, the number of interconnect trees to be studied becomes tremendous. It is a challenging task to evaluate the full EM reliability of on-chip interconnects with good accuracy as well as fast speed. Further application of machine learning techniques deserves more attention to accelerate the evaluation speed and to improve the models' prediction capability. On the other hand, since Cu is predicted to be replaced by novel materials such as Co and Ru to ensure interconnects' reliability at sub-10 nm technology nodes, more efforts need to be made to novel modeling methodologies which suitably estimate the aging of new material-based interconnects and networks.

Author Contributions: Writing—Original Draft Preparation, R.Z. and W.-S.Z.; Writing—Review and Editing, W.-S.Z. and D.-W.W. All authors have read and agreed to the published version of the manuscript.

Funding: This research was funded by the Zhejiang Provincial Natural Science Foundation under Grants LXR22F040001 and LD22F040003, and the Natural Science Foundation of China (NSFC) under Grants 61874038 and 62101170.

Conflicts of Interest: The authors declare no conflict of interest.

References

1. Parihar, N.; Sharma, U.; Southwick, R.G.; Wang, M.; Stathis, J.H.; Mahapatra, S. Ultrafast measurements and physical modeling of NBTI stress and recovery in RMG FinFETs under diverse DC–AC experimental conditions. *IEEE Trans. Electron Devices* **2017**, *65*, 23–30. [CrossRef]
2. Zhang, R.; Liu, T.; Yang, K.; Milor, L. Modeling of the Reliability Degradation of a FinFET-based SRAM Due to Bias Temperature Instability, Hot Carrier Injection, and Gate Oxide Breakdown. In Proceedings of the 2017 IEEE International Integrated Reliability Workshop (IIRW), South Lake Tahoe, CA, USA, 8–12 October 2017; pp. 1–4.
3. Yang, K.; Liu, T.; Zhang, R.; Kim, D.; Milor, L. Front-end of line and middle-of-line time-dependent dielectric breakdown reliability simulator for logic circuits. *Microelectron. Reliab.* **2017**, *76*, 81–86. [CrossRef]
4. Yang, K.; Liu, T.; Zhang, R.; Milor, L. Circuit-level reliability simulator for front-end-of-line and middle-of-line time-dependent dielectric breakdown in FinFET technology. In Proceedings of the 2018 IEEE 36th VLSI Test Symposium (VTS), San Francisco, CA, USA, 22–25 April 2018; pp. 1–6.
5. Parihar, N.; Goel, N.; Mukhopadhyay, S.; Mahapatra, S. BTI Analysis Tool-Modeling of NBTI DC, AC Stress and Recovery Time Kinetics, Nitrogen Impact, and EOL Estimation. *IEEE Trans. Electron Devices* **2018**, *65*, 392–403. [CrossRef]
6. Thirunavukkarasu, A.; Amrouch, H.; Joe, J.; Goel, N.; Parihar, N.; Mishra, S.; Dabhi, C.K.; Chauhan, Y.S.; Henkel, J.; Mahapatra, S. Device to circuit framework for activity-dependent nbti aging in digital circuits. *IEEE Trans. Electron Devices* **2018**, *66*, 316–323. [CrossRef]
7. Mishra, S.; Amrouch, H.; Joe, J.; Dabhi, C.K.; Thakor, K.; Chauhan, Y.S.; Henkel, J.; Mahapatra, S. A simulation study of nbti impact on 14-nm node FinFET technology for logic applications: Device degradation to circuit-level interaction. *IEEE Trans. Electron Devices* **2018**, *66*, 271–278. [CrossRef]
8. Zhang, R.; Yang, K.; Liu, T.; Milor, L. Impact of front-end wearout mechanisms on the performance of a ring oscillator-based thermal sensor. In Proceedings of the 2019 IEEE 8th International Workshop on Advances in Sensors and Interfaces (IWASI), Otranto, Italy, 13–14 June 2019; pp. 258–263.
9. Zhang, R.; Liu, Z.; Yang, K.; Liu, T.; Cai, W.; Milor, L. Impact of front-end wearout mechanisms on FinFET SRAM soft error rate. *Microelectron. Reliab.* **2019**, *100–101*, 113487. [CrossRef]

10. Zhang, R.; Liu, Z.; Yang, K.; Liu, T.; Cai, W.; Milor, L. A library based on deep neural networks for modeling the degradation of FinFET SRAM performance metrics due to aging. *Microelectron. Reliab.* **2019**, *100–101*, 113486. [CrossRef]
11. Choudhury, N.; Parihar, N.; Goel, N.; Thirunavukkarasu, A.; Mahapatra, S. Modeling of DC-AC NBTI Stress-Recovery Time Kinetics in P-Channel Planar Bulk and FDSOI MOSFETs and FinFETs. *IEEE J. Electron Devices Soc.* **2020**, *8*, 1281–1288. [CrossRef]
12. Zhang, R.; Liu, T.; Yang, K.; Chen, C.C.; Milor, L. SRAM Stability Analysis and Performance-Reliability Tradeoff for Different Cache Configurations. *IEEE Trans. Very Large Scale Integr. (VLSI) Syst.* **2020**, *28*, 620–633. [CrossRef]
13. Choudhury, N.; Samadder, T.; Tiwari, R.; Zhou, H.; Southwick, R.G.; Wang, M.; Mahapatra, S. Analysis of Sheet Dimension (W, L) Dependence of NBTI in GAA-SNS FETs. In Proceedings of the 2021 IEEE International Reliability Physics Symposium (IRPS), Monterey, CA, USA, 21–25 March 2021; pp. 1–8.
14. Chung, E.; Nam, K.; Nakanishi, T.; Park, S.; Yang, H.; Kauerauf, T.; Jiao, G.; Kim, D.; Hwang, K.H. Investigation of hot carrier degradation in bulk FinFET. In Proceedings of the 2017 IEEE International Reliability Physics Symposium (IRPS), Monterey, CA, USA, 2–6 April 2017; pp. XT-6.1–XT-6.4.
15. Sharma, U.; Mahapatra, S. A SPICE compatible compact model for hot-carrier degradation in MOSFETs under different experimental conditions. *IEEE Trans. Electron Devices* **2018**, *66*, 839–846. [CrossRef]
16. van Santen, V.M.; Thomann, S.; Pasupuleti, C.; Genssler, P.R.; Gangwar, N.; Sharma, U.; Henkel, J.; Mahapatra, S.; Amrouch, H. Bti and hcd degradation in a complete 32× 64 bit sram array–including sense amplifiers and write drivers–under processor activity. In Proceedings of the 2020 IEEE International Reliability Physics Symposium (IRPS), Dallas, TX, USA, 28 April–30 May 2020; pp. 1–7.
17. Kumar, S.; Anandkrishnan, R.; Parihar, N.; Mahapatra, S. A stochastic framework for the time kinetics of interface and bulk oxide traps for BTI, SILC, and TDDB in MOSFETs. *IEEE Trans. Electron Devices* **2020**, *67*, 4741–4748. [CrossRef]
18. Qian, Y.; Gao, Y.; Shukla, A.K.; Wu, T.; Wei, X.; Lu, K.; Yang, Y. Modeling of Hot Carrier Injection on Gate-Induced Drain Leakage in PDSOI nMOSFET. In Proceedings of the 2021 IEEE International Conference on Integrated Circuits, Technologies and Applications (ICTA), Zhuhai, China, 24–26 November 2021; pp. 239–240.
19. Crespo-Yepes, A.; Nasarre, C.; Garsot, N.; Martin-Martinez, J.; Rodriguez, R.; Barajas, E.; Aragones, X.; Mateo, D.; Nafria, M. Combined effects of BTI, HCI and OFF-State MOSFETs Aging on the CMOS Inverter Performance. In Proceedings of the 2021 Joint International EUROSOI Workshop and International Conference on Ultimate Integration on Silicon (EuroSOI-ULIS), Caen, France, 1–3 September 2021; pp. 1–5.
20. Tang, Q.; Kim, C.H. Characterizing the impact of RTN on logic and SRAM operation using a dual ring oscillator array circuit. *IEEE J. Solid-State Circuits* **2017**, *52*, 1655–1663. [CrossRef]
21. Wirth, G. Modeling and Simulation of Charge Trapping in 1/f Noise, RTN and BTI: From Devices to Circuits. In Proceedings of the 2021 28th International Conference on Mixed Design of Integrated Circuits and System, Lodz, Poland, 24–26 June 2021; pp. 14–19.
22. Yang, H.; Robitaille, M.; Chen, X.; Elgabra, H.; Wei, L.; Kim, N.Y. Random Telegraph Noise of a 28-nm Cryogenic MOSFET in the Coulomb Blockade Regime. *IEEE Electron Device Lett.* **2021**, *43*, 5–8. [CrossRef]
23. González-Cordero, G.; González, M.B.; Zabala, M.; Kalam, K.; Tamm, A.; Jiménez-Molinos, F.; Campabadal, F.; Roldán, J.B. Study of RTN signals in resistive switching devices based on neural networks. *Solid-State Electron.* **2021**, *183*, 108034. [CrossRef]
24. Zhang, R.; Yang, K.; Liu, Z.; Liu, T.; Cai, W.; Milor, L. A Comprehensive Framework for Analysis of Time-Dependent Performance-Reliability Degradation of SRAM Cache Memory. *IEEE Trans. Very Large Scale Integr. (VLSI) Syst.* **2021**, *29*, 857–870. [CrossRef]
25. da Silva, M.B.; Both, T.H.; Wirth, G.I. Random Telegraph Noise Modeling for Circuit Analysis: RTN in Ring Oscillators. *IEEE J. Electron Devices Soc.* **2022**. [CrossRef]
26. Yang, K.; Liu, T.; Zhang, R.; Milor, L. A comparison study of time-dependent dielectric breakdown for analog and digital circuit's optimal accelerated test regions. In Proceedings of the 2017 32nd Conference on Design of Circuits and Integrated Systems (DCIS), Barcelona, Spain, 22–24 November 2017; pp. 1–6.
27. Peng, S.; Zhou, H.; Kim, T.; Chen, H.B.; Tan, S.X.D. Physics-Based Compact TDDB Models for Low-k BEOL Copper Interconnects With Time-Varying Voltage Stressing. *IEEE Trans. Very Large Scale Integr. (VLSI) Syst.* **2017**, *26*, 239–248. [CrossRef]
28. Yang, K.; Zhang, R.; Liu, T.; Kim, D.; Milor, L. Optimal accelerated test regions for time-dependent dielectric breakdown lifetime parameters estimation in FinFET technology. In Proceedings of the 2018 Conference on Design of Circuits and Integrated Systems (DCIS), Lyon, France, 14–16 November 2018; pp. 1–6.
29. Yang, K.; Liu, T.; Zhang, R.; Milor, L. A Comprehensive Time-Dependent Dielectric Breakdown Lifetime Simulator for Both Traditional CMOS and FinFET Technology. *IEEE Trans. Very Large Scale Integr. (VLSI) Syst.* **2018**, *26*, 2470–2482. [CrossRef]
30. Peng, S.; Demircan, E.; Shroff, M.D.; Tan, S.X.D. Full-chip wire-oriented back-end-of-line TDDB hotspot detection and lifetime analysis. *Integration* **2020**, *70*, 90–98. [CrossRef]
31. International Technology Roadmap for Semiconductors 2.0 (ITRS 2.0) (2015). Available online: http://www.itrs2.net/itrs-reports.html (accessed on 1 April 2022).
32. Blech, I.A. Electromigration in Thin Aluminum Films on Titanium Nitride. *J. Appl. Phys.* **1976**, *47*, 1203–1208. [CrossRef]
33. Black, J.R. Electromigration-A Brief Survey and Some Recent Results. *IEEE Trans. Electron Devices* **1969**, *16*, 338–347. [CrossRef]
34. Tan, S.; Sun, Z.; Sadiqbatcha, S. Interconnect electromigration modeling and analysis for nanometer ICs: From physics to full-chip. *IPSJ Trans. Syst. LSI Des. Methodol.* **2020**, *13*, 42–55. [CrossRef]

35. Sukharev, V.; Najm, F.N. Electromigration check: Where the design and reliability methodologies meet. *IEEE Trans. Device Mater. Reliab.* **2018**, *18*, 498–507. [CrossRef]
36. Mishra, V.; Sapatnekar, S.S. The impact of electromigration in copper interconnects on power grid integrity. In Proceedings of the 50th Annual Design Automation Conference, Austin, TX, USA, 29 May 2013–7 June 2013; pp. 1–6.
37. Huang, X.; Yu, T.; Sukharev, V.; Tan, S. Physics-based electromigration assessment for power grid networks. In Proceedings of the 51st Annual Design Automation Conference, San Francisco, CA, USA, 1–5 June 2014; pp. 1–6.
38. Mishra, V.; Sapatnekar, S.S. Circuit delay variability due to wire resistance evolution under AC electromigration. In Proceedings of the 2015 IEEE International Reliability Physics Symposium, Monterey, CA, USA, 19–23 April 2015; p. 3D-3.
39. Mishra, V.; Sapatnekar, S.S. Predicting electromigration mortality under temperature and product lifetime specifications. In Proceedings of the 53rd Annual Design Automation Conference, Austin, TX, USA, 5–9 June 2016; pp. 1–6.
40. Mishra, V.; Sapatnekar, S.S. Probabilistic wire resistance degradation due to electromigration in power grids. *IEEE Trans. Comput.-Aided Des. Integr. Circuits Syst.* **2016**, *36*, 628–640. [CrossRef]
41. Chatterjee, S.; Sukharev, V.; Najm, F.N. Fast physics-based electromigration checking for on-die power grids. In Proceedings of the IEEE/ACM International Conference on Computer-Aided Design (ICCAD), Austin, TX, USA, 7–10 November 2016; pp. 1–8.
42. Huang, X.; Kteyan, A.; Tan, S.; Sukharev, V. Physics-based electromigration models and full-chip assessment for power grid networks. *IEEE Trans. Comput.-Aided Des. Integr. Circuits Syst.* **2016**, *35*, 1848–1861. [CrossRef]
43. Mishra, V.; Jain, P.; Marella, S.K.; Sapatnekar, S.S. Incorporating the role of stress on electromigration in power grids with via arrays. In Proceedings of the ACM/EDAC/IEEE Design Automation Conference (DAC), Austin, TX, USA, 18–22 June 2017; pp. 1–6.
44. Chatterjee, S.; Sukharev, V.; Najm, F.N. Power grid electromigration checking using physics-based models. *IEEE Trans. Comput. Aided Des. Integr. Circuits Syst.* **2017**, *37*, 1317–1330. [CrossRef]
45. Tan, S.; Amrouch, H.; Kim, T.; Sun, Z.; Cook, C.; Henkel, J. Recent advances in EM and BTI induced reliability modeling, analysis and optimization. *Integration* **2018**, *60*, 132–152. [CrossRef]
46. Sapatnekar, S.S. Electromigration-Aware Interconnect Design. In Proceedings of the 2019 International Symposium on Physical Design, San Francisco, CA, USA, 14–17 April 2019; pp. 83–90.
47. Tan, S.; Tahoori, M.; Kim, T.; Wang, S.; Sun, Z.; Kiamehr, S. *VLSI Systems Long-Term Reliability–Modeling, Simulation and Optimization*; Springer Publishing: Boston, MA, USA, 2019.
48. Najm, F.N.; Sukharev, V. Efficient simulation of electromigration damage in large chip power grids using accurate physical models. In Proceedings of the IEEE International Reliability Physics Symposium (IRPS), Monterey, CA, USA, 31 March–4 April 2019; pp. 1–10.
49. Zhang, R.; Liu, T.; Yang, K.; Milor, L. CacheEM: For Reliability Analysis on Cache Memory Aging Due to Electromigration. *IEEE Trans. Comput.-Aided Des. Integr. Circuits Syst.* **2021**. [CrossRef]
50. Korhonen, M.A.; Børgesen, P.; Tu, K.N.; Li, C.Y. Stress evolution due to electromigration in confined metal lines. *J. Appl. Phys.* **1993**, *73*, 3790–3799. [CrossRef]
51. Sukharev, V.; Kteyan, A.; Huang, X. Postvoiding stress evolution in confined metal lines, *IEEE Trans. Device Mater. Reliab.* **2016**, *16*, 50–60. [CrossRef]
52. Zhang, L. Effects of Scaling and Grain Structure on Electromigration Reliability of cu Interconnects. Ph.D. Dissertation, The University of Texas at Austin, Austin, TX, USA, 2010.
53. Zahedmanesh, H.; Pedreira, O.V.; Wilson, C.; Tőkei, Z.; Croes, K. Copper electromigration; prediction of scaling limits. In Proceedings of the IEEE International Interconnect Technology Conference, Brussels, Belgium, 2–6 June 2019.
54. Zahedmaesh, H.; Pedreira, O.V.; Tokei, Z.; Croes, K. Electromigration limits of copper nano-interconnects. In Proceedings of the 2021 IEEE International Reliability Physics Symposium (IRPS), Monterey, CA, USA, 21–25 March 2021; pp. 1–6.
55. Zhang, D.; Xu, J.; Mao, S.; Zhao, C.; Wang, G.; Luo, X.; Li, J.; Li, Y.; Wang, W.; Chen, D.; et al. Co-sputtering Co–Ti alloy as a single barrier/liner for Co interconnects and thermal stability enhancement using TiN metal capping. *J. Mater. Sci. Mater. Electron.* **2019**, *30*, 10579–10588. [CrossRef]
56. Beyne, S.; Pedreira, O.V.; Oprins, H.; De Wolf, I.; Tőkei, Z.; Croes, K. Electromigration activation energies in alternative metal interconnects. *IEEE Trans. Electron Devices* **2019**, *66*, 5278–5283. [CrossRef]
57. Zahedmanesh, H.; Vanstreels, K. Mechanical integrity of back-end-of-line with Ru nanowires and airgaps. *Microelectron. Reliab.* **2020**, *110*, 113700. [CrossRef]
58. Zhang, D.; Du, A.; Zhao, C.; Gao, J.; Mao, S.; Xu, J.; Gu, S.; Liu, Y.; Luo, X.; Li, Y.; et al. Investigation of barrier property of amorphous co–ti layer as single barrier/liner in local co interconnects. *IEEE Trans. Electron Devices* **2020**, *67*, 2076–2081. [CrossRef]
59. Kotsugi, Y.; Han, S.M.; Kim, Y.H.; Cheon, T.; Nandi, D.K.; Ramesh, R.; Yu, N.K.; Son, K.; Tsugawa, T.; Ohtake, S.; et al. Atomic Layer Deposition of Ru for Replacing Cu-Interconnects. *Chem. Mater.* **2021**, *33*, 5639–5651. [CrossRef]
60. Vega-Gonzalez, V.; Montero, D.; Versluijs, J.; Pedreira, O.V.; Jourdan, N.; Puliyalil, H.; Chehab, B.; Peissker, T.; Haider, A.; Batuk, D.; et al. Process Integration of High Aspect Ratio Vias with a Comparison between Co and Ru Metallizations. In Proceedings of the 2021 IEEE International Interconnect Technology Conference (IITC), Kyoto, Japan, 6–9 July 2021; pp. 1–3.
61. Croes, K.; Adelmann, C.; Wilson, C.J.; Zahedmanesh, H.; Pedreira, O.V.; Wu, C.; Leśniewska, A.; Oprins, H.; Beyne, S.; Ciofi, I.; et al. Interconnect metals beyond copper: Reliability challenges and opportunities. In Proceedings of the 2018 IEEE International Electron Devices Meeting (IEDM), San Francisco, CA, USA, 1–5 December 2018; pp. 531–534.

62. Pedreira, O.V.; Stucchi, M.; Gupta, A.; Gonzalez, V.V.; van der Veen, M.; Lariviere, S.; Wilson, C.J.; Tőkei, Z. Metal reliability mechanisms in Ruthenium interconnects. In Proceedings of the 2020 IEEE International Reliability Physics Symposium (IRPS), Dallas, TX, USA, 28 April–30 May 2020; pp. 1–7.
63. Pedreira, O.V.; Lofrano, M.; Zahedmanesh, H.; Roussel, P.J.; van der Veen, M.; Simons, V.; Chery, E.; Ciofi, I.; Croes, K. Assessment of critical Co electromigration parameters. In Proceedings of the 2022 IEEE International Reliability Physics Symposium (IRPS), Dallas, TX, USA, 27–31 March 2022; p. 8C-2.
64. Pedreira, O.V.; Croes, K.; Zahedmanesh, H.; Vandersmissen, K.; Van Der Veen, M.H.; Gonzalez, V.V.; Dictus, D.; Zhao, L.; Kolies, A.; Tőkei, Z. Electromigration and thermal storage study of barrierless Co vias. In Proceedings of the 2018 IEEE International Interconnect Technology Conference (IITC), Santa Clara, CA, USA, 4–7 June 2018; pp. 48–50.
65. Ceric, H.; Selberherr, S.; Zahedmanesh, H.; de Orio, R.L.; Croes, K. Modeling Methods for Analysis of Electromigration Degradation in Nano-Interconnects. *ECS J. Solid State Sci. Technol.* **2021**, *10*, 035003. [CrossRef]
66. Posser, G.; Mishra, V.; Jain, P.; Reis, R.; Sapatnekar, S.S. A systematic approach for analyzing and optimizing cell-internal signal electromigration. In Proceedings of the IEEE/ACM International Conference on Computer-Aided Design (ICCAD), San Jose, CA, USA, 2–6 November 2014; pp. 486–491.
67. Posser, G.; Mishra, V.; Jain, P.; Reis, R.; Sapatnekar, S.S. Impact on performance, power, area and wirelength using electromigration-aware cells. In Proceedings of the IEEE International Conference on Electronics, Circuits, and Systems (ICECS), Cairo, Egypt, 6–9 December 2015; pp. 129–132.
68. Posser, G.; Paris, L.D.; Mishra, V.; Jain, P.; Reis, R.; Sapatnekar, S.S. Reducing the signal electromigration effects on different logic gates by cell layout optimization. In Proceedings of the IEEE 6th Latin American Symposium on Circuits and Systems (LASCAS), Montevideo, Uruguay, 24–27 February 2015; pp. 1–4.
69. Jain, P.; Sapatnekar, S.S.; Cortadella, J. A retargetable and accurate methodology for logic-IP-internal electromigration assessment. In Proceedings of the Asia and South Pacific Design Automation Conference, Chiba, Japan, 19–22 January 2015; pp. 346–351.
70. Jain, P.; Sapatnekar, S.S.; Cortadella, J. Stochastic and topologically aware electromigration analysis for clock skew. In Proceedings of the IEEE International Reliability Physics Symposium, Monterey, CA, USA, 19–23 April 2015; p. 3D-4.
71. Jain, P.; Cortadella, J.; Sapatnekar, S.S. A fast and retargetable framework for logic-IP-internal electromigration assessment comprehending advanced waveform effects. *IEEE Trans. Very Large Scale Integr. (VLSI) Syst.* **2016**, *24*, 2345–2358. [CrossRef]
72. Jain, P.; Mishra, V.; Sapatnekar, S.S. Fast stochastic analysis of electromigration in power distribution networks. *IEEE Trans. Very Large Scale Integr. (VLSI) Syst.* **2017**, *25*, 2512–2524. [CrossRef]
73. Chhabria, V.A.; Sapatnekar, S.S. Impact of self-heating on performance and reliability in FinFET and GAAFET designs. In Proceedings of the IEEE International Symposium on Quality Electronic Design (ISQED), Santa Clara, CA, USA, 6–7 March 2019; pp. 235–240.
74. Fawaz, M.; Chatterjee, S.; Najm, F.N. A vectorless framework for power grid electromigration checking. In Proceedings of the IEEE/ACM International Conference on Computer-Aided Design (ICCAD), San Jose, CA, USA, 18–21 November 2013; pp. 553–560.
75. Chatterjee, S.; Fawaz, M.; Najm, F.N. Redundancy-aware electromigration checking for mesh power grids. In Proceedings of the IEEE/ACM International Conference on Computer-Aided Design (ICCAD), San Jose, CA, USA, 18–21 November 2013; pp. 540–547.
76. Chatterjee, S.; Fawaz, M.; Najm, F.N. Redundancy-aware power grid electromigration checking under workload uncertainties. *IEEE Trans. Comput.-Aided Des. Integr. Circuits Syst.* **2015**, *34*, 1509–1522. [CrossRef]
77. Huang, X.; Sukharev, V.; Choy, J.; Chen, H.; Tlelo-Cuautle, E.; Tan, S. Full-chip electromigration assessment: Effect of cross-layout temperature and thermal stress distributions. In Proceedings of the International Conference on Synthesis, Modeling, Analysis and Simulation Methods and Applications to Circuit Design (SMACD), Istanbul, Turkey, 7–9 September 2015; pp. 1–4.
78. Huang, X.; Sukharev, V.; Choy, J.; Chew, M.; Kim, T.; Tan, S. Electromigration assessment for power grid networks considering temperature and thermal stress effects. *Integration* **2016**, *55*, 307–315. [CrossRef]
79. He, K.; Huang, X.; Tan, S. EM-based on-chip aging sensor for detection and prevention of counterfeit and recycled ICs. In Proceedings of the IEEE/ACM International Conference on Computer-Aided Design (ICCAD), Austin, TX, USA, 2–6 November 2015; pp. 146–151.
80. Cook, C.; Sun, Z.; Kim, T.; Tan, S. Finite difference method for electromigration analysis of multi-branch interconnects. In Proceedings of the International Conference on Synthesis, Modeling, Analysis and Simulation Methods and Applications to Circuit Design (SMACD), Lisbon, Portugal, 27–30 June 2016; pp. 1–4.
81. Kim, T.; Liu, Z.; Tan, S. Dynamic reliability management based on resource-based EM modeling for multi-core microprocessors. *Microelectron. J.* **2018**, *74*, 106–115. [CrossRef]
82. Sun, Z.; Demircan, E.; Shroff, M.D.; Cook, C.; Tan, S. Fast electromigration immortality analysis for multisegment copper interconnect wires. *IEEE Trans. Comput.-Aided Des. Integr. Circuits Syst.* **2018**, *37*, 3137–3150. [CrossRef]
83. Sun, Z.; Sadiqbatcha, S.; Zhao, H.; Tan, S. Saturation-Volume Estimation for Multisegment Copper Interconnect Wires. *IEEE Trans. Very Large Scale Integr. (VLSI) Syst.* **2019**, *27*, 1666–1674. [CrossRef]
84. Sun, Z.; Yu, S.; Zhou, H.; Liu, Y.; Tan, S. EMSpice: Physics-based electromigration check using coupled electronic and stress simulation. *IEEE Trans. Device Mater. Reliab.* **2020**, *20*, 376–389. [CrossRef]

85. Zhao, H.; Tan, S. Multi-physics-based FEM analysis for post-voiding analysis of electromigration failure effects. In Proceedings of the International Conference on Computer-Aided Design, San Diego, CA, USA, 5–8 November 2018; pp. 1–8.
86. Zhao, H.; Tan, S. Postvoiding fem analysis for electromigration failure characterization. *IEEE Trans. Very Large Scale Integr. (VLSI) Syst.* **2018**, *26*, 2483–2493. [CrossRef]
87. Choy, J.; Sukharev, V.; Chatterjee, S.; Najm, F.N.; Kteyan, A.; Moreau, S. Finite-difference methodology for full-chip electromigration analysis applied to 3D IC test structure: Simulation vs experiment. In Proceedings of the International Conference on Simulation of Semiconductor Processes and Devices (SISPAD), Kamakura, Japan, 7–9 September 2017; pp. 41–44.
88. Moudallal, Z.; Sukharev, V.; Najm, F.N. Power grid fixing for electromigration-induced voltage failures. In Proceedings of the IEEE/ACM International Conference on Computer-Aided Design (ICCAD), Westminster, CO, USA, 4–7 November 2019; pp. 1–8.
89. Issa, A.; Sukharev, V.; Najm, F.N. Electromigration checking using a stochastic effective current model. In Proceedings of the IEEE/ACM International Conference on Computer Aided Design (ICCAD), San Diego, CA, USA, 2–5 November 2020; pp. 1–8.
90. Najm, F.N. Equivalent circuits for electromigration. *Microelectron. Reliab.* **2021**, *123*, 114200. [CrossRef]
91. Torosyan, S.; Kteyan, A.; Sukharev, V.; Choy, J.; Najm, F.N. Novel physics-based tool-prototype for electromigration assessment in commercial-grade power delivery networks. *J. Vac. Sci. Technol. B Nanotechnol. Microelectron. Mater. Processing Meas. Phenom.* **2021**, *39*, 013203. [CrossRef]
92. Sukharev, V.; Kteyan, A.; Najm, F.N.; Yi, Y.H.; Kim, C.H.; Choy, J.; Torosyan, S.; Zhu, Y. Experimental Validation of a Novel Methodology for Electromigration Assessment in On-chip Power Grids. *IEEE Trans. Comput.-Aided Des. Integr. Circuits Syst.* **2021**. [CrossRef]
93. Zahedmanesh, H.; Varela Pedreira, O.; Tőkei, Z.; Croes, K. Investigating the electromigration limits of Cu nano-interconnects using a novel hybrid physics-based model. *J. Appl. Phys.* **2019**, *126*, 055102. [CrossRef]
94. Nair, S.M.; Bishnoi, R.; Tahoori, M.B.; Zahedmanesh, H.; Croes, K.; Garello, K.; Kar, G.S.; Catthoor, F. Variation-aware physics-based electromigration modeling and experimental calibration for VLSI interconnects. In Proceedings of the 2019 IEEE International Reliability Physics Symposium (IRPS), Monterey, CA, USA, 31 March–4 April 2019; pp. 1–6.
95. Nair, S.M.; Bishnoi, R.; Tahoori, M.B.; Zahedmanesh, H.; Croes, K.; Garello, K.; Kar, G.S.; Catthoor, F. Physics based modeling of bimodal electromigration failure distributions and variation analysis for VLSI interconnects. In Proceedings of the 2020 IEEE International Reliability Physics Symposium (IRPS), Dallas, TX, USA, 28 April–30 May 2020; pp. 1–5.
96. Zahedmanesh, H.; Ciofi, I.; Zografos, O.; Croes, K.; Badaroglu, M. System-Level Simulation of Electromigration in a 3 nm CMOS Power Delivery Network: The Effect of Grid Redundancy, Metallization Stack and Standard-Cell Currents. In Proceedings of the 2022 IEEE International Reliability Physics Symposium (IRPS), Dallas, TX, USA, 27–31 March 2022; pp. 1–7.
97. Zahedmanesh, H.; Ciofi, I.; Zografos, O.; Badaroglu, M.; Croes, K. A Novel System-Level Physics-Based Electromigration Modelling Framework: Application to the Power Delivery Network. In Proceedings of the ACM/IEEE International Workshop on System Level Interconnect Prediction (SLIP), Munich, Germany, 4 November 2021; pp. 1–7.
98. Wang, S.; Sun, Z.; Cheng, Y.; Tan, S.; Tahoori, M.B. Leveraging recovery effect to reduce electromigration degradation in power/ground TSV. In Proceedings of the IEEE/ACM International Conference on Computer-Aided Design (ICCAD), Irvine, CA, USA, 13–16 November 2017; p. 811.
99. Wang, S.; Zhao, H.; Tan, S.; Tahoori, M.B. Recovery-aware proactive TSV repair for electromigration in 3D ICs. In Proceedings of the Design, Automation and Test in Europe Conference and Exhibition (DATE), Lausanne, Switzerland, 27–31 March 2017; pp. 220–225.
100. Wang, S.; Kim, T.; Sun, Z.; Tan, S.; Tahoori, M.B. Recovery-aware proactive TSV repair for electromigration lifetime enhancement in 3-D ICs. *IEEE Trans. Very Large Scale Integr. (VLSI) Syst.* **2017**, *26*, 531–543. [CrossRef]
101. Sukharev, V.; Huang, X.; Tan, S. Electromigration induced stress evolution under alternate current and pulse current loads. *J. Appl. Phys.* **2015**, *118*, 034504. [CrossRef]
102. Huang, X.; Sukharev, V.; Kim, T.; Chen, H.; Tan, S. Electromigration recovery modeling and analysis under time-dependent current and temperature stressing. In Proceedings of the 21st Asia and South Pacific Design Automation Conference (ASP-DAC), Macao, China, 25–28 January 2016; pp. 244–249.
103. Huang, X.; Sukharev, V.; Kim, T.; Tan, S. Dynamic electromigration modeling for transient stress evolution and recovery under time-dependent current and temperature stressing. *Integration* **2017**, *58*, 518–527. [CrossRef]
104. Chen, H.; Tan, S.; Sukharev, V.; Huang, X.; Kim, T. Interconnect reliability modeling and analysis for multi-branch interconnect trees. In Proceedings of the 52nd Annual Design Automation Conference, San Francisco, CA, USA, 7–11 June 2015; pp. 1–6.
105. Chen, H.; Tan, S.; Huang, X.; Kim, T.; Sukharev, V. Analytical modeling and characterization of electromigration effects for multibranch interconnect trees. *IEEE Trans. Comput.-Aided Des. Integr. Circuits Syst.* **2016**, *35*, 1811–1824. [CrossRef]
106. Chen, H.; Tan, S.; Huang, X.; Sukharev, V. New electromigration modeling and analysis considering time-varying temperature and current densities. In Proceedings of the Asia and South Pacific Design Automation Conference, Chiba, Japan, 19–22 January 2015; pp. 352–357.
107. Chen, H.; Tan, S.; Peng, J.; Kim, T.; Chen, J. Analytical modeling of electromigration failure for VLSI interconnect tree considering temperature and segment length effects. *IEEE Trans. Device Mater. Reliab.* **2017**, *17*, 653–666. [CrossRef]
108. Wang, X.; Yan, Y.; He, J.; Tan, S.; Cook, C.; Yang, S. Fast physics-based electromigration analysis for multi-branch interconnect trees. In Proceedings of the IEEE/ACM International Conference on Computer-Aided Design (ICCAD), Irvine, CA, USA, 13–16 November 2017; pp. 169–176.

109. Wang, X.; Wang, H.; He, J.; Tan, S.; Cai, Y.; Yang, S. Physics-based electromigration modeling and assessment for multi-segment interconnects in power grid networks. In Proceedings of the Design, Automation and Test in Europe Conference and Exhibition (DATE), Lausanne, Switzerland, 27–31 March 2017; pp. 1727–1732.
110. Wang, X.; Ma, S.; Tan, S.; Cook, C.; Chen, L.; Yang, J.; Yu, W. Fast Physics-Based Electromigration Analysis for Full-Chip Networks by Efficient Eigenfunction-Based Solution. *IEEE Trans. Comput.-Aided Des. Integr. Circuits Syst.* **2020**, *40*, 507–520. [CrossRef]
111. Chen, L.; Tan, S.; Sun, Z.; Peng, S.; Tang, M.; Mao, J. Fast analytic electromigration analysis for general multisegment interconnect wires. *IEEE Trans. Very Large-Scale Integr. (VLSI) Syst.* **2019**, *28*, 421–432. [CrossRef]
112. Chen, L.; Tan, S.; Sun, Z.; Peng, S.; Tang, M.; Mao, J. A Fast Semi-Analytic Approach for Combined Electromigration and Thermomigration Analysis for General Multisegment Interconnects. *IEEE Trans. Comput.-Aided Des. Integr. Circuits Syst.* **2020**, *40*, 350–363. [CrossRef]
113. Kavousi, M.; Chen, L.; Tan, S. Electromigration Immortality Check considering Joule Heating Effect for Multisegment Wires. In Proceedings of the IEEE/ACM International Conference on Computer Aided Design (ICCAD), San Diego, CA, USA, 2–5 November 2020; pp. 1–8.
114. Kavousi, M.; Chen, L.; Tan, S. Fast Electromigration Stress Analysis Considering Spatial Joule Heating Effects. In Proceedings of the 27th Asia and South Pacific Design Automation Conference (ASP-DAC), Taipei, Taiwan, 17–20 January 2022; pp. 208–213.
115. Al, S.; Abdullah, M.; Chhabria, V.A.; Sapatnekar, S.S. A New, Computationally Efficient "Blech Criterion" for Immortality in General Interconnects. In Proceedings of the ACM/IEEE Design Automation Conference (DAC), San Francisco, CA, USA, 5–9 December 2021; pp. 913–918.
116. Al, S.; Abdullah, M.; Chhabria, V.A.; Evmorfopoulos, N.; Sapatnekar, S.S. Analytical Modeling of Transient Electromigration Stress based on Boundary Reflections. In Proceedings of the IEEE/ACM International Conference on Computer Aided Design (ICCAD), Munich, Germany, 1–4 November 2021; pp. 1–8.
117. Chatterjee, S.; Sukharev, V.; Najm, F.N. Fast physics-based electromigration assessment by efficient solution of linear time-invariant (LTI) systems. In Proceedings of the IEEE/ACM International Conference on Computer-Aided Design (ICCAD), Irvine, CA, USA, 13–16 November 2017; pp. 659–666.
118. Cook, C.; Sun, Z.; Demircan, E.; Shroff, M.D.; Tan, S. Fast electromigration stress evolution analysis for interconnect trees using Krylov subspace method. *IEEE Trans. Very Large Scale Integr. (VLSI) Syst.* **2018**, *26*, 969–980. [CrossRef]
119. Jin, W.; Sadiqbatcha, S.; Sun, Z.; Zhou, Z.; Tan, S. Em-gan: Data-driven fast stress analysis for multi-segment interconnects. In Proceedings of the IEEE 38th International Conference on Computer Design (ICCD), Hartford, CT, USA, 18–21 October 2020; pp. 296–303.
120. Jin, W.; Chen, L.; Sadiqbatcha, S.; Peng, S.; Tan, S. EMGraph: Fast Learning-Based Electromigration Analysis for Multi-Segment Interconnect Using Graph Convolution Networks. In Proceedings of the ACM/IEEE Design Automation Conference (DAC), San Francisco, CA, USA, 5–9 December 2021; pp. 919–924.
121. Zhou, H.; Jin, W.; Tan, S. Gridnet: Fast data-driven EM-induced IR drop prediction and localized fixing for on-chip power grid networks. In Proceedings of the IEEE/ACM International Conference on Computer Aided Design (ICCAD), San Diego, CA, USA, 2–5 November 2020; pp. 1–9.
122. Zhou, H.; Sun, Y.; Sun, Z.; Zhao, H.; Tan, S. Electromigration-lifetime constrained power grid optimization considering multi-segment interconnect wires. In Proceedings of the 23rd Asia and South Pacific Design Automation Conference (ASP-DAC), Jeju, Korea, 22–25 January 2018; pp. 399–404.
123. Zhou, H.; Sun, Z.; Sadiqbatcha, S.; Chang, Z.; Tan, S. EM-aware and lifetime-constrained optimization for multisegment power grid networks. *IEEE Trans. Very Large Scale Integr. (VLSI) Syst.* **2019**, *27*, 940–953. [CrossRef]
124. Zhou, H.; Yu, S.; Sun, Z.; Tan, S. Reliable power grid network design framework considering EM immortalities for multi-segment wires. In Proceedings of the 25th Asia and South Pacific Design Automation Conference (ASP-DAC), Beijing, China, 13–16 January 2022; pp. 74–79.
125. Zhou, H.; Chen, L.; Tan, S. Robust power grid network design considering EM aging effects for multi-segment wires. *Integration* **2021**, *77*, 38–47. [CrossRef]
126. Sun, Z.; Sadiqbatcha, S.; Zhao, H.; Tan, S. Accelerating electromigration aging for fast failure detection for nanometer ICs. In Proceedings of the 23rd Asia and South Pacific Design Automation Conference (ASP-DAC), Jeju, Korea, 22–25 January 2018; pp. 623–630.
127. Sadiqbatcha, S.; Cook, C.; Sun, Z.; Tan, S. Accelerating electromigration wear-out effects based on configurable sink-structured wires. In Proceedings of the International Conference on Synthesis, Modeling, Analysis and Simulation Methods and Applications to Circuit Design (SMACD), Prague, Czech Republic, 2–5 July 2018; pp. 21–24.
128. Sadiqbatcha, S.; Sun, Z.; Tan, S. Accelerating electromigration aging: Fast failure detection for nanometer ICs. *IEEE Trans. Comput.-Aided Des. Integr. Circuits Syst.* **2019**, *39*, 885–894. [CrossRef]
129. Guan, Z.; Marek-Sadowska, M.; Nassif, S. SRAM bit-line electromigration mechanism and its prevention scheme. In Proceedings of the IEEE International Symposium on Quality Electronic Design (ISQED), Santa Clara, CA, USA, 4–6 March 2013; pp. 286–293.
130. Guan, Z.; Marek-Sadowska, M.; Nassif, S. Statistical analysis of process variation induced SRAM electromigration degradation. In Proceedings of the IEEE International Symposium on Quality Electronic Design (ISQED), Santa Clara, CA, USA, 3–5 March 2014; pp. 700–707.

131. Guan, Z.; Marek-Sadowska, M. Incorporating process variations into SRAM electromigration reliability assessment using atomic flux divergence. *IEEE Trans. Very Large Scale Integr. (VLSI) Syst.* **2015**, *24*, 2195–2207. [CrossRef]
132. Zhang, R.; Yang, K.X.; Liu, T.Z.; Milor, L. Modeling of FinFET SRAM array reliability degradation due to electromigration. *Microelectron. Reliab.* **2019**, *100*, 113485. [CrossRef]

micromachines

Article

Wideband Substrate Integrated Waveguide Chip Filter Using Spoof Surface Plasmon Polariton

Dongzhe Pan [1], Bin You [1,*], Xuan Wen [2] and Xungen Li [1]

[1] The Key Laboratory of RF Circuits and Systems of Ministry of Education, Microelectronics CAD Center, School of Electronics and Information, Hangzhou Dianzi University, Hangzhou 310018, China; pdz@hdu.edu.cn (D.P.); lixg@hdu.edu.cn (X.L.)

[2] The School of Information Engineering, Hangzhou Dianzi University, Hangzhou 310018, China; fyrrrr@163.com

* Correspondence: youbin@hdu.edu.cn

Abstract: This article presents a novel wideband bandpass filter based on the integration of a substrate integrated waveguide (SIW) and a spoof surface plasmon polariton (SSPP). An SIW cavity with periodic arrays of meander-slot units is etched on the top metallic layer to achieve the characteristics of a multi-order filter with good performance. The passbands can be flexibly selected by varying the geometric parameters of the SIW and SSPP to adjust the lower and upper sidebands independently. Using a redistribution layer (RDL) process, a novel 3D capacitive interconnection called a through-dielectric capacitor (TDC) is proposed and collaboratively designed with an interdigital capacitor to achieve capacitive source-load cross-coupling. The proposed filter has a center frequency of 60 GHz, with a wide 3-dB fractional bandwidth of about 45.8%. The improved simulated sideband suppression has a 30 dB rejection at 40 GHz and 75.4 GHz, corresponding to a 30-dB rectangular coefficient of 1.28.

Keywords: substrate integrated waveguide (SIW); spoof surface plasmon polaritons (SSPPs); integrated passive device; through-dielectric capacitor (TDC)

Citation: Pan, D.; You, B.; Wen, X.; Li, X. Wideband Substrate Integrated Waveguide Chip Filter Using Spoof Surface Plasmon Polariton. *Micromachines* **2022**, *13*, 1195. https://doi.org/10.3390/mi13081195

Academic Editor: Stefano Mariani

Received: 13 June 2022
Accepted: 27 July 2022
Published: 28 July 2022

Publisher's Note: MDPI stays neutral with regard to jurisdictional claims in published maps and institutional affiliations.

Copyright: © 2022 by the authors. Licensee MDPI, Basel, Switzerland. This article is an open access article distributed under the terms and conditions of the Creative Commons Attribution (CC BY) license (https://creativecommons.org/licenses/by/4.0/).

1. Introduction

With the availability of 60 GHz unlicensed frequency bands and millimeter-wave spectrum for fast data transmission, wireless systems are increasingly operating at higher frequencies. Meanwhile, filters are key front-end components in communication systems, and higher transmission rates must be supported by wider bandwidths and higher frequencies. High-quality and compact BPFs are essential for the improvement of the overall performance of an RF receiver [1]. Acoustic-wave resonators (AWR) such as surface acoustic wave (SAW) [2] and film bulk acoustic resonators (FBAR) [3] exhibit remarkable features in the field of integrated circuit design, with excellent quality factor (Q) and compact size. However, the confined electromechanical coupling coefficient of piezoelectric material limits their bandwidth and center frequency expansion, and starting from C-band, the performance of SAW/FBAR decreases sharply in broadband and high frequency [4–6], making it unsuitable for high frequency bands.

The substrate integrated waveguide (SIW) is popular, with high-pass characteristics, and it exhibits many advantages, such as low insertion loss, high quality factor, and ease of integration. However, its applications in mmW frequency are limited by the harmonics in stopband and bulk size. Frequency selectivity can be improved by introducing transmission zeros (TZs) into the SIW filter. The filter in [7] employs a source-load coupling to improve the slope of the sideband. The conventional folded coupling configuration in [8] introduces a TZ on each side of the passband. An SIW filter with higher-order mode resonators is proposed in order to achieve modal bypass coupling [9]. Multi-order filters with couplings require complicated coupling structures that are larger in size and create some inevitable losses. Miniaturization is not possible because cascaded multiple resonators prohibit full miniaturization. The use of 3D-stacked structures to achieve miniaturization has also become a popular method in recent years [10–16]; however, it introduces additional losses.

To avoid the deficiencies mentioned above, a new design structure to implement a quasi-elliptic SIW filter is proposed. Coplanar waveguides (CPW) [17] and complementary split-ring resonators (CSRR) [18–20] have been embedded in the SIW cavities to achieve miniaturization, but they result in bad insertion loss. The spoof surface plasmon polariton (SSPP) waveguides that support the surface wave transmission on thin, planar corrugated metals [21,22] have attracted the interest of researchers due to their exceptional properties, such as high confinement [23], low loss [24], potential for solving severe on-chip signal integrity and interference issues [25], as well as the possibility for minimizing the circuit area [26,27]. By embedding the SSPP directly in the SIW, it has been proven to further minimize the filter and improve the transmission performance [28,29]. However, only the upper sideband is improved, due to the characteristics of the SSPPs.

This article proposes a novel SIW-based bandpass filter with periodic arrays of meander-slot etching on the top metallic layer. In [30], an advanced RDL (redistribution layer) process was adopted to design passive components, and it proved to be feasible. In this article, we design the filter using a silicon substrate with 4 RDLs. The dispersion and transmission characteristics are numerically studied by simulation. Due to the distinctive structure of the SSPPs, the asymptotic frequency is lower than that of the original groove structure; only one SIW cavity can achieve the characteristic of multi-order filter, and the size of BPF is greatly diminished. The low-cutoff and high-cutoff frequencies can be flexibly adjusted with the variation of size of the SIW and SSPPs, respectively. With the RDL process, an improved TDC structure is proposed and analyzed. After obtaining the equivalent lump model, a TDC is designed as a part of the coupling circuit to achieve capacitive source-load cross-coupling, and a left transmission zero (TZ) appears successfully. The location of the TZ id can be selected by tuning the TDC and interdigital capacitor. The filter simulation results show that the sideband suppression and stopband suppression are significantly improved, TZ is generated in 37.92 GHz with the rejection of −45 dB, and the 30-dB rectangular coefficient is 1.28.

2. Design of the Proposed BPF

This work adopts a redistribution layer (RDL) process whose stack-up is illustrated in Figure 1a. All circuit structures are designed in a four-layer dielectric system interconnected by through-dielectric vias (TDV). The material of the dielectric layer is polyimide (PI-HD4100) with a relative permittivity of 3.2. As shown in Figure 1b, two dielectric layers (P3, P4) are utilized as the SIW cavity. The device layer (M4) is employed to etch the SSPPs grooves and M2 is used as a ground plane. The M1 layer is used to design interdigital capacitors. The low-cutoff frequency is determined by the size of the SIW cavity, and the high-cutoff is finally acquired from the dispersion curve of the SSPPs. All electromagnetic (EM) simulations were carried out using the finite-element method (FEM) of the HFSS 3D simulator.

The vertical view of the proposed bandpass filter is shown in Figure 2., An array of periodic curved slots is etched on the top surface metal of the SIW cavity along the direction of electromagnetic wave propagation. The top layer comprises three parts: the GCPW input (region I), the SIW–SSPP transition (region II), and the periodic array part for the propagation of the SSPP. The GCPW transmission line with lower loss and better heat dissipation is more appropriate for mmW integrated circuits; the filter is excited by two 50-Ω GCPW transmission lines with a pair of quarter wavelength coupling slots. Three curved slot cells of different lengths (L_{s1}, L_{s2}, L_{s3}), which are gradually increased in length, constitute region II. The increase is optimized to achieve smooth transition and mode matching between the SIW and SSPP, and then the SSPP mode is effectively excited. The dimensions of the BPF are calculated below.

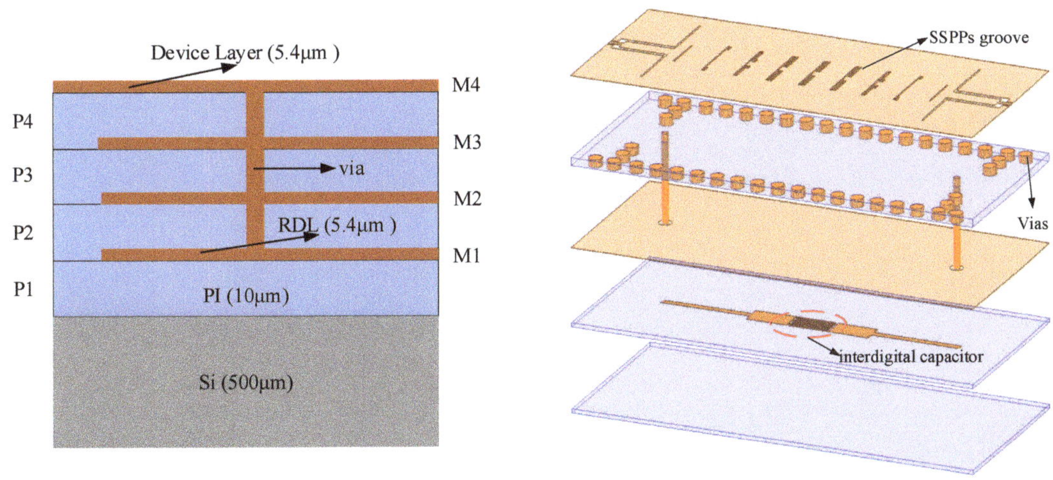

(**a**) (**b**)

Figure 1. (**a**) Stack-up of the RDL process. (**b**) Configuration of the proposed filter (without Si layer).

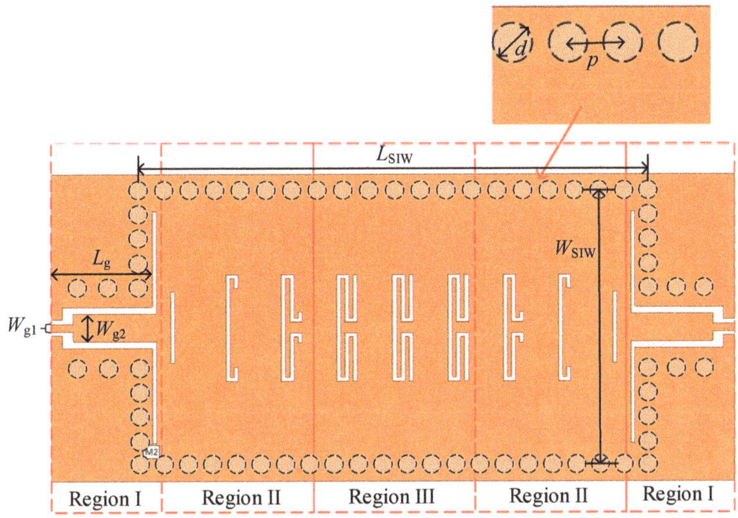

Figure 2. Vertical view of the proposed SIW BPF based on SSPP.

The TE101 mode resonant frequency of the SIW cavity is adopted as the low-cutoff frequency, and the dimensions are determined as in [31]:

$$f_0 = \frac{c_0}{2\sqrt{\varepsilon_r}}\sqrt{\frac{1}{W_{SIW}^2} + \frac{1}{L_{SIW}^2}} \tag{1}$$

where W_{SIW} and L_{SIW} are the width and length of the equivalent rectangular waveguide, calculated as:

$$W_{SIW} = w - 1.08\frac{d^2}{p} + 0.1\frac{d^2}{w} \tag{2}$$

$$L_{SIW} = l - 1.08\frac{d^2}{p} + 0.1\frac{d^2}{l} \tag{3}$$

where w and l represent the width and length of the SIW cavity, respectively, d is the diameter of Cu via, and p is the center-to-center pitch between the adjacent via holes. c_0 is the light velocity in vacuum, and ε_r is the relative permittivity. In order to make the resonance frequency $f_0 = 47$ GHz, $W_{SIW} = 2$ mm and $L_{SIW} = 3.8$ mm is finally determined by the above equations.

The schematic diagram of the unit cell of the SSPP is illustrated in Figure 3. According to [32], the dispersion curve for the SSPP mode propagated in the metallic grove array can be expressed as

$$k = k_0\sqrt{1 + \frac{W^2}{C^2}tan^2(k_0 L)} \quad (4)$$

where $k_0 = 2\pi/\lambda$ refers to the propagation constant in free space, C is the width of the unit cell, W and L represent the width and depth of the grooves, respectively.

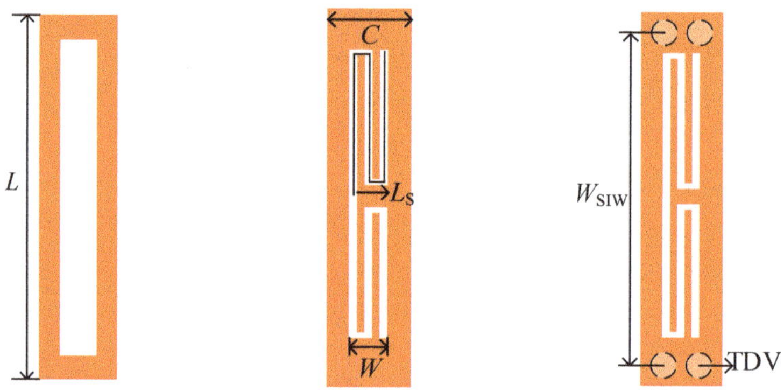

Figure 3. The vertical view of (**a**) rectangular groove unit cell, (**b**) meander–slot unit cell, (**c**) SIW with meander-slot unit cell.

When the width of the groove W and the width of the unit cell C is fixed, the depth of the groove L is the main factor affecting the dispersion. Meanwhile, to keep the occupied area of the groove unchanged, a meander-slot structure is proposed. Figure 3a,b show a traditional groove unit and a meander-slot unit with the same height (L). Figure 3c shows the unit cell consisting of an SIW and meander-slot units. The orange and white parts represent the metal surface on the top layer and the slot line, respectively. The geometric parameters are set as shown in Table 1.

Table 1. The geometric parameters of the SSPPs unit cell.

C	L	L_s	W	W_{SIW}
0.4 mm	0.39 mm	1 mm	0.075 mm	2 mm

To visualize the relationship between the dimension and frequency of the SSPP, the dispersion curves are calculated using the eigenmode solver of commercial electromagnetic software. In the simulations, the dispersion relation was obtained by calculating the eigenfrequency of the SSPPs unit. Figure 4 shows the dispersion curves of fundamental SSPP modes, k represents propagation constant, and k is swept from 0° to 180° between the period boundaries in the propagation direction. It is clear that the two dispersion curves have similar frequency trends, increasing as k increases. It is obvious that these two dispersion curves have similar frequency variation trends, but the asymptotic frequency of the groove unit is 97 GHz when the asymptotic frequency of meander-slot unit is 73.7 GHz. This means that adopting a meander-slot SSPP in the same occupied area can reduce the available asymptotic frequency, which is effective in minimizing the device. Figure 4 demonstrates that the SIW can be combined with SSPPs to achieve bandpass characteristics, starting from the cutoff frequency, and its dispersion curve behaves like the SIW in the low frequency range and like SSPPs in the high frequency range.

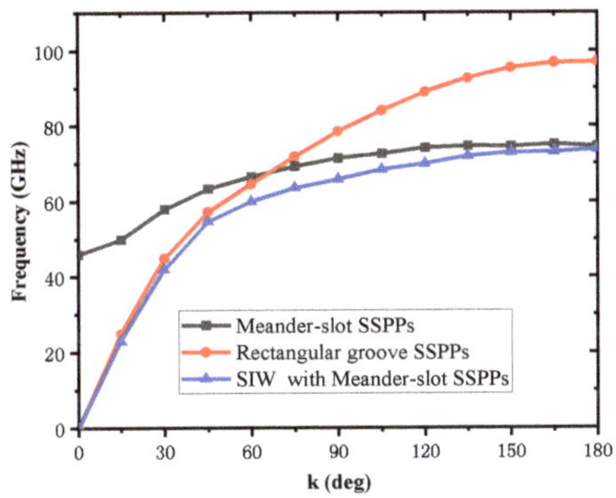

Figure 4. Dispersion curve of the unit cell.

To confirm the passband alteration characteristics of the proposed filter, a parameter inquiry was conducted. The simulated results of transmission coefficients are demonstrated in Figure 5 with different geometric parameters. It can be inferred that the high-cutoff frequency of the proposed filter is determined by the SSPPs length L_S, and the low-cutoff frequency is determined by the SIW width W_{SIW}. Figure 5a shows that when the value of L_S increases gradually, the upper sideband frequency of the filter will move to the left, while the lower sideband remains constant; the bandwidth becomes narrower accordingly. It can be attributed to the gradual increase of the propagation constant and momentum, and the decrease of the asymptotic frequency, with the increase of the geometric length L_S of the curved slot. In addition, as shown in Figure 5b, the W_{SIW} decreases and the lower sideband shifts to the right because the resonant frequency of the SIW is determined by the SIW width W_{SIW}. Therefore, the bandwidth of the filter can be flexibly controlled by adjusting the size of the SIW cavity and the SSPP slot. Finally, the dimensions of the BPF are determined by the low-cutoff frequency (47 GHz) and high-cutoff frequency (74 GHz), as shown in Table 2.

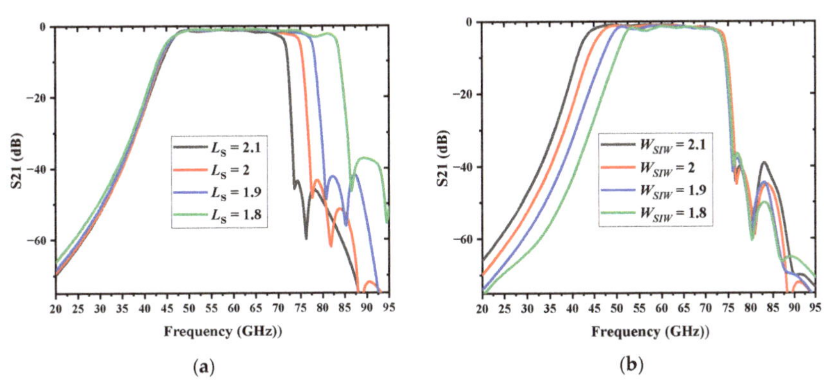

Figure 5. Simulated results of the proposed bandpass filter with different (**a**) SSPPs length L_S. (**b**) SIW width W_{SIW}.

Table 2. The dimensions of the BPF.

L_{SIW}	W_{SIW}	L_g	W_{g1}	W_{g2}	L_{s1}	L_{s2}	L_{s3}
4 mm	2 mm	0.76 mm	0.055 mm	0.2 mm	0.26 mm	0.49 mm	0.78 mm

The SIW filter using SSPP without source-load coupling was simulated, and the result is shown in Figure 6. Good bandpass features and high-efficiency propagation are obtained. The BPF operates from 46.13 GHz to 74 GHz; the 30-dB bandwidth is from 37.5 GHz to 76.5 GHz; and the corresponding rectangular coefficient is 1.4.

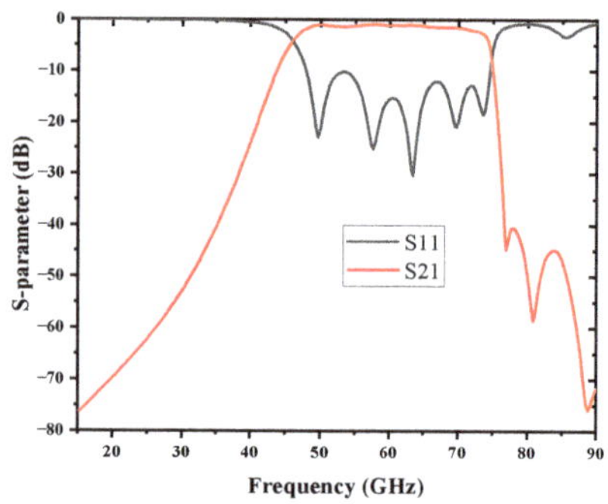

Figure 6. Simulation results of the BPF based on SIW and SSPP.

With the multi-layer stacked structure, 3D interconnections can be applied. To further improve the performance of the filter, source-load cross-coupling was employed. The sectional drawing of the proposed filter with source-load cross-coupling is shown in Figure 7a. With the adoption of the TDC, signals can be transmitted vertically through different layers. There are two microstrip lines and an interdigital capacitor in the M1 layer. The interdigital capacitor is more suitable for applications where low values of capacitance are required. Letting the finger width (X = 0.025 mm) equal the slot width to achieve maximum capacitance density, the expression [33] for estimation of capacitance of the interdigital capacitor can be given by

$$C_i = (\varepsilon_r + 1)\frac{L_i}{W_i}[(n-3)A_1 + A_2] \tag{5}$$

where n is the number of fingers and C is in pF. $A_1 = 0.75$ and $A_2 = 0.175$ are determined by the value of $\frac{T}{X}$ refer to in [33], where T is the height of the substrate ($T = 0.01$ mm). As shown in Figure 7b, the complete coupling circuit consists of two TDCs, two microstrips and an interdigital capacitor which are lumped models. The collaborative design method of the TDC and the coupling circuit can be used to minimize the influence of the interconnection structure. The series connection of the interdigital capacitor and TDC can also reduce the circuit capacitance and increase the adjustment range of the interdigital capacitor. The admittance of the equivalent coupling circuit with the TDC in Figure 7b can be expressed as

$$Z = Z_t + R_m + j(\omega L_m - \frac{1}{\omega C_i}) \tag{6}$$

where Z_t is the impedance of the TDC structure, which is analyzed below.

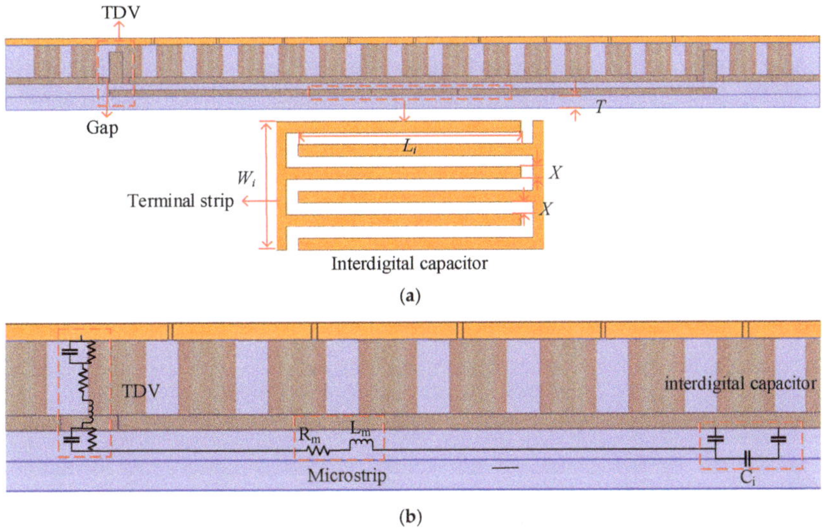

Figure 7. (**a**) Cross-section schematic view of the filter; (**b**) Equivalent lump model of this proposed coupling circuit (**left**).

Figure 8a shows the structure and the main parasitic components of the TDC in the designed filter. Between the two metal layers, two polyimide bonding layers act as capacitive coupling media, while one is filled with copper through the dielectric via. The equivalent lump model of the TDC can be referred to [34], as shown in Figure 8b.

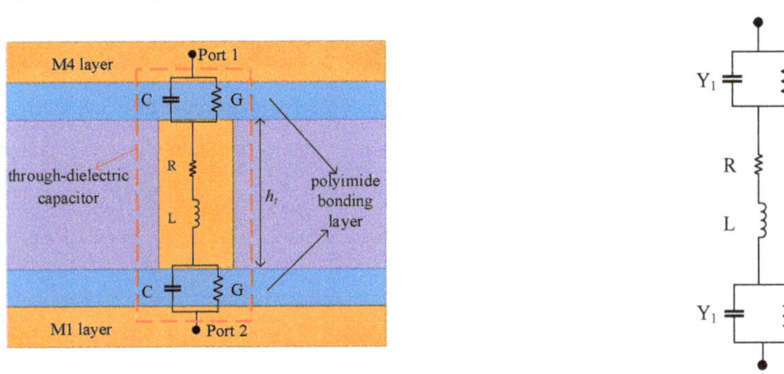

(**a**) (**b**)

Figure 8. Structure of TDC in the designed filter: (**a**) Cross-section schematic view with the main parasitic components; (**b**) Equivalent lump model.

The impedance Z_t can be approximately expressed by

$$Z_t = \frac{2}{Y_1} + j\omega L + R \tag{7}$$

$$Y_1 = G + j\omega C \tag{8}$$

where Y_1 is the admittance of the coupling medium, R and L represent via resistance and inductance, respectively, when C and G represent the parasitic capacitance and conductance, respectively.

To accurately verify the compatibility and analyze the electrical performance of the TDC, Ansys Q3D was used to extract all the parasitic components of TDC. Substituting all the parasitic components into the proposed lump model then enables the transmission characteristics to be obtained by the ADS. The comparison of transmission characteristics between the TDC simulation result by HFSS and the proposed lumped model is shown in Figure 9. Two experimental cases with different heights of through dielectric via (h_t) are used in simulation. The lumped models of the two cases both match well with the HFSS simulations. The insertion loss S_{21} of TDC increases with frequency below 10 GHz, which implies that the coupling behavior of the TDC is mainly capacitive coupling, and slightly decreases in high frequency. The capacitive TDC effect is dominant in the lower frequency range, and the inductive TDV effect becomes dominant in the higher frequency range [34]; the inductance of the TSV channel starts to affect the insertion loss over 10 GHz. The above analysis shows that the proposed TDC can be regarded as a CGRL lumped model, collaboratively designed with the coupling circuit. In addition, it will not seriously influence the circuit during the working frequency.

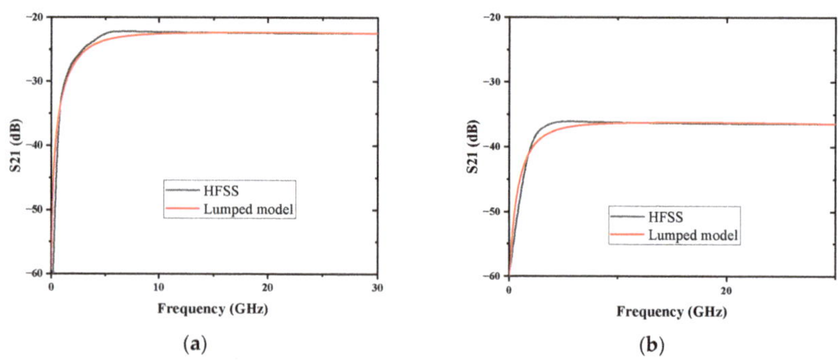

Figure 9. Comparison of S parameters between TDC and model with different h_t: (**a**) $h_t = 0.024$ mm; (**b**) $h_t = 0.08$ mm.

To further verify the lumped model of the whole coupling circuit, the S-parameter is extracted from Figure 6 and simulated with the above coupling circuit model, and the result is compared with the HFSS simulation of the proposed filter, as shown in Figure 10. The frequency responses of the equivalent circuit and filter simulations both exhibit a transmission zero at 37.6 GHz, and their simulation results are in good agreement.

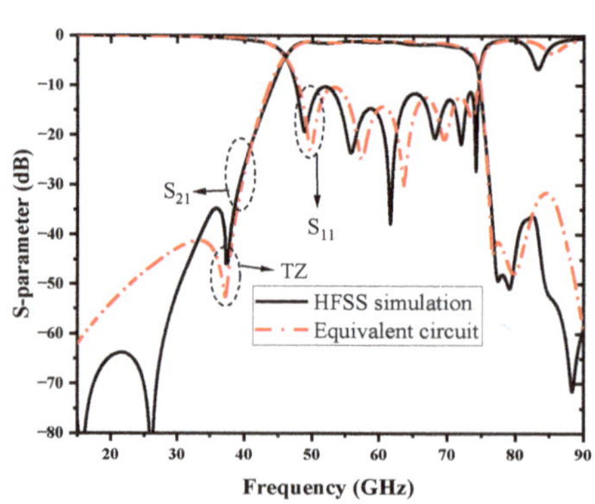

Figure 10. Comparison of equivalent circuit and filter simulations.

According to the above analysis and verification, the proposed TDC structure can be applied to transmit signals vertically through different layers, and it is possible to design a circuit with a lumped model. With the adoption of the previous lumped model, capacitive source-load cross-coupling is investigated.

The capacitance of the interdigital capacitor will change over L_i. By adjusting the capacitance of the interdigital capacitor, the location of the TZ is moved, as shown in Figure 11, and the suppression deteriorates as TZ approaches the passband. To optimize the performance of the filter, a transmission zero is finally generated at 37.92 GHz, and the height of the through dielectric via $h_t = 0.0246$ mm and the radius of the via $r_t = 0.04$ mm are finally determined.

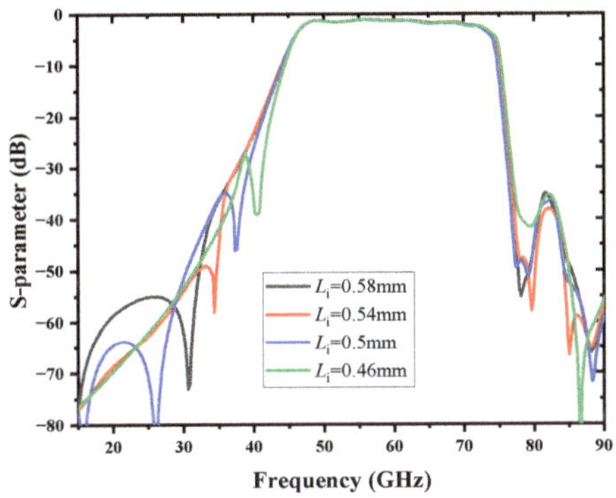

Figure 11. Simulated results with different interdigital capacitor length L_i.

3. Results and Analysis

The simulated results of the S-parameters with reflection coefficients (S11) and transmission coefficients (S21) are demonstrated in Figure 12. An SIW cavity is employed to provide multi-transmission poles in order to achieve a passband. The compact BPF operates at 46.1–73.7 GHz with a wide 3 dB FBW of 45.8%. Good bandpass features and high-efficiency propagation are obtained; the minimum insertion loss is 1.08 dB, and the return loss is better than −10 dB in the passband. Due to the steep upper sideband, the 30-dB bandwidth is from 40 GHz to 75.4 GHz, and the corresponding rectangular coefficient is 1.28 when the BPF without TZ is 1.4. Meanwhile, the stopband rejection is better than 30 dB up to 125 GHz ($2f_c$). The final core size of the BPF is 2 mm × 4.4 mm ($0.74\lambda_g \times 1.63\lambda_g$). It is notable that the proposed BPF has the merits of competitive wideband performance, broad stopband, and compact size. The performance of the proposed filter is compared with other work in Table 3; all the data are based on the simulation results.

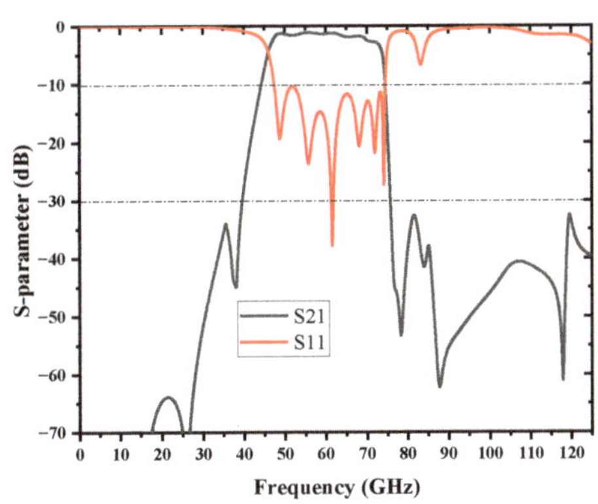

Figure 12. Simulation result of the bandpass filter with source-load coupling.

Table 3. Performance comparisons of BPFs operating above millimeter-wave. (All the data are based on the simulation results).

Ref.	f_c (GHz)	Size ($\lambda_g \times \lambda_g$)	MIN. IL (dB)	$\frac{BW_{30dB}}{BW_{3dB}}$	FBW (%)
[4]	27	0.09 × 0.09	0.84	2.6	20
[6]	3.5	2.04 × 0.85	1.139	1.25	56
[7]-2	9.1	0.218 × 0.218	0.84	2.5	19.8
[9]	93	2.31 × 1.57	4	1.59	3.5
[29]	236.5	1.16 × 0.37	2	1.86	12
This work	60	1.63 × 0.74	1.08	1.28	45.8

4. Conclusions

In this article, an SIW-based 46.1 GHz to 73.7 GHz bandpass filter is proposed, which is designed on the RDL process. The simulation results show low insertion loss, good frequency selectivity, and wideband harmonic suppression. The etching of periodic arrays of meander slot units on the top metallic layer enables the substrate integrated waveguide (SIW) to attain bandpass characteristics with high-efficiency and strongly confined microwave SSPP transmission. In addition, its asymptotic frequency can be significantly reduced compared to the conventional groove SSPP. This means that the propagation of the structure can occupy a smaller area with the benefit of lower cost, especially for the process of integrating a passive device, and the leakage loss will decrease as the gap area reduces. The simulated results show that the bandwidth of the proposed filter can be flexibly elected by tuning the geometric parameters of the SIW and SSPPs. To improve the performance of the upper sideband, a novel 3D capacitive interconnection is proposed and investigated, the lump model of the TDC is obtained, and a collaborative design with interdigital capacitance is adopted to achieve a transmission zero. By adjusting the height of the metal via (h_t) and the capacitance of the interdigital capacitor (C_i), the location of the TZ can be selected. The proposed filter based on one SIW resonator is the same as the multi-order filter with coupling, realizing the miniaturization with good performance.

Author Contributions: Conceptualization, D.P. and B.Y.; Data curation, D.P., X.W. and X.L.; Formal analysis, D.P. and X.W.; Funding acquisition, B.Y.; Methodology, D.P., X.W. and X.L.; Validation, B.Y. and X.L.; Writing—original draft, D.P.; Writing—review & editing, D.P., B.Y. and X.W. All authors have read and agreed to the published version of the manuscript.

Funding: This research was funded by the Natural Science Foundation of Zhejiang Province under Grant LZ22F010006 and National Natural Science Foundation of China under Grant 61671195.

Conflicts of Interest: The authors declare no conflict of interest.

References

1. Chaturvedi, S.; Bozanic, M.; Sinha, S. Millimeter wave passive bandpass filters. *Microw. J.* **2017**, *60*, 98–108.
2. Cai, T.; Chen, C.; Zhang, X.; Lin, F.; Zhang, H. A Hybrid Transmission-Line/SAW-Resonator Analog Signal-Interference Dual-Band Bandpass Filter. *IEEE Microw. Wirel. Compon. Lett.* **2020**, *30*, 27–30. [CrossRef]
3. Warder, P.; Link, A. Golden age for filter design: Innovative and proven approaches for acoustic filter, duplexer, and multiplexer design. *IEEE Microw. Mag.* **2015**, *16*, 60–72. [CrossRef]
4. Shen, G.; Che, W.; Feng, W.; Shi, Y.; Xu, F.; Xue, Q. Ultra-low-loss millimeter-wave LTCC bandpass filters based on flexible design of lumped and distributed circuits. *IEEE Trans. Circuits Syst. II Exp. Briefs* **2021**, *68*, 1123–1127. [CrossRef]
5. Zeng, X.; Bi, X.; Cao, Z.; Wan, R.; Xu, Q. High selectivity dual-wideband balun filter utilizing a multimode t-line loaded middle-shorted CSRR. *IEEE Trans. Circuits Syst. II Exp. Briefs* **2020**, *67*, 2447–2451. [CrossRef]
6. Wu, H.; Wu, Y.; Lai, Z.; Wang, W.; Yang, Q. A Hybrid Filter With Extremely Wide Bandwidth and High Selectivity Using FBAR Network. *IEEE Trans. Circuits Syst. II Exp. Briefs.* **2020**, *68*, 1123–1127. [CrossRef]
7. Wang, X.; Zhu, X.; Jiang, Z.; Hao, Z.; Wu, Y.; Hong, W. Analysis of eighth-mode substrate-integrated waveguide cavity and flexible filter design. *IEEE Trans. Microw. Theory Tech.* **2019**, *67*, 2701–2712. [CrossRef]

8. Zhou, K.; Zhou, C.; Wu, W. Dual-mode characteristics of half-mode SIW rectangular cavity and applications to dual-band filters with widely separated passbands. *IEEE Trans. Microw. Theory Tech.* **2018**, *66*, 4820–4829. [CrossRef]
9. Xiao, Y.; Shan, P.; Zhao, Y.; Sun, H.; Yang, F. Design of a W-Band GaAs-Based SIW Chip Filter Using Higher Order Mode Resonances. *IEEE Microw. Wirel. Compon. Lett.* **2019**, *29*, 104–106. [CrossRef]
10. Iqbal, A.; Tiang, J.-J.; Kee, C.-K.; Lee, B.-M. Tunable substrate integrated waveguide diplexer with high isolation and wide stopband. *IEEE Microw. Wirel. Compon. Lett.* **2019**, *29*, 456–458. [CrossRef]
11. Li, P.; Chu, H.; Zhao, D.; Chen, R.-S. Compact dual-band balanced SIW bandpass filter with improved common-mode suppression. *IEEE Microw. Wirel. Compon. Lett.* **2017**, *27*, 347–349. [CrossRef]
12. Ho, M.-H.; Tang, K.-H. Miniaturized SIW cavity tri-band fil- ter design. *IEEE Microw. Wirel. Compon. Lett.* **2020**, *30*, 589–592. [CrossRef]
13. Deng, H.-W.; Han, Y.-K.; Sun, L.; Zhu, J.-M.; Xing, S.-B. Multilayer dual-mode balanced SIW filter utilizing PEC–PMC characteristic for common-mode suppression. *IEEE Microw. Wirel. Compon. Lett.* **2020**, *30*, 865–868. [CrossRef]
14. Jia, D.; Feng, Q.; Xiang, Q.; Wu, K. Multilayer substrate integrated waveguide (SIW) filters with higher-order mode suppression. *IEEE Microw. Wirel. Compon. Lett.* **2016**, *26*, 678–680. [CrossRef]
15. Liu, X.; Zhu, Z.; Liu, Y.; Lu, Q.; Yin, X.; Yang, Y. Wideband Substrate Integrated Waveguide Bandpass Filter Based on 3-D ICs. *IEEE Trans. Compon. Packag. Manuf. Technol.* **2019**, *9*, 728–735. [CrossRef]
16. Shen, W.; Zhu, H.-R. Vertically stacked trisection SIW filter with controllable transmission zeros. *IEEE Microw. Wirel. Compon. Lett.* **2020**, *30*, 237–240. [CrossRef]
17. Chu, P.; Hong, W.; Dai, L.; Tang, H.; Chen, J.; Hao, Z.; Zhu, X.; Wu, K. A planar bandpass filter implemented with a hybrid structure of substrate integrated waveguide and coplanar waveguide. *IEEE Trans. Microw. Theory Tech.* **2014**, *62*, 266–274. [CrossRef]
18. Muchhal, N.; Srivastava, S. Design of Miniaturized Diamond Shaped Substrate Integrated Waveguide CSRR Band Pass Filter for X Band Applications. In Proceedings of the International Conference on Signal Processing and Communication (ICSC), Noida, India, 7–9 March 2019.
19. Jin, B.; Zhang, P.; Mu, J.; Zhang, M.; Li, M. A Miniaturized Bandpass Filter Basing on HMSIW Loaded Dual-mode CSRR. In Proceedings of the 2021 IEEE MTT-S International Wireless Symposium (IWS), Nanjing, China, 23–26 May 2021.
20. Pradhan, N.C.; Subramanian, K.S.; Barik, R.K.; Cheng, Q.-S. Design of compact substrate integrated waveguide based triple- and quad-band power dividers. *IEEE Microw. Wirel. Compon. Lett.* **2021**, *31*, 365–368. [CrossRef]
21. Shen, X.; Cui, T.-J. Planar plasmonic metamaterial on a thin film with nearly zero thicknes. *Appl. Phys. Lett.* **2013**, *102*, 211909-1–211909-4. [CrossRef]
22. Shen, X.; Cui, T.-J.; Martin-Cano, D.F.; Garcia-Vidal, J. Conformal surface plasmons propagating on ultrathin and flexible films. *Proc. Nat. Acad. Sci. USA* **2013**, *110*, 40–45. [CrossRef]
23. Pendry, J.B.; Martín-Moreno, L.; Garcia-Vidal, F.J. Mimicking surface plasmons with structured surfaces. *Science* **2004**, *305*, 847–848. [CrossRef] [PubMed]
24. Martin-Cano, D.; Nesterov, M.L.; Fernandez-Dominguez, A.I.; Garcia-Vidal, F.J.; Martin-Moreno, L.; Moreno, E. Domino plasmons for subwavelength terahertz circuitry. *Opt. Exp.* **2010**, *18*, 754–764. [CrossRef] [PubMed]
25. Zhang, H.-C.; Cui, T.-J.; Zhang, Q.; Fan, Y.; Fu, X. Breaking the challenge of signal integrity using time-domain spoof surface plasmon polaritons. *ACS Photon.* **2015**, *2*, 1333–1340. [CrossRef]
26. Unutmaz, M.A.; Unlu, M. Modelling of the effective dielectric constant of planar spoof surface plasmon polariton waveguides. *Electron. Lett.* **2017**, *53*, 1418–1420. [CrossRef]
27. Unutmaz, M.A.; Unlu, M. Fixed physical length spoof surface plasmon polariton delay lines for a 2-bit phase shifter. *J. Opt. Soc. Amer. B Opt. Phys.* **2020**, *37*, 1116–1121. [CrossRef]
28. Guan, D.F.; You, P.; Zhang, Q.; Xiao, K.; Yong, S.-W. Hybrid spoof surface plasmon polariton and substrate integrated waveguide transmission line and its application in filter. *IEEE Trans. Microw. Theory Tech.* **2017**, *65*, 4925–4932. [CrossRef]
29. Zhu, H.; Zhang, Y.; Ye, L.; Li, Y.; Xu, Y.; Xu, R. On-chip Terahertz Bandpass Filter based on Substrate Integrated Plasmonic Waveguide. *Results Phys.* **2021**, *104553*, 2211–3797. [CrossRef]
30. Xu, J.; Sun, Y.; Liu, J.; Wei, Y.-D.; Zhao, W.-S.; Wang, D.-W. Fabrication and high-frequency characterization of low-cost fan-in/out WLP technology with RDL for 2.5D/3D heterogeneous integration. *Microelectron. J.* **2022**, *119*, 105332. [CrossRef]
31. Zhou, K.; Zhou, C.-X.; Wu, W. Resonance Characteristics of Substrate-Integrated Rectangular Cavity and Their Applications to Dual-Band and Wide-Stopband Bandpass Filters Design. *IEEE Trans. Microw. Theory Tech.* **2017**, *65*, 1511–1524. [CrossRef]
32. Liu, H.; Wang, Z.; Zhang, Q.; Ma, H.; Ren, B.; Wen, P. Design wideband differential bandpass filter using slotline surface plasmon polariton. *IEEE Access* **2019**, *7*, 44212–44218. [CrossRef]
33. Alley, G.D. Interdigital Capacitors and Their Application to Lumped-Element Microwave Integrated Circuits. *IEEE Trans. Microw. Theory Tech.* **1970**, *18*, 1028–1033. [CrossRef]
34. Kim, J.; Pak, J.S.; Cho, J.; Song, E.; Cho, J.; Kim, H.; Song, T.; Lee, J.; Lee, H.; Park, K.; et al. High-Frequency Scalable Electrical Model and Analysis of a Through Silicon Via (TSV). *IEEE Trans. Compon. Packag. Manuf. Technol.* **2011**, *1*, 181–195.

micromachines MDPI

Article

Antenna Current Calculation Based on Equivalent Transmission Line Model

Shusheng Wei [1] and Wusong Wen [2,*]

1 Department of Electrical Engineering, Tsinghua University, Beijing 100084, China; weiss13@mails.tsinghua.edu.cn
2 Key Laboratory of Military Special Power Supply, Army Engineering University of PLA, Chongqing 400035, China
* Correspondence: wenwusong@163.com

Abstract: This paper provides a new way for spatial current/field profiles for frequency-selective surface analytical approximation. It confirms that the per unit length radiation resistance of an equivalent transmission line model for line antenna has little influence on the normalized current distribution. The two-wire equivalent transmission line model (typically used for transmitting line antenna) is applied to the receiving line antenna. In this case, the corresponding incident field is decomposed into odd and even mode for asymmetric distribution. A one-wire equivalent transmission line model is then introduced for any antenna composed of relative narrow strips. The incident field does not need to be decomposed. According to the simulation, the transmission line loss has little influence on the current distribution.

Keywords: antenna current; transmission line model; frequency-selective surface analytical approximation

Citation: Wei, S.; Wen, W. Antenna Current Calculation Based on Equivalent Transmission Line Model. *Micromachines* **2022**, *13*, 714. https://doi.org/10.3390/mi13050714

Academic Editor: Wensheng Zhao

Received: 8 April 2022
Accepted: 27 April 2022
Published: 30 April 2022

Publisher's Note: MDPI stays neutral with regard to jurisdictional claims in published maps and institutional affiliations.

Copyright: © 2022 by the authors. Licensee MDPI, Basel, Switzerland. This article is an open access article distributed under the terms and conditions of the Creative Commons Attribution (CC BY) license (https://creativecommons.org/licenses/by/4.0/).

1. Introduction

Numerical methods, such as method of moment (MoM), are usually applied to calculate the transmitting current of dipole antenna. However, the current distribution calculated is in numerical form, which cannot be directly used for frequency-selective surface (FSS), where the analytical current distribution is needed [1–7]. Two promising approximate analytical methods for FSS are the periodic method of moment (PMM) and multimodal network approach. However, it is necessary for both of them to obtain the current/field distribution in the scatterer. PMM solves this problem using MoM [8–10], while the multimodal network approach only considers simple rectangular scatters [11–13]. For a rectangular patch (or aperture) under oblique incidence in the principle scan plane of the structure, approximate closed-form expressions for aperture field and current profiles of patch surface were obtained. However, to the authors' knowledge, the approximate closed-form expressions for other structures have not been reported. Therefore, the application of the above approximate analytical methods depends on the development of current/field distribution in the scatterer of other types. In [14], the spatial current/field profile is calculated using a full-wave simulation, at a single- and low-frequency value. It is then assumed to be independent of the frequency in the considered range of interest. However, this method is restrictive, treating structures that are relatively simple and incidence angles that are relatively small. Besides, the numerical preparation is time consuming and means the analytical method is complex.

The calculation of spatial current/field profile for an FSS element is similar to the analysis of current distribution for line antenna [15,16]. The equivalent transmission line model was often adopted for the study of input impedance and mutual coupling of transmitting line antenna [17–19]. It was usually a two-wire transmission line model with lossy lumped feed. For single antennas, radiation lossy can be considered easily, as can the mutual coupling lossy. A two-wire straight antenna refers to two collinear or equidistant antennas, which can be one or more segments. The segments may or may not be parallel. The two-wire straight antenna can only be equivalent to the two-wire transmission line model, and the requirement is that the antenna structures are symmetrical. Besides, the tangential component of the incident electric field on the two antennas should be even or odd symmetrical. For the asymmetric distribution of the incident field, the incident field can be decomposed into odd and even symmetrical. The total current in the antenna is then the superposition of currents in the above two modes.

Schelkunoff mainly used a one-wire transmission line model in [17] to study the input impedance, mutual coupling, antenna impedance, and the approximate solution of transmitting/receiving antenna current. Compared to the two-wire transmission line model, the one-wire transmission model is more flexible. The former makes it easier to study the dipole oscillator antenna. For the one-wire transmission line model of complex structures, the main problem is that the equations and parameters are restrictive in approximating the lossless case. Schelkunoff also studied the current distribution on the receiving antenna (actually a reflector antenna, which is a straight continuous fine wire antenna, which is exposed to the incident field and the corresponding tangential component is uniform) using a two-wire transmission line model. The model is restrictive for vertical incidence (the tangential field of incident field is uniform on the antenna). For oblique incidence, the tangential electric field of the incident field on the antenna varies with the spatial position. It results in the induced electromotive force distributed asymmetrically about the center of the antenna. The currents on the lines are not equivalent and reverse, so the two-wire model cannot be used.

In this paper, the current distributions of transmitting and receiving antenna, based on the equivalent transmission line method, are analyzed. Both the two-wire and one-wire model are studied, together with the radiation lossy. An iterative scheme is then introduced to further improve the accuracy. The current distribution calculated is in analytic form, which can be directly used to approximate the analytical method for FSS.

2. Transmitting Antenna

Equivalent Circuit of the Transformers

Symmetric diploes transmitting antenna (Figure 1) can be equivalent to the two-wire transmission line model (Figure 2a). For the lossless equivalent transmission line corresponding to equidistance symmetric diploes antenna (Figure 1a), the average characteristic impedance is evaluated using [14,15]

$$Z_0 = \frac{120}{\sqrt{\varepsilon_r}} \ln \frac{d}{\rho}. \tag{1}$$

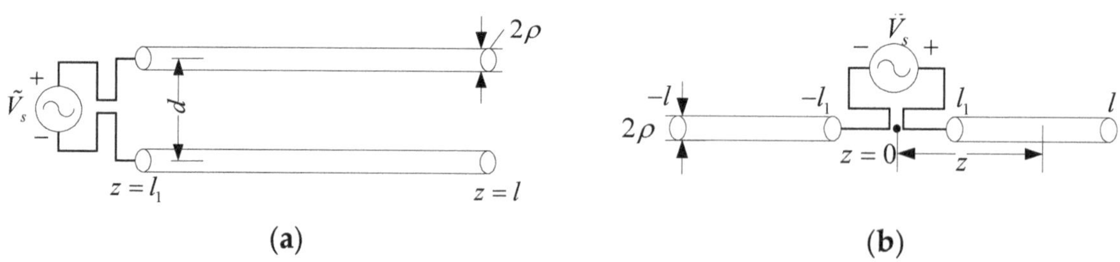

Figure 1. Symmetric diploes transmitting line antenna (**a**) equidistance; (**b**) collinear.

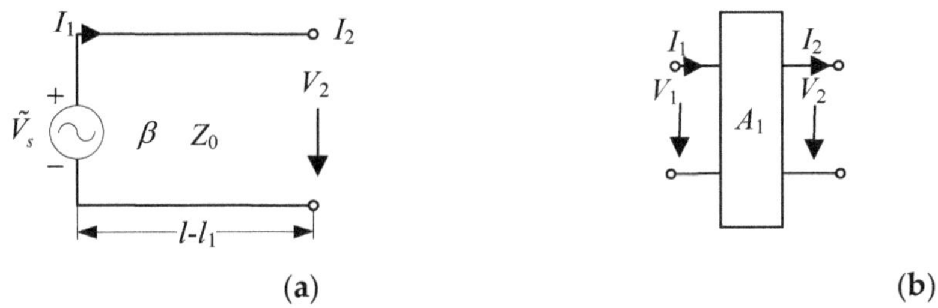

Figure 2. Two-wire equivalent transmission line model of Figure 1. (**a**) Equivalent transmission line; (**b**) cascaded two-port network.

For the lossless equivalent transmission line corresponding to collinear symmetric diploes antenna (Figure 1b), the average characteristic impedance is calculated as [17,18]

$$Z_0 = \frac{1}{l - l_1} \int_{l_1}^{l} \frac{120}{\sqrt{\varepsilon_r}} \ln \frac{2(z - l_1)}{\rho} dz. \tag{2}$$

The per-unit-length inductance and capacity can then be obtained

$$L = \sqrt{\mu_0 \varepsilon_0 \varepsilon_r} Z_0,\ C = \frac{\sqrt{\mu_0 \varepsilon_0 \varepsilon_r}}{Z_0}. \tag{3}$$

The propagation constant is $r_1 = j\beta = j\omega\sqrt{LC} = j\omega\sqrt{\mu_0 \varepsilon_0 \varepsilon_r}$.

According to the theory of the transmission line [20,21], the equivalent transmission line model in Figure 2a can be equivalent to the cascaded two-port network in Figure 2b. V_i and I_i denote the corresponding port voltage and current, respectively. A voltage source $V_1 = \widetilde{V}_s$ is connected at the left-hand-side port and the right-hand-side is an open circuit ($I_2 = 0$). The transmission matrix A_1 can be expressed as

$$A_1 = \begin{bmatrix} \cos[\beta(l - l_1)] & -jZ_0 \sin[\beta(l - l_1)] \\ \frac{\sin[\beta(l - l_1)]}{jZ_0} & \cos[\beta(l - l_1)] \end{bmatrix}. \tag{4}$$

According to the microwave network theory, we obtain

$$\begin{bmatrix} V_2 \\ I_2 \end{bmatrix} = A_1 \begin{bmatrix} V_1 \\ I_1 \end{bmatrix}. \tag{5}$$

Substituting the boundary condition $V_1 = \widetilde{V}_s$ and $I_2 = 0$ into Equation (5), we can obtain the coefficients of the voltages and currents

$$V_2 = \frac{\widetilde{V}_s}{\cos[\beta(l - l_1)]},\ I_1 = j\tan[\beta(l - l_1)]\frac{\widetilde{V}_s}{Z_0}. \tag{6}$$

Based on the above port voltages and currents, the currents distribution of the equivalent transmission lines (Figure 1) can be written as

$$I(z) = I_1 \cos[\beta(z - l_1)] + \frac{V_1}{jZ_0} \sin[\beta(z - l_1)]. \tag{7}$$

The simulated normalized current distribution of the proposed method and MoM are shown in Figure 3. It can be observed that the results comply well with each other when $l = 0.05\lambda$. However, they differ from each other when $l = 0.25\lambda$. In fact, there is deviation for MoM when $\varepsilon_r \neq 1.0$. It should be pointed out that this method can be extended to multi-section line antenna. In this case, the equivalent transmission line model and two-port network are cascaded by the corresponding multi sections. It would affiliate the analysis of the complicated line antenna.

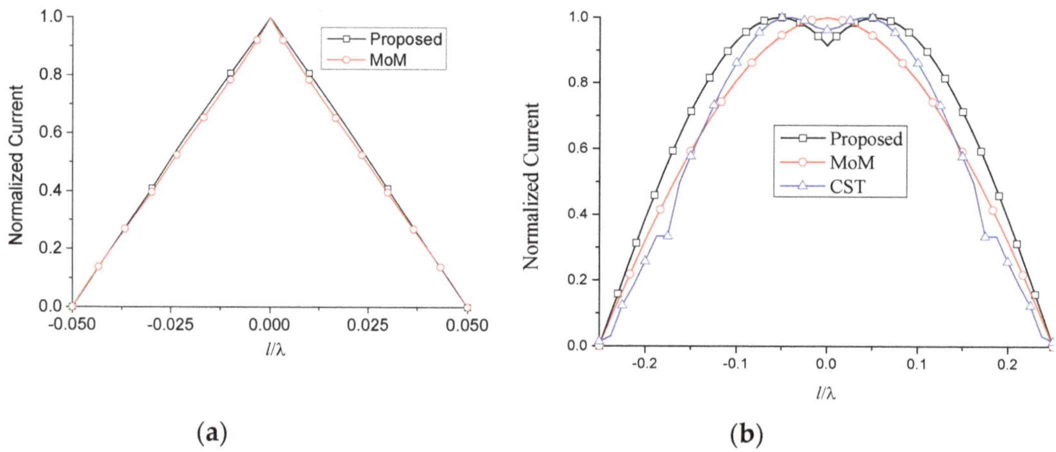

Figure 3. Normalized current distribution of the proposed method and MoM. (**a**) $l = 0.05\lambda$; (**b**) $l = 0.25\lambda$.

3. Two-Wire Transmission Line Model of Receiving Antenna

As discussed in the introduction, the necessary condition for a single straight antenna to be equivalent to a two-wire transmission line model is that the incident field of the antenna is symmetrical. Mode decomposition can be applied to asymmetric situations. In this section, the equivalent transmission model is applied to the analysis of receiving antenna, shown in Figure 4.

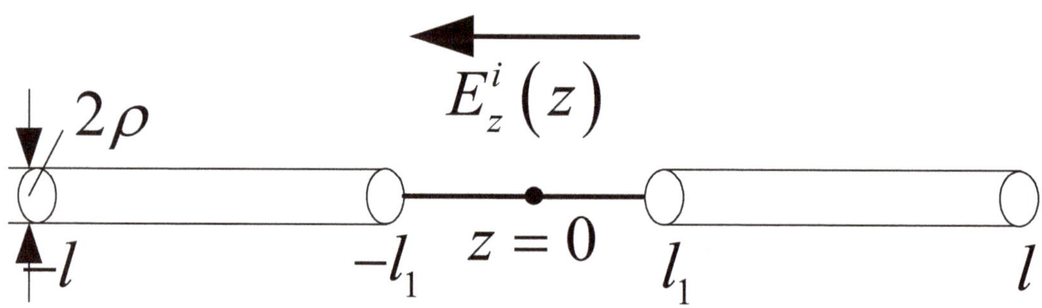

Figure 4. Two-wire straight antenna exposed to incident electric field.

The two-wire lossless transmission line model of Figure 4 is shown in Figure 5, where $V(z)$ is the voltage between the two lines. The voltage and current on the line are governed by the transmission line equations

$$\frac{\partial V(z)}{\partial z} = -jwL \cdot I(z) + 2E_z^i(z) \quad (8)$$

$$\frac{\partial I(z)}{\partial z} = -jwC \cdot V(z) \quad (9)$$

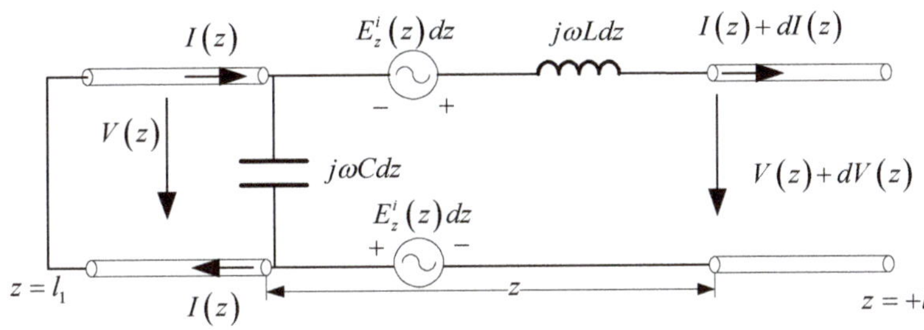

Figure 5. Two-wire transmission line model of Figure 4.

The corresponding transmission line parameters can be calculated following Section 2. The boundary condition is

$$I(-l) = I(+l) = 0. \tag{10}$$

Substituting (10) into (8) and (9) yields the general expression of potential and current

$$\begin{aligned} V(z) &= Ae^{-r_1 z} + Be^{r_1 z} + \varphi_v(z) \\ I(z) &= \tfrac{1}{Z_0}\left[Ae^{-r_1 z} - Be^{r_1 z} + \varphi_i(z)\right] \end{aligned} \tag{11}$$

$\varphi_v(z)$ and $\varphi_i(z)$ are the terms corresponding to $E_z^i(z)$. The coefficients A and B can be obtained according to the boundary condition of the transmission line. Suppose the incident field on the antenna is

$$E^i(z) = E_0 \cos\theta \cdot e^{-j\beta_0 \sin\theta z}. \tag{12}$$

where $\beta_0 = j\omega\sqrt{\mu_0\varepsilon_0}$, θ is the incident angle. Since the incident field is asymmetrically distributed, mode decomposition is required.

$$E_e^i(z) = \frac{E^i(z) + E^i(-z)}{2} = E_0 \cos\theta \cdot \cos(\beta_0 \sin\theta z) \tag{13}$$

$$E_o^i(z) = \frac{E^i(z) - E^i(-z)}{2} = -jE_0 \cos\theta \cdot \sin(\beta_0 \sin\theta \cdot z) \tag{14}$$

The total current on the antenna is then obtained by the superposition of the above two modes. The corresponding simulated results are shown in Figure 6.

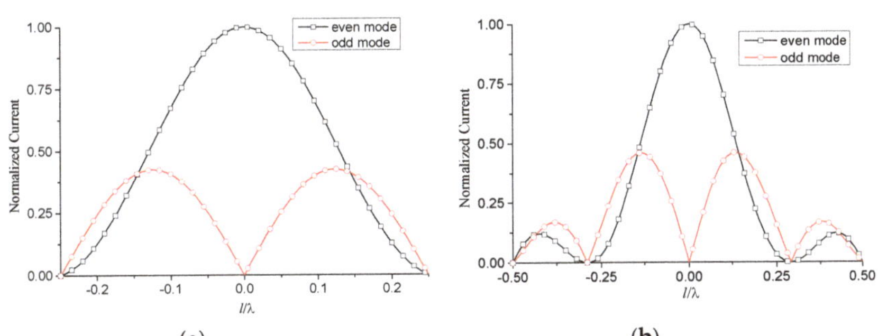

Figure 6. Normalized current distribution of two-wire transmission line model. $l_1 = 0$, $\rho = 1 \times 10^{-7}$ m, $f = 10$ GHz, $\varepsilon_r = 1.6$, $\theta = 0^{\circ}$. (**a**) $l = 0.25\lambda$; (**b**) $l = 0.5\lambda$.

4. One-Wire Transmission Line Model of Receiving Antenna

In this section, the one-wire transmission line model for a single straight antenna is analyzed. In this case, the mode decomposition of the incident field is eliminated. A single straight antenna is a continuous straight-line antenna, which could be one section or multiple non-parallel sections. It can be equivalent to the one-wire equivalent transmission line model. Each section corresponds to a potential and a current equation. A single straight antenna with one section is illustrated in Figure 7, and $E_z^i(z)$ is the tangential component in the incident electric field. The corresponding one-wire equivalent transmission line model is shown in Figure 8. For the transmission line excited by the distribution voltage source, the electric potential and current on the line are governed by the transmission line equations.

$$\frac{\partial V(z)}{\partial z} = -jwL \cdot I(z) + E_z^i(z) \tag{15}$$

$$\frac{\partial I(z)}{\partial z} = -jwC \cdot V(z) \tag{16}$$

$V(z)$ and $I(z)$ are potential and current, respectively. The boundary condition is

$$I(-l) = I(+l) = 0 \tag{17}$$

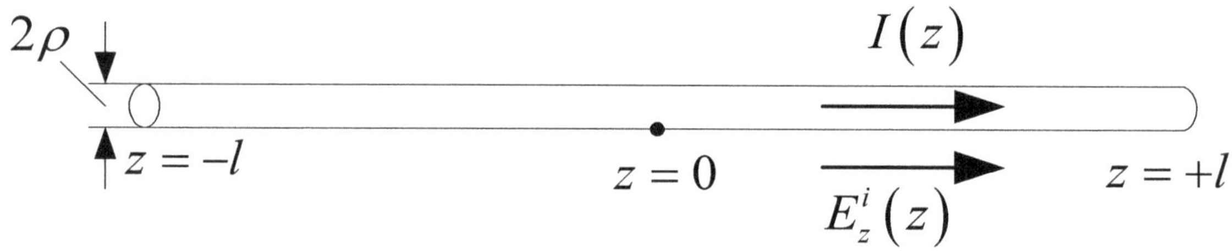

Figure 7. Single straight antenna exposed to incident electric field.

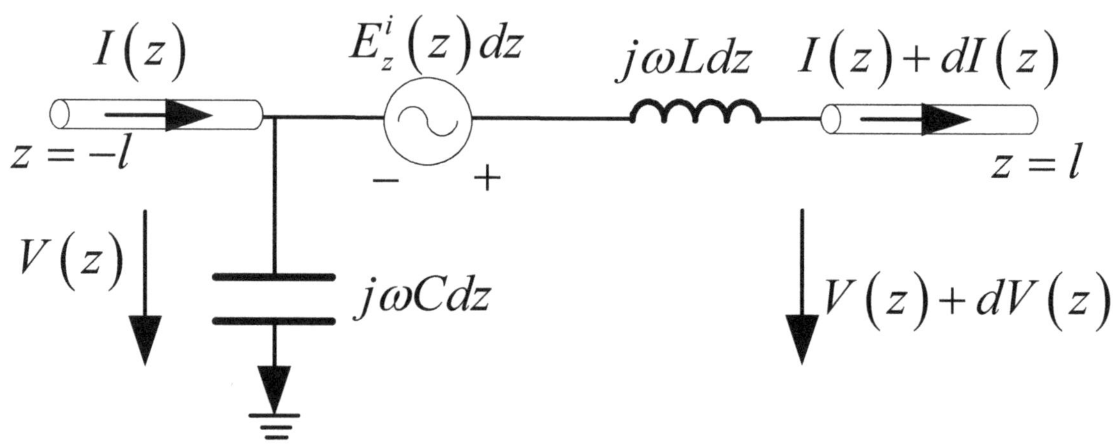

Figure 8. One-wire equivalent transmission line model of Figure 7.

Compared with the two-wire equivalent transmission line model, the one-wire model is a more general method. It can be used for any antenna composed of relatively narrow strips or slots. This will benefit the analysis of FSS, where the element currents are usually unknown. The normalized current distribution of Figure 7 is shown in Figure 9, together with the results of the two-wire model.

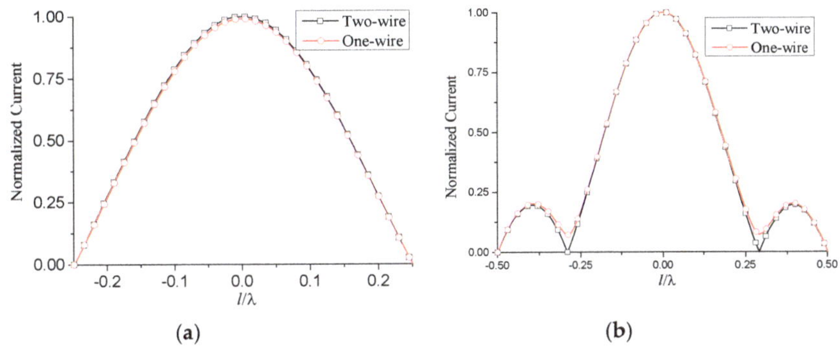

Figure 9. Normalized current distribution of one-wire transmission line model. $l_1 = 0$, $\rho = 1 \times 10^{-7}$ m, $f = 10$ GHz, $\varepsilon_r = 1.6$, $\theta = 0^{\circ}$. (**a**) $l = 0.25\lambda$; (**b**) $l = 0.5\lambda$.

5. Loss of Equivalent Transmission Line

The equivalent transmission line models discussed above are lossless. In this section, the antenna lossy is analyzed. In this case, the current in the receiving antenna needs to be obtained first and the per unit length radiation resistance can then be calculated. However, the current is practically an unknown quantity. One way to obtain it is to approximate the current in the form of sine function. However, it is restricted to some special cases (e.g., electrically small antenna and vertical incident). An alternative way is to use the lossless receiving current $I_0(z)$ as the approximation, which has been discussed in Sections 3 and 5. $I_0(z)$ produces the scattered fields, namely the re-radiated field. The radiation energy is the radiation lossy of the antenna. Consequently, the radiation resistance, potential and current of the equivalent transmission line model can then be solved. Through the iterative algorithm of the above process, a more accurate current distribution of the receiving antenna can be obtained.

Suppose r is parallel to $r_1, \theta = \theta_1$. When calculating the magnitude of the field, assume $r_1 \approx r$. When calculating the phase of a field, assume $r_1 \approx r - z\cos\theta$. The far-field radiation produced by the current element is

$$dE_\theta \approx j\frac{30\beta}{\sqrt{\varepsilon_r}}\frac{I(z)dz}{r}\sin\theta e^{-j\beta(r-z\cos\theta)} \tag{18}$$

The far-field radiation electric field of the antenna is the integral of (18) on the whole antenna

$$E_\theta = \int_{-l}^{l} dE_\theta dz = j\frac{30\beta}{\sqrt{\varepsilon_r}}\frac{e^{-j\beta r}}{r}\sin\theta\int_{-l}^{l} I(z)e^{j\beta z\cos\theta}dz. \tag{19}$$

According to the approximation property of the plane wave in the far field, the active power of the antenna is

$$P_r = \frac{1}{2}\oint_S\left[\vec{E}\times\vec{H}^*\right]\cdot d\vec{s} = \frac{1}{2}\int_0^{2\pi}\int_0^{\pi}\frac{1}{Z_w}|E_\theta|^2 r^2\sin\theta d\theta d\varphi \tag{20}$$

where $Z_w = 120\pi/\sqrt{\varepsilon_r}$. Substituting (19) into (20) yields

$$P_r = \frac{5\pi\beta^2}{\sqrt{\varepsilon_r}}\int_0^{\pi}\left|\int_{-l}^{l} I(z)e^{j\beta z\cos\theta}dz\right|^2\sin^3\theta d\theta \tag{21}$$

The total lossy on the transmission line (Figure 10, the average actual power) is

$$P_r{}' = \frac{1}{2}\int_{-l}^{l}|I(z)|^2 R_1 dz \tag{22}$$

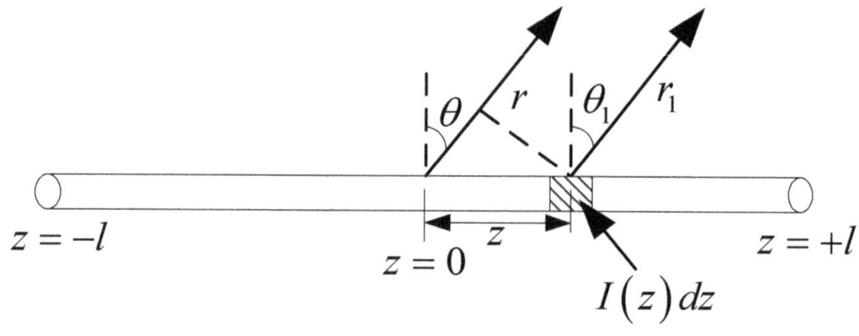

Figure 10. The radiation field of the receiving antenna current.

Suppose that all radiation lossy of an antenna are equivalent to the total power dissipated on the per unit length radiation resistance R_1 of the transmission line, we can solve R_1

$$R_1 = \frac{\frac{15\beta_r^2}{\sqrt{\varepsilon_r}} \int_0^{\pi} \left| \int_{-l}^{l} I(z) e^{j\beta z \cos\theta} dz \right|^2 \sin^3\theta d\theta}{\int_{-l}^{l} |I(z)| dz} \tag{23}$$

Consequently, the equivalent transmission line model of Figures 2, 5 and 8 should be revised. Taking Figure 7 as an example, the lossy is shown in Figure 11. The corresponding transmission line equations and parameters should be revised as well

$$\frac{\partial V(z)}{\partial z} = -(jwL + R_1) \cdot I(z) + E_z^i(z) \tag{24}$$

$$\frac{\partial I(z)}{\partial z} = jwC \cdot V(z) \tag{25}$$

$$Z_c = \sqrt{(jwL + R_1)/jwC} \tag{26}$$

$$r = j\beta_r = \sqrt{(jwL + R_1) j\omega C} = j\omega\sqrt{\mu_0 \varepsilon_0 \varepsilon_r} \tag{27}$$

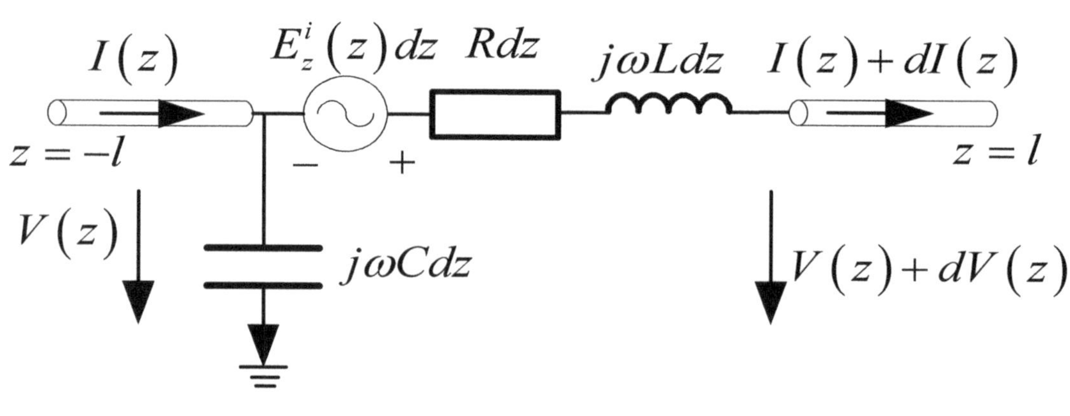

Figure 11. One-wire equivalent lossy transmission line model of Figure 7.

The simulation results show that the lossy has little influence on the normalized current distribution. The per unit length radiation resistance R_1 is shown in Figure 12.

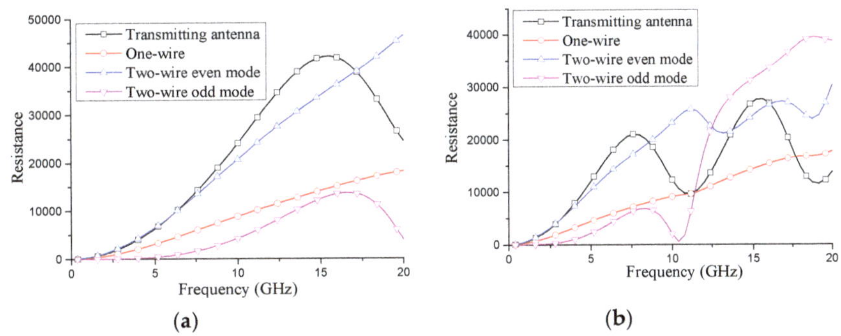

Figure 12. Per unit length radiation resistance. (**a**) $l = 0.25\lambda$; (**b**) $l = 0.5\lambda$.

6. Conclusions

The equivalent transmission line model is established for the current analysis of line antenna. The two-wire model is discussed for transmitting antenna and receiving antenna (the incident field should be decomposed for asymmetric case). The one-wire model is introduced for any receiving antenna composed of narrow strips. The corresponding mode decomposition is eliminated. The transmission line radiation resistance hardly effects the normalized current distribution.

Author Contributions: Conceptualization, S.W. and W.W.; methodology, S.W.; software, S.W.; validation, W.W.; formal analysis, W.W.; investigation, W.W.; resources, S.W.; data curation, S.W.; writing—original draft preparation, S.W.; writing—review and editing, S.W.; visualization, S.W.; supervision, W.W.; project administration, W.W.; funding acquisition, W.W. All authors have read and agreed to the published version of the manuscript..

Funding: This work was supported by the key project of Science and Technology Innovation Program of Army Engineering University of PLA under Grant KYCQJQZL2119.

Conflicts of Interest: The authors declare no conflict of interest.

References

1. Orta, R.; Tascone, R.; Zich, R. A unified formulation for the analysis of general frequency selective surfaces. *Electromagnetics* **2007**, *5*, 307–329. [CrossRef]
2. Pelletti, C.; Bianconi, G.; Mittra, R.; Monorchio, A. Analysis of finite conformal frequency selective surfaces via the characteristic basis function method and spectral rotation approaches. *IEEE Antennas Wirel. Propag. Lett.* **2013**, *12*, 1404–1407. [CrossRef]
3. Florencio, R.; Boix, R.R.; Encinar, J.A.; Toso, G. Optimized periodic MoM for the analysis and design of dual polarization multilayered reflectarray antennas made of dipoles. *IEEE Trans. Antennas Propag.* **2017**, *65*, 3623–3637. [CrossRef]
4. Bozzi, M.; Perregrini, L. Efficient analysis of thin conductive screens perforated periodically with arbitrarily shaped apertures. *Electron. Lett.* **1999**, *35*, 1085–1087. [CrossRef]
5. Dong, J.; Ma, Y.; Li, Z.; Mo, J. A Miniaturized Quad-Stopband Frequency Selective Surface with Convoluted and Interdigitated Stripe Based on Equivalent Circuit Model Analysis. *Micromachines* **2021**, *12*, 1027. [CrossRef] [PubMed]
6. Lv, X.; Withayachumnankul, W.; Fumeaux, C. Single-FSS-Layer Absorber with Improved Bandwidth-Thickness Trade-Off Adopting Impedance Matching Superstrate. *IEEE Antennas Wirel. Propag. Lett.* **2019**, *18*, 916–920. [CrossRef]
7. Luo, K.; Meng, J.; Zhu, D.; Ge, S.; Han, J. Approximate analytical method for hexagonal slot frequency selective surface analysis. *Int. J. RF Microw. Comput. Eng.* **2021**, e22776. [CrossRef]
8. Munk, B.A.; Burrell, G.A. Plane wave expansion for arrays of arbitrarily oriented piecewise linear elements and its application in determining the impedance of a single linear antenna in a lossy half-space. *IEEE Trans. Antennas Propag.* **1979**, *34*, 331–343. [CrossRef]
9. Munk, B.A. *Frequency Selective Surfaces: Theory and Design*; Wiley: New York, NY, USA, 2000.
10. Chen, K.; Du, P.A.; Nie, B.L.; Ren, D. An improved MOM approach to determine the shielding properties of a rectangular enclosure with a doubly periodic array of apertures. *IEEE Trans. Electromagn. Compat.* **2016**, *58*, 1456–1464. [CrossRef]
11. Medina, F.; Mesa, F.; Marqués, R. Extraordinary transmission through arrays of electrically small holes from a circuit theory perspective. *IEEE Trans. Microw. Theory Tech.* **2008**, *56*, 3108–3120. [CrossRef]
12. Rodriguez-Berral, R.; Mesa, F.; Medina, F. Analytical multimodal network approach for 2-D arrays of planar patches/apertures embedded in a layered medium. *IEEE Trans. Antennas Propag.* **2015**, *63*, 1969–1983. [CrossRef]

13. Molero, C.; Rodríguez-Berral, R.; Mesa, F.; Medina, F.; Yakovlev, A.B. Wideband analytical equivalent circuit for one-dimensional periodic stacked arrays. *Phys. Rev. E* **2016**, *93*, 013306. [CrossRef] [PubMed]
14. Mesa, F.; García-Vigueras, M.; Medina, F.; Rodríguez-Berral, R.; Mosig, J.R. Circuit-model analysis of frequency selective surfaces with scatterers of arbitrary geometry. *IEEE Antennas Wirel. Propag. Lett.* **2015**, *14*, 135–138. [CrossRef]
15. Carter, P.S. Circuit relations in radiating systems and applications to antenna problems. *Proc. Inst. Radio Eng.* **1932**, *20*, 1004–1041. [CrossRef]
16. Brown, G.H.; Woodward, O.M. Experimentally determined impedance characteristics of cylindrical antennas. *Proc. IRE* **1945**, *33*, 257–262. [CrossRef]
17. Schelkunoff, S.A.; Friis, H.T. *Antennas Theory and Practice*; Wiley: New York, NY, USA, 1952.
18. Li, B.Y. *Antenna Principle and Application*; Lanzhou University Press: Lanzhou, China, 1993.
19. Luo, K.; Yi, Y.; Zong, Z.Y.; Chen, B.; Zhou, X.L.; Duan, Y.T. Approximate Analysis Method for Frequency Selective Surface Based on Kirchhoff Type Circuit. *IEEE Trans. Antennas Propag.* **2018**, *66*, 6076–6085. [CrossRef]
20. Liao, C.E. *Foundations for Microwave Technology*; Xi'an University Press: Xi'an, China, 1979.
21. Pozar, D.M. *Microwave Enginnering*, 3rd ed.; Wiley: Hoboken, NJ, USA, 2005.

micromachines

Article

Broadband Frequency Selective Rasorber Based on Spoof Surface Plasmon Polaritons

Jin Bai [1], Qingzhen Yang [1], Yichao Liang [2,*] and Xiang Gao [1]

1 School of Power and Energy, Northwestern Polytechnical University, Xi'an 710129, China
2 School of Aerospace Engineering, Xiamen University, Xiamen 361005, China
* Correspondence: echo.chao@outlook.com

Abstract: A broadband frequency selective rasorber (FSR) based on spoofsurface plasmon polaritons (SSPP) is proposed. The FSR is composed of a multi-layer structure comprising frequency selective surface (FSS)-polyresin (PR)-indium tin oxide (ITO)-PR-FSS and placed vertically on a metal base plate. A periodic square cavity structure is formed. The transmission characteristics of the FSR are studied by full-wave simulation and equivalent circuit method. The simulation results demonstrate that under normal incidence, the absorption rate of the structure remains 95% in the 5–30 GHz band, and the absorption rate is also 80% in the 3.5–5 GHz band. As the incident angle of the electromagnetic wave increases to 40°, the absorption rate in the 15–20 GHz band decreases to 70% in the transverse electric (TE) mode, and the absorption rate in the transverse magnetic (TM) mode is almost the same as that of vertical incidence. The transmission response of the structure is measured in an anechoic chamber. The measurement results agree well with the simulation results, proving the reliability of the design and fabrication. The structure is less sensitive to the incident angle of magnetic waves and has a better broadband absorbing ability.

Keywords: frequency selective rasorber (FSR); spoof surface plasmon polaritons (SSPP); frequency selective surface (FSS); broadband

Citation: Bai, J.; Yang, Q.; Liang, Y.; Gao, X. Broadband Frequency Selective Rasorber Based on Spoof Surface Plasmon Polaritons. *Micromachines* **2022**, *13*, 1969. https://doi.org/10.3390/mi13111969

Academic Editor: Wensheng Zhao

Received: 25 September 2022
Accepted: 8 November 2022
Published: 13 November 2022

Publisher's Note: MDPI stays neutral with regard to jurisdictional claims in published maps and institutional affiliations.

Copyright: © 2022 by the authors. Licensee MDPI, Basel, Switzerland. This article is an open access article distributed under the terms and conditions of the Creative Commons Attribution (CC BY) license (https://creativecommons.org/licenses/by/4.0/).

1. Introduction

With the rapid development of communication systems, stealth and anti-stealth technology have increasingly become decisive factors in military electronic information warfare. It is more and more essential to improve aircraft stealth. For single-station radar detection, the detection wave can be reflected to other angular domains through the shape structure design to achieve radar stealth. However, the expanding receiving angular territory for multi-station radar detection is far from enough to reflect the radar detection wave. Frequency Selective Rasorber (FSR) [1–3] directly absorbs the electromagnetic waves in the working frequency band, and the reflectivity is minimal. Therefore, FSR with a broadband wave absorption ability demonstrates an excellent application prospect.

FSR is a periodic array composed of metal patches or aperture units, which exhibit band-pass or band-stop spectral filtering characteristics for electromagnetic waves of different frequencies. It is widely used in radome [4], electromagnetic compatibility and electromagnetic shielding [5], satellite communication [6,7], and other fields. The traditional metal-dielectric Frequency Selective Surface (FSS) [8,9] is based on the scattering properties of metal resonator elements. When the electromagnetic wave is incidental on the surface of the FSS, an induced current is generated, thus generating a scattered field. The scattered and incident fields are superimposed to form an entire field with spatial filtering characteristics. This kind of FSS is generally a "sandwich" structure, which consists of a metal pattern on the top layer, a dielectric substrate in the middle, and a metal plate on the bottom layer, which is easy to process and low-cost. However, the disadvantage is that the absorption range is narrow, and the absorption rate is sensitive to the polarization mode and incidence angle of electromagnetic waves. Many scholars have researched and proposed various schemes to widen the absorbing bandwidth of FSS and reduce polarization sensitivity, such as designing a multi-mode resonant unit structure or fractal structure [10–12], using a multi-layer structure to realize the superposition of the absorbing frequency band [13] and loading lumped elements to realize bandwidth expansion [14]. They are adding active controllable devices such as PIN diode or varactor diode in the FSS [15,16] and using Spoof surface plasmon polaritons (SSPP) [17–19] to confine the incident wave to the interface and then dissipate it. In [20], the dispersive properties and subwavelength field confinement properties of artificial surfaces in low-frequency bands such as microwave and terahertz waves, which are similar to electrical surface plasmons in the optical frequency band, were studied for the first time. Since then, scholars have widely used SSPP in waveguide design [21], filters [22], notch filters [23], leaky-wave antennas [24,25], and other directions. At the same time, the designed FSR based on SSPP exhibits high rejection performance in the design stopband and achieves steep cutoff and high permeability in the working passband [26–28].

SSPP is based on Surface Plasmonic Polaritons (SPPs) by etching periodic structures on the metal surface or arranging periodic metal patches on the dielectric layer. When the electromagnetic wave is incident on the interface between the metal and the medium, the free electrons in the metal conductor oscillate collectively, generating SPPs in the microwave frequency band. The electromagnetic field strength peaks at the interface between the metal and the medium. The energy propagates along the structure's surface and is completely bound near the structure's surface, thereby achieving the wave-absorbing effect. Compared with SPPs, SSPP has more vital electromagnetic wave confinement ability, and the field strength decays exponentially in the vertical direction, reducing the interference to adjacent structures and being conducive to the miniaturized design of the FSR. In contrast, the unit structure of loading lumped components makes the FSR challenging to process and has poor practicability; if active devices are loaded, an external circuit needs to be introduced, which is slightly inconvenient. This paper uses a multi-layer stacking method to couple the SSPP to design the FSR.

2. Presentation of the Proposed Structure

The FSR designed in this paper is shown in Figure 1. It consists of a periodic square cavity array composed of a metal base plate and a multi-layer FSS cross-arranged. The depth of the square cavity unit is represented as a, and the length, width, and depth dimensions are consistent. The multi-layer FSS is shown in Figure 2a. The structure is symmetrically distributed concerning the indium tin oxide (ITO) layer. The dielectric is placed on the sub-top layer, and the sub-bottom layer is composed of polyresin (PR). The thickness of the dielectric layer is represented as h, and the dielectric constant is $\varepsilon = 4.3, \tan\delta = 0.025$. The ITO layer is placed in the middle, where the square resistance of ITO is represented as R, its unit is $\Omega/\square$, which characterizes the resistance value of a conductive material per square centimeter. Moreover, the thickness of ITO in the calculation model is 0. In reality, ITO is often formed by adhering to the surface of polyethylene glycol terephthalate (PET). Therefore, the ITO layer of the FSR is flanked by a PET layer with a thickness of d, and the dielectric constant of PET is $\varepsilon = 3.0, \tan\delta = 0.18$. The bottom and top layers are right-angle metal patch arrays. As shown in Figure 2b, the metal line width of the patch array is represented as w, the thickness is represented as c, and the distance between adjacent metal lines is also represented as w. These parameters are as follows: $R = 200\ \Omega/\square$, $a = 10$ mm, $b = 8$ mm, $c = 35$ μm, $d = 10$ μm, $w = 125$ μm, and $h = 1$ mm.

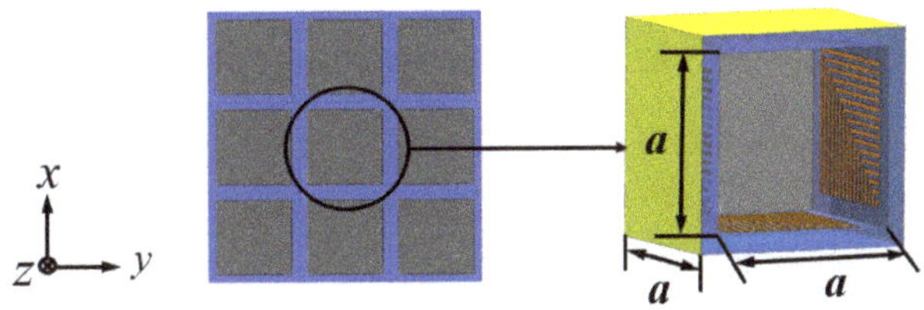

Figure 1. The structure of the FSR.

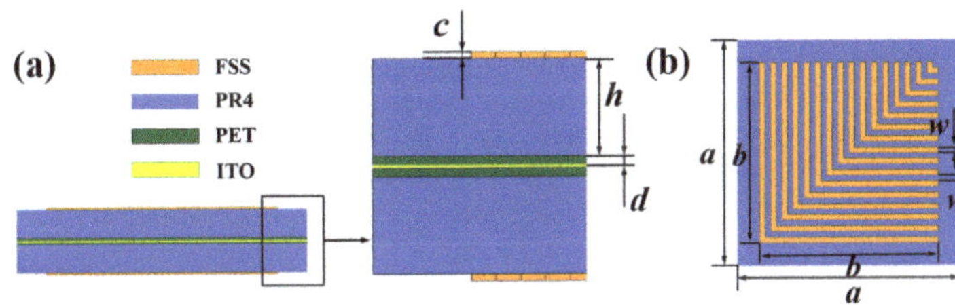

Figure 2. (**a**) Multi-layer FSS Sectional View. (**b**) The FSS patch array.

3. Design Principles

Compared with linear dipoles, the advantages of using L-shaped metal patches in this design are mainly reflected in two aspects. First, the same resonance characteristics can be obtained in both transverse electric (TE) and transverse magnetic (TM) modes. Second, one side of the L-shaped metal patch is equivalent to an inductor, and the other is equivalent to a capacitor between the adjacent patch. At the same time, the L-shaped metal patches array can achieve the most significant number with the most negligible length gradient so that the center frequency distribution of the FSS is tighter, which is more conducive to the realization of the effect of broadband wave absorption. The structure of the FSR designed in this paper is complex and cannot be directly analyzed as a whole utilizing an equivalent circuit. However, the FSR can be regarded as a multi-layer FSS with a cross-distribution in the *xoz* and *yoz* planes. According to the different incident angles of electromagnetic waves, there are two main mechanisms for the absorption principle of the FSR.

When the electromagnetic wave is vertically incident on the bottom surface of the absorber along the *z* direction, the wave vector direction is parallel to the wall surface of the square cavity. At this time, the wall surface of the FSR is equivalent to the SSPP transmission line. According to the SSPP theory, the electromagnetic wave in TE mode cannot induce polarization charges on the interface due to the absence of an electric field component perpendicular to the interface. Therefore, SSPP transmission lines only work in TM mode for specific surfaces. The FSR designed in this paper is a cross-distributed multi-layer FSS structure. For any set of mutually perpendicular multi-layer FSS, when the multi-layer FSS in the *xoz* plane is in TM mode, the multi-layer FSS in the *yoz* plane is in the TE mode, and vice versa. Therefore, relative to the whole FSR, whether it is TE mode or TM mode, the FSR can support SSPP. For a one-dimensional periodic groove array, when the groove width is much smaller than the incident wavelength ($w \ll \lambda$), the dispersion relation can be expressed as Equation (1) [19]. Where k_z is the wave number along the *z* direction of the conductor surface, the period of the groove array is represented as D, the width of the groove interval is represented as A, the depth of the groove is represented as H, and k_0 is the wave vector in the free space. Equation (2) is the electromagnetic wave's attenuation constant along the wave vector's propagation direction, which characterizes the attenuation ability of the transmission line to the electromagnetic field. From Equations (1) and (2), it can be concluded that in order to increase the attenuation capability of SSPP to electromagnetic waves, k_z should be as large as possible so that the cut-off angular frequency (w_p) of SSPP can be obtained as shown in Equation (3). The FSR designed in this paper is no longer a simple one-dimensional groove array. Due to the reflection effect of the metal base plate and the non-negligible coupling effect between adjacent multi-layer FSS, its internal working mechanism will be more complicated.

$$k_z{}^2 = k_0{}^2(1 + (\frac{A}{D})^2\tan^2(k_0H)) \tag{1}$$

$$\alpha_T = \sqrt{k_x{}^2 - k_0{}^2} = k_0\frac{A}{D}\tan(k_0H) \tag{2}$$

$$\omega_p = \frac{\pi c_0}{2H} \tag{3}$$

When the electromagnetic wave is vertically incident on the FSS wall of the FSR along the direction x or y, the wave vector direction is perpendicular to the wall of the square cavity. At this time, the working mechanism of the filter is similar to that of the traditional multi-layer FSS. The equivalent circuit of the formed FSS is shown in Figure 3. The upper and lower surfaces of the multi-layer FSS are air, and its impedance is $Z_0 = 377\ \Omega$. The PR layer with a thickness of h can be equivalent to a transmission line with a length of h, and its impedance is represented by Z_1, $Z_1 = Z_0/\sqrt{\varepsilon_r}$. The PET layer of thickness d can be equivalent to a transmission line of length d. Due to its small thickness, the impedance on the corresponding transmission line can be ignored. The ITO of the center layer is equivalent to a parallel resistance, which is represented by R. The FSS of the upper and lower layers is equivalent to two LC series branches, and the inductances of the branches interact with each other. Suppose the mutual inductance value and the impedance of the dielectric layer are ignored. In that case, the equivalent impedance of the multi-layer FSS can be obtained as in Equation (4). Where L is the equivalent inductance, C is the equivalent capacitance. The square resistance of the ITO layer is R, $R = 200\ \Omega/\square$. The inductance corresponding to each metal patch in the FSS array is shown in Equation (5). Where b is half of the total length of the metal wire, w is the width of the metal wire, and μ_0 is the vacuum permeability. The capacitance of the metal patch is shown in Equation (6), ε_0 is the vacuum conductivity, the effective dielectric constant of the dielectric layer is shown in Equation (7) with ε_{eff}, and ε_r is the relative dielectric constant of the medium. When the interaction between adjacent metal patches is not considered, it is not difficult to find the transmission pole corresponding to a single metal patch through $Z = \infty$. By analogy, all the poles of the FSS can be found. The distance between adjacent metal patches is minimal, and their interaction cannot be ignored. At the same time, the FSR is a square cavity structure, and the interaction between adjacent walls and the reflection effect of the metal base plate cannot be ignored.

$$Z = \frac{R\cdot(j\omega L_2 + \frac{1}{j\omega C_2})\cdot(j\omega L_2 + \frac{1}{j\omega C_2})}{R\cdot(j\omega L_1 + \frac{1}{j\omega C_1}) + R\cdot(j\omega L_2 + \frac{1}{j\omega C_2}) + (j\omega L_1 + \frac{1}{j\omega C_1})\cdot(j\omega L_2 + \frac{1}{j\omega C_2})} \tag{4}$$

$$L = -\mu_0\frac{b}{\pi}\ln\left[\sin(\frac{\pi w}{4b})\right] \tag{5}$$

$$C = -\varepsilon_0\varepsilon_{eff}\frac{4b}{\pi}\ln\left[\sin(\frac{\pi w}{4b})\right] \tag{6}$$

$$\varepsilon_{eff} = \frac{1}{\sqrt{(\varepsilon_r + 1)/2}} \tag{7}$$

In fact, due to the complex structure of the FSR, the SSPP action mechanism and the traditional FSS action mechanism exist at any incident angle of electromagnetic waves, and the transmission characteristics of the absorber appear more complicated. In order to study the transmission characteristics of the absorber more accurately, we use the CST studio software to analyze the transmission characteristics of the whole waveband and finally make the FSR and test its transmission characteristics in the microwave anechoic chamber.

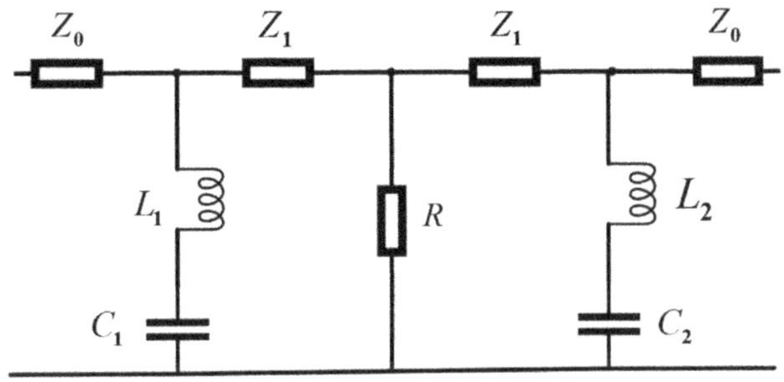

Figure 3. Equivalent circuit diagram of multi-layer FSS.

4. Analysis and Discussion

The working frequency band of the designed FSR should cover the S, C, X, Ka, and K bands as much as possible. Therefore, the cavity size of the FSR is optimized first, and then the ITO layer directly affects the circuit structure of the multilayer FSS. Therefore, the influence of the ITO square resistance on the transmission characteristics of the absorber is studied. The PR is a lossy medium, and the thickness of the dielectric layer also has a specific influence on the performance of the FSR. Finally, the optimization analysis of the width of the FSS metal patch is carried out. When the electromagnetic wave is vertically incident on the absorber along the z direction, the FSR is symmetrical about the xoz plane and the yoz plane. The transmission characteristics in the TE mode and TM mode are consistent, so only the transmission characteristics of the absorber in TE mode are studied. Figure 4 shows the absorptivity distribution of FSR with different sizes in the 0–30 GHz band when electromagnetic waves are incident vertically. It can be observed that when the incident frequency is less than 9 GHz, at the same frequency, with the decrease of a, the absorption rate of the FSR gradually decreases. When the incident frequency is greater than 9 GHz, the fluctuation characteristics of the absorption rate are enhanced. Except for the $a = 10$ mm model, the absorption rates of other models have unstable fluctuations in different bands. When a is greater than 10 mm, the absorption rate gradually increases with the decrease of a. When the incident frequency is more than 18 GHz, the absorption rate remains unchanged with the decrease of a, and remains above 95%. In a comprehensive comparison, although the absorption rate of the $a = 10$ mm model is lower than that of the $a = 15$ mm and $a = 20$ mm models when the incident frequency is less than 5GHz, the reduction is not too noticeable. However, the model with $a = 10$ mm maintains the absorption rate above 95% in the 5–30 GHz band, which other models unmatch.

Figure 5 shows the absorption rate distribution of the model with the ITO layer and the model without the ITO layer in the 0–30 GHz band when the electromagnetic wave is vertically incident. It can be observed that when the incident frequency is less than 12 GHz, the FSR absorption rate is greatly improved due to the addition of the ITO layer. When the incident frequency is greater than 12 GHz, the ITO layer also dramatically reduces the fluctuation of the absorption rate of the FSR. In order to more accurately study the impact of the ITO layer on FSR's inhalation transmission characteristics, this paper studies the distribution of FSR electric fields under the resonance frequency. Figure 6 shows the FSR electric field distribution when the incident frequency is 9 GHz. Due to the addition of the ITO layer, the strength of the electric field motivated on the FSR surface is significantly enhanced; this is mainly because the conductive characteristics of the ITO layer promote the coupling of the FSS on both sides of the square wall surface. To a certain extent, the electric field can penetrate the PR medium, thereby increasing FSS resonance intensity. This paper studies the effect of different square resistances of the ITO layer on the FSR performance. It can be observed from Figure 7 that, except for the $R = 100\ \Omega/\square$ model, when the incident frequency is less than 5 GHz, the absorptivity gradually decreases with the increasing R. When the incident frequency is more than 5 GHz, except when the absorption rate of the $R = 100\ \Omega/\square$ and $R = 800\ \Omega/\square$ models decreases obviously, the changes in the absorption rates of other models are small.

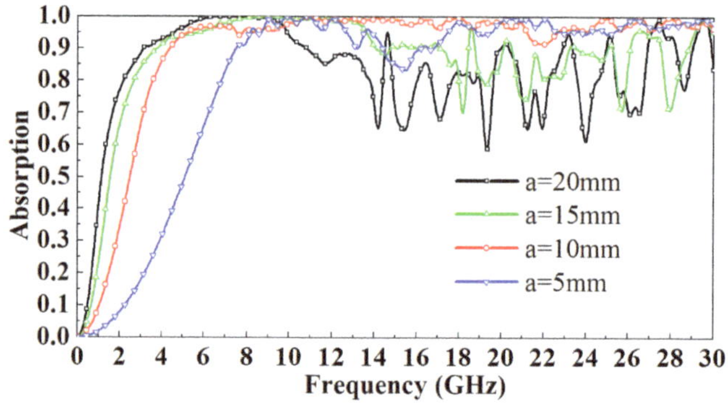

Figure 4. Effect of square cavity size on FSR absorption rate under normal incidence.

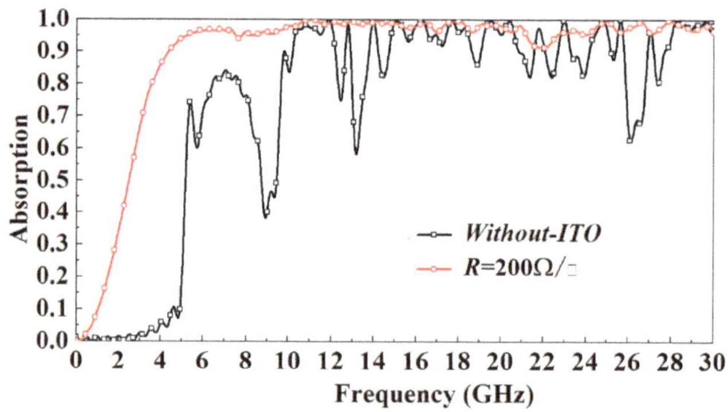

Figure 5. Effect of ITO layer on FSR absorption rate under normal incidence.

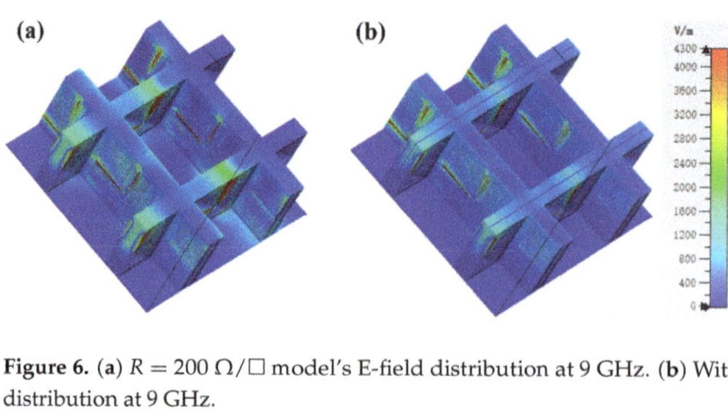

Figure 6. (**a**) $R = 200\ \Omega/\square$ model's E-field distribution at 9 GHz. (**b**) Without-ITO model's E-field distribution at 9 GHz.

Figure 8 shows the distribution of the influence of the thickness of the PR medium on the absorptivity. It can be observed that when the incident frequency is in the 3–7 GHz band, as h increases gradually, the absorptivity increases. When the incident frequency is in the 7–22 GHz band, the absorptivity is basically independent of h. When the incident frequency is in the 22–29 GHz band, the absorptivity of the $h = 0.8$ mm model and $h = 0.6$ mm model fluctuates wildly, and the absorption rate of the $h = 0.6$ mm model drops to a minimum of 75%. Overall, the absorption rate increased gradually with the increase of h to more than 95%. However, when the incident frequency is more than 29 GHz, the absorptivity of the $h = 1.2$ mm model drops to 80%, while the absorptivity of the $h = 1$ mm model remains above 95%. After careful consideration, the PR medium thickness of the designed FSR is set to 1mm. This paper also studies the influence of the width of the FSS metal patch on the absorption rate. It can be observed from Figure 9 that the influence of the width of the metal patch within a specific range on the absorption rate is minimal. Except for the $w = 500$ μm mode and $w = 250$ μm model, the absorption rate is slightly lower than other models in some frequency bands, the absorption rate change is small, and both are above 90%. Considering the miniaturization design and processing difficulty of FSR, the $w = 125$ μm model is finally selected.

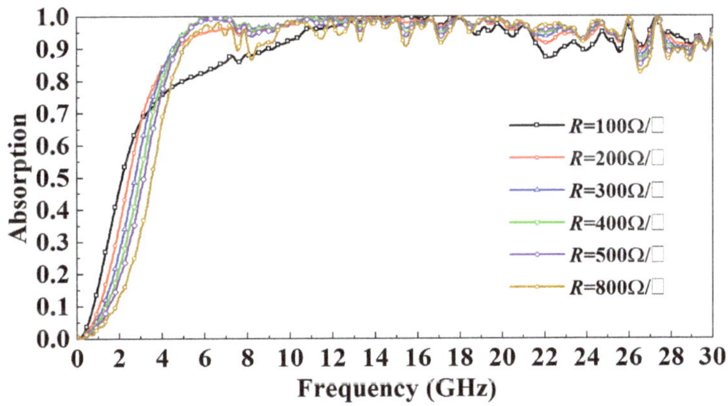

Figure 7. Effect of square resistance of the ITO layer on FSR absorption rate under normal incidence.

Finally, the absorptivity of the FSR under different incident angles is compared and analyzed. In TE mode, as the incidence angle increases, the magnetic field direction always has an angle with the metal line on the FSR wall, and the component of the incident wave parallel to the metal line decreases with the increase in the incidence angle. The absorbing principle of the FSR designed in this paper is to use the SSPP to bind the spatial electromagnetic wave to the wall of the FSR, convert it into a plane wave, and finally dissipate it. As the incidence angle increases, the electromagnetic wave component bound to the wall of the FSR will become smaller, so the absorption rate of the FSR will decrease. It can be observed from Figure 10 that in the TE mode, with the increase in the incident angle of the electromagnetic wave, the most sensitive band to the incident angle is mainly in the 15–20 GHz band. When the incident angle increases to 20°, this band's lowest point of absorptivity drop to 85%. When the incident angle increases to 40°, this band's lowest point of absorptivity drops to 70%. The absorption rate in other bands is highly insensitive to the incident angle, and the absorption rate is above 90%. In TM mode, the direction of the magnetic field is parallel to part of the metal wire. As the incidence angle increases, the metal wire in the absorption cavity will be directly irradiated by the incident wave, thus increasing the utilization rate of the metal wire, so the absorption rate increases. Nevertheless, the incidence angle within a specific range has little influence on the absorption rate. It can be observed from Figure 11 that in the TM mode, when the incident angle increases to 40°, the lowest point of the absorption rate is up to around 95%.

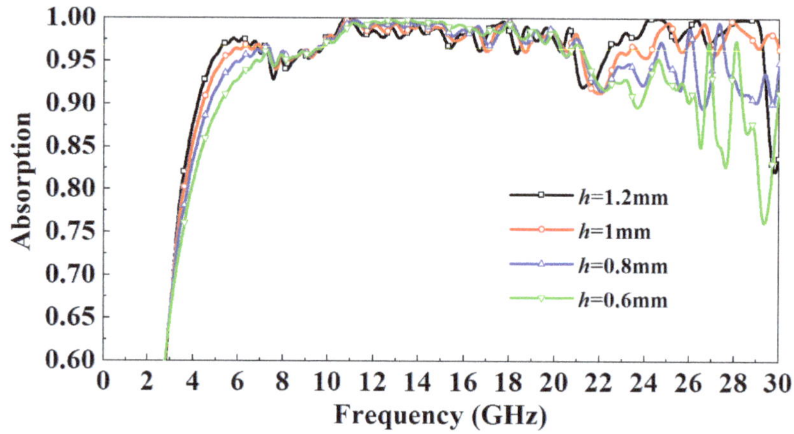

Figure 8. Effect of PR thickness on FSR absorption rate under normal incidence.

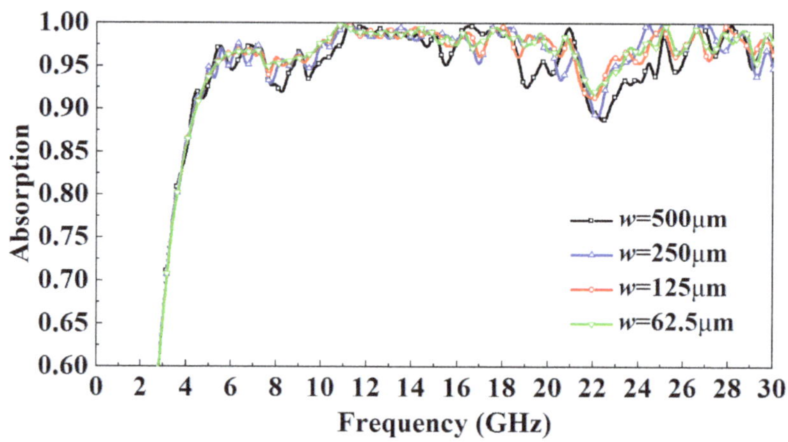

Figure 9. Effect of the width of the metal patch on FSR absorption rate under normal incidence.

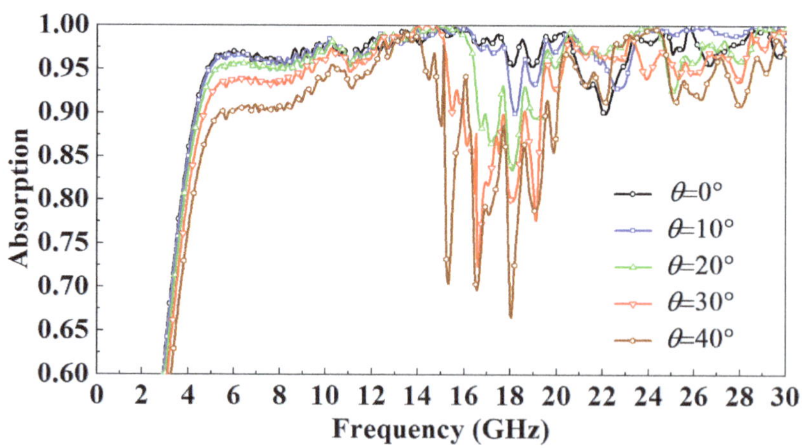

Figure 10. Effect of incident angle on FSR absorption rate in TE mode.

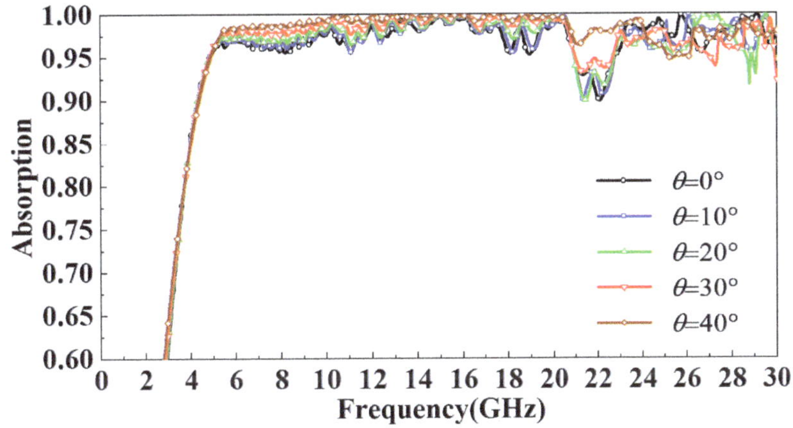

Figure 11. Effect of incident angle on FSR absorption rate in TM mode.

5. Fabrication and Measurement

In order to verify the simulation results, the FSR based on SSPP is fabricated and processed using printed circuit board technology. The FSR size is 244 mm × 244 mm × 10 mm, consisting 20 × 20 square cavity cell arrays, and the processed model is shown in Figure 12. The FSS on the inner wall of the square cavity is made by a printed circuit board. The ITO layer outside the square cavity is a dense ITO film formed on the PET substrate by magnetron sputtering under high vacuum conditions. The ITO-PET layer and the PR layer are tightly bonded together by hot pressing technology.

The experimental test adopts the stepped-frequency test system, and the schematic diagram of the system is shown in Figure 13. Agilent E8363A VNA (Agilent Technologies, Palo Alto, CA, USA) generates the stepped-frequency signal. Compared to point-frequency continuous test systems, the stepped-frequency test system does not require complex hardware cancellers. The signal sent by the test system is sent by the standard gain antenna aiming at the target, and the reflected echo signal is received by another standard gain antenna and sent to the vector network analyzer. Then, the vector network analyzer is used for time domain cancellation. Limited by the laboratory antenna specifications, the FSR is tested in the 1–18 GHz band, the test angle range is −40∼40° , and the test angle interval is 0.1°.

Figure 12. The photo of the FSR composed of 20 × 20 square cavity units, and the enlarged view of the FSS metal patch.

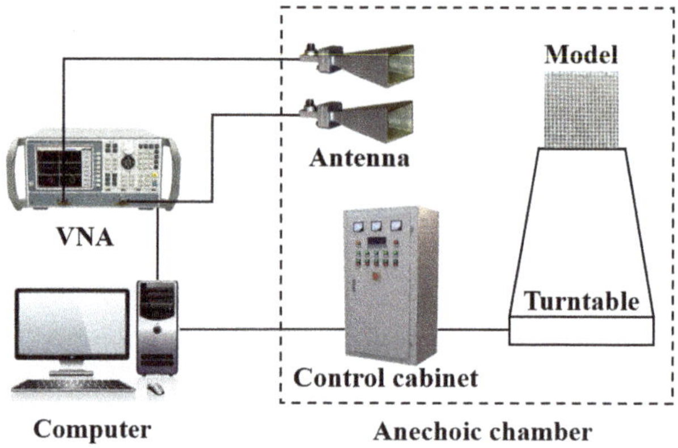

Figure 13. Stepped-frequency test system.

Figure 14 is the ISAR image of the metal plate and FSR. It can be observed that the FSR has a strong absorption of incident electromagnetic waves. Figure 15 compares simulated and measured results of FSR absorption rates in the TE mode under normal incidence. In the 1–5 GHz band, the absorption rate gradually increases and decreases with the increase of the incident frequency. At the same incident frequency, the absorption rate measured is slightly larger than the simulated results by about 10%. In the 5–8 GHz band, the absorption rate of FSR is basically about 90%, which is slightly smaller than the simulated results. In the 8–18 GHz band, FSR's absorption rate is above 95%, consistent with the simulated results. Figure 16 compares the measured results of different incident angles in TE mode and the simulated results. It can be observed that with the increase in the incident angle of electromagnetic waves, the most sensitive bands to the incident angle are mainly in the 4.5–7 GHz band and the 13–18 GHz band. In these two bands, the absorptivity decreases roughly gradually with the increase of the incident angle. In the 4.5–7 GHz band, the fluctuation of the absorption rate is relatively small. When the incident angle increases to 40°, the lowest point of the absorption rate drops to around 77%. In the 13–18 GHz band, the absorptivity fluctuates wildly. When the incident angle increases to 20°, this band's lowest point of absorptivity drop to around 83%. When the incident angle increases to 40°, this band's lowest point of absorptivity drop to around 66%. Furthermore, absorptivity in other bands is less sensitive to the incident angle. Figure 17 compares the measured results of different incident angles in the TM mode and the simulated results. It can be observed that with the increase in the incident angle of electromagnetic waves, the absorption rate of FSR changed little, and the measured results are consistent with the simulation results.

By comparison, it is found that the experimental results are consistent with the numerical simulation results, proving that the FSR designed in this paper has excellent reliability.

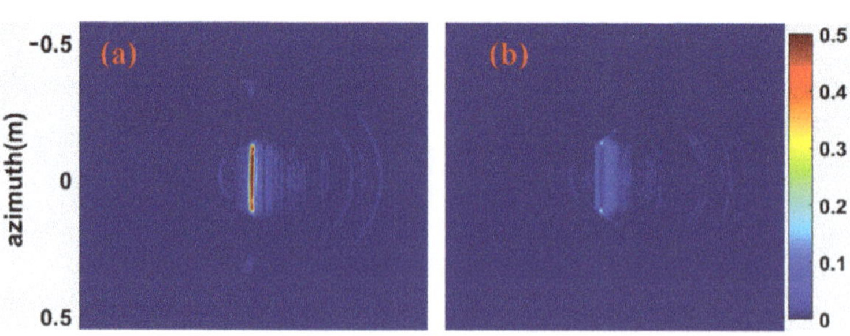

Figure 14. (**a**) ISAR imaging of sheet metal. (**b**) ISAR imaging of FSR.

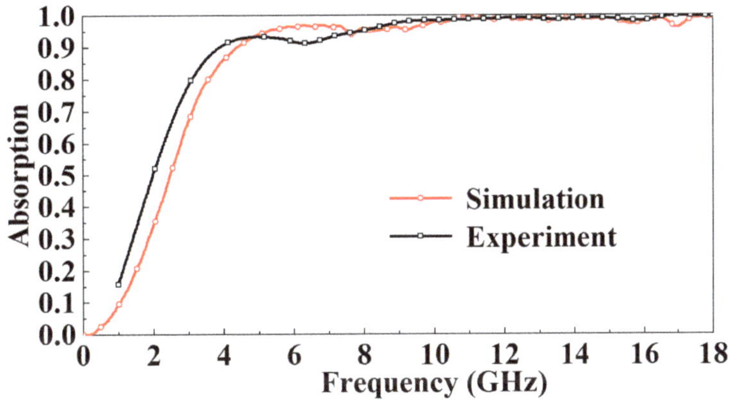

Figure 15. Comparison of simulated results and measured results of FSR absorption rate in TE mode under normal incidence.

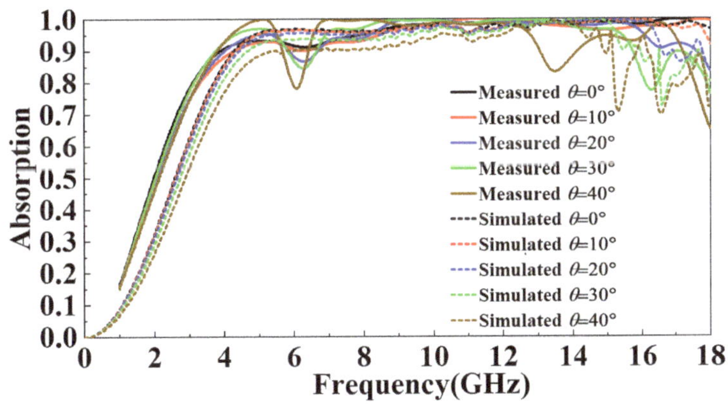

Figure 16. Comparison of the measured results of different incident angles in TE mode and the simulated results.

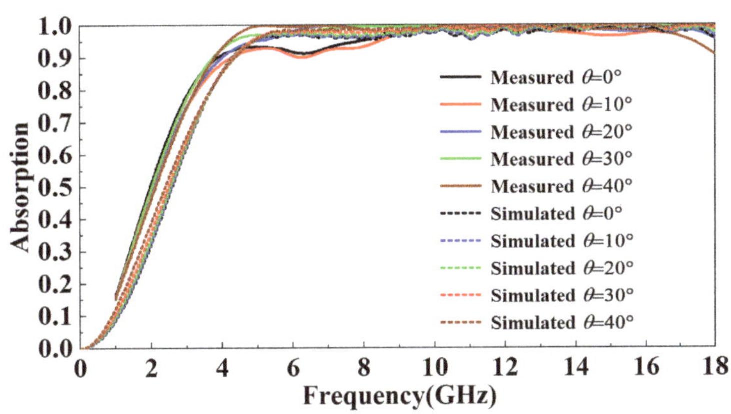

Figure 17. Comparison of the measured results of different incident angles in TM mode and the simulated results.

Table 1 compares the proposed FSR with the reported FSR in terms of the working frequency band, band continuity, maximum incident angle, and size of the absorbing unit. The comparison shows that the proposed FSR has a wider absorption bandwidth, better band continuity, and weaker sensitivity to the incident angle. Furthermore, the smaller size of the absorbing unit, which is very conducive to the miniaturization design of FSR, makes the application prospect of FSR wider.

Table 1. Feature comparison between the proposed FSR and the reported FSR.

FSR in Reference	Bandwidth/GHz	Band Continuity	Maximum Angle of Incidence	Cell Size/mm
[2]	7.99–11.97	Yes	Not reported	9 × 9 × 13.5
[3]	2.5–4.7; 7.8–14.6	No	40°	20 × 20 × 20
This work	3.5–30	Yes	40°	10 × 10 × 10

6. Conclusions

In this paper, a broadband FSR based on SSPP is designed, which is composed of a square cavity array composed of multiple layers of FSS. The numerical simulation results demonstrate that the designed FSR has an excellent absorption effect. When the electromagnetic wave is incident vertically, the absorption rate of the FSR in the 5–30 GHz band is above 95%, and in the 3.5–5 GHz band, the absorption rate is also above 80%. At the same time, FSR is highly insensitive to the incident angle. When the incident angle increases by 40°, the absorption rate fluctuates wildly only in the 15–20 GHz band, and the absorption rate drops to about 70% at the lowest. In TM mode, the incidence angle has a little effect on the absorption rate of FSR. The FSR designed in this paper is fabricated, and a microwave anechoic chamber measures its transmission characteristics. The experimental results demonstrate that the FSR absorption rate is consistent with the numerical simulation results, which proves that the FSR designed in this paper is less sensitive to the incident angle of magnetic waves. It has excellent broadband absorbing ability and outstanding reliability and robustness. The proposed FSR has a high absorption rate and little reflectivity of electromagnetic waves in its operating frequency band, so that it can be applied to the stealth design for dual-station radar detection. At the same time, the structural strength of the FSR is also excellent, which can realize the integrated design of structure and stealth and has a good application prospect in the stealth design of aircraft and other targets.

Author Contributions: Conceptualization, J.B. and Y.L.; methodology, J.B.; software, J.B.; validation, J.B., Y.L. and Q.Y.; formal analysis, Q.Y.; investigation, X.G.; resources, Q.Y.; data curation, J.B.; writing—original draft preparation, J.B.; writing—review and editing, J.B., Y.L. and Q.Y.; visualization, J.B. and Y.L.; funding acquisition, X.G. All authors have read and agreed to the published version of the manuscript.

Funding: This research was funded by the Natural Science Foundation of Shanxi Province under Grant 2022JQ052.

Data Availability Statement: The data that support the findings of this study are available from the corresponding author (Y.L.), upon reasonable request.

Conflicts of Interest: The authors declare no conflict of interest.

References

1. Wang, Z.; Fu, J.; Zeng, Q.; Song, M.; Denidni,T. Wide-band Transmissive Frequency-Selective Absorber. *IEEE Antennas Wirel. Propag. Lett.* **2019**, *18*, 1443–1447. [CrossRef]
2. Yu, W.; Luo, G.; Yu, Y.; Liao, Z.; Shen, Z. Broadband Band-Absorptive Frequency-Selective Rasorber With a Hybrid 2-D and 3-D Structure. *IEEE Antennas Wirel. Propag. Lett.* **2019**, *18*, 1701–1705. [CrossRef]

3. Xia, J.; Wei, J.; Liu, Y.; Zhang, Y.; Guo, S.; Li, C.; Bie, S.; Jiang, J. Design of a Wideband Absorption Frequency Selective Rasorber Based on Double Lossy Layers. *IEEE Trans. Antennas Propag.* **2020**, *68*, 5718–5723. [CrossRef]
4. Lu, G.; Zhang, J.; Yang, J.; Zhang, T.; Kou, Y. Comprehensive Survey for the Frontier Disciplines Status and Development of Frequency Selective Surface Radome. *Acta Phys. Sin.* **2013**, *62*, 9–18.
5. Sun, Z.; Chen, Q.; Guo,M.; Yu, H.; Fu, Y. Frequency Selective Rasorber and Reflector With Two-Sided Absorption Bands. *IEEE Access* **2018**, *7*, 6025–6031. [CrossRef]
6. Alwareth, H.; Ibrahim, I.M.; Zakaria, Z.; Al-Gburi, A.J.A.; Ahmed, S.; Nasser, Z.A. A Wideband High-Gain Microstrip Array Antenna Integrated with Frequency-Selective Surface for Sub-6 GHz 5G Applications. *Micromachines* **2022**, *13*, 1215. [CrossRef]
7. Ruan, J.; Meng, Z.; Zou, R.; Cai, F.; Pan, S. Miniaturized Frequency Selective Surface for 6G Communication. *Micromachines* **2022**, *13*, 427. [CrossRef]
8. Hu, F.; Liu, F.; Zhu, M.; Bao, J. Design of Miniaturized and Polarization-Insensitive Double-Layer Frequency Selective Surface Based on Meander Lines. *Int. J. Antennas Propag.* **2021**, *2021* , 9576494. [CrossRef]
9. De Sabata, A.; Matekovits, L.; Buta, A.; Dassano, G.; Silaghi, A. Frequency Selective Surface for Ultra-Wide Band Filtering and Shielding. *Sensors* **2022**, *22*, 1896. [CrossRef]
10. Yu, Q.; Liu, S.; Kong, X.; Bian, B. A Broadband Miniaturized Ultra-Thin Tri-Band Bandpass FSS with Triangular Layout. *Int. J. RF Microw. Comput.-Aided Eng.* **2019**, *29*, 21738. [CrossRef]
11. Xu, C.; Qu, S.; Wang, J.; Yan, M.; Pang, Y.; Wang, W.; Wang, A.; Fan, Q. Dual Atmospheric Windows Infrared Stealth Research Based On Frequency Selective Surface. *J. Infrared Millim. Waves* **2019**, *38*, 315–319.
12. Fan, Y.; Wang, J.; Ma,H.; Li, Y.; Feng, M.; Qu, S. Ultra-Thin and Light-Weight Spoof Surface Plasmon Polariton Coupler Achieved by Broadside Coupled Split Ring Resonators. *Chin. Phys. B* **2018**, *27*, 179–184. [CrossRef]
13. Kim, Y.J.; Yoo, Y.J.; Hwang, J.S.; Lee, Y.P. Ultra-Broadband Microwave Metamaterial Absorber Based on Resistive Sheets. *J. Opt. UK* **2017**, *19*, 015103. [CrossRef]
14. Lan, J.; Cao, X.; Gao, J.; Cong, L.; Wang, S.; Yang, H.H. Design of Miniaturized Wideband Microwave Absorber Loaded with Lumped Resistance. *Radioengineering* **2018**, *27*, 746–752. [CrossRef]
15. Xiao, S.; Wang, T.; Liu, T.; Zhou, C.; Zhang, J. Active Metamaterials and Metadevices: a Review. *J. Phys. D Appl. Phys.* **2020**, *53*, 503002. [CrossRef]
16. Yang, J.; Wang, J.; Zhang, J.; Jiang, W.; Qu, S. Malposed Spoof Surface Plasmon Structure with Enhanced Microwave Absorption and Compressive Performances Realized by Carbon-Based Foams. *Mater. Sci. Eng. B* **2020**, *262*, 114787. [CrossRef]
17. Niknam, S.; Yazdi, M.; Amlashi, S.B.; Khalily, M. Numerical Performance Analysis of Terahertz Spectroscopy Using an Ultra-Sensitive Resonance-Based Sensor. *Front. Phys.* **2020**, *8*, 19. [CrossRef]
18. Zhu, R.; Wang, J.; Sui, S.; Meng, Y.; Qu, S. Wideband Absorbing Plasmonic Structures via Profile Optimization Based on Genetic Algorithm. *Front. Phys.* **2020**, *8*, 231. [CrossRef]
19. Martin-Cano, D.; Nesterov, M.L.; Fernandez-Dominguez, A.I.; Garcia-Vidal, F.J.; Martin-Moreno, L.; Moreno, E. Domino Plasmons for Subwavelength Terahertz Circuitry. *Opt. Express* **2010**, *18*, 754–764. [CrossRef]
20. Pendry, J.B.; Martín-Moreno, L.; Garcia-Vidal, F.J. Mimicking Surface Plasmons with Structured Surfaces. *Science* **2004**, *305*, 847–848. [CrossRef]
21. Ye, L.; Chen, Y.; Zhou, J.; Feng, H.; Zhang, Y.; Liu, Q. High-Performance Spoof Surface Plasmon Polariton Waveguides and Splitters Based on Greek-Cross Fractal Units. *J. Phys. D Appl. Phys.* **2020**, *53*, 235502. [CrossRef]
22. Ye, L.; Chen, Y.; Xu, K.; Li, W.; Zhang, Y. Substrate Integrated Plasmonic Waveguide for Microwave Bandpass Filter Applications. *IEEE Access* **2019**, *7*, 75957–75964. [CrossRef]
23. Ye, L.; Chen, Y.; Wang, Z.; Zhu, C.; Zhuo, J.; Liu, Q. Compact Spoof Surface Plasmon Polariton Waveguides and Notch Filters Based on Meander-Strip Units. *IEEE Photonics Technol. Lett.* **2021**, *33*, 135–138. [CrossRef]
24. Jackson, D.R.; Caloz, C.; Itoh, T. Leaky-Wave Antennas. *Proc. IEEE* **2012**, *100*, 2194–2206. [CrossRef]
25. Karmokar, D.K.; Chen, S.; Thalakotuna, D.; Qin, P.; Guo, Y. Continuous Backward-to-Forward Scanning 1-D Slot-Array Leaky-Wave Antenna With Improved Gain. *IEEE Antennas Wirel. Propag. Lett.* **2019**, *19*, 89–93. [CrossRef]
26. Wang, C.; Li, Y.; Han, Y.; Feng, M.; Wang, J.; Ma, H.; Zhang, J.; Qu, S. Design of Triple-Band-Pass Frequency Selective Structure Based on Spoof Surface Plasmon Polariton. *AIP Adv.* **2018**, *67*, 368–375. [CrossRef]
27. Pan, D.; You, B.; Wen, X.; Li, X. Wideband Substrate Integrated Waveguide Chip Filter Using Spoof Surface Plasmon Polariton. *Micromachines* **2022**, *13*, 1195. [CrossRef]
28. Zhang, Y.; Li, B.; Zhu, L.; Tang, Y.; Chang, Y.; Bo, Y. Frequency Selective Rasorber With Low Insertion Loss and Dual-Band Absorptions Using Planar Slotline Structures. *IEEE Antennas Wirel. Propag. Lett.* **2018**, *17*, 633–636. [CrossRef]

micromachines

MDPI

Article

Dynamic Enhancement for Dual Active Bridge Converter with a Deadbeat Current Controller

Chengfu Tian [1], Shusheng Wei [2,*], Jiayu Xie [1] and Tainming Bai [1]

1 Unit 91977 of the Chinese People's Liberation Army, Beijing 100036, China
2 School of Automotive Engineering, Wuhan University of Technology, Wuhan 430070, China
* Correspondence: weishusheng13@163.com

Abstract: This paper investigates the deadbeat current controllers for isolated bidirectional dual-active-bridge dc-dc converter (IBDC), including the peak current mode (PCM) and middle current mode (MCM). The controller uses an enhanced single phase shift (ESPS) modulation method by exploiting pulse width as an extra control variable in addition to phase shift ratio. The control variables for PCM controllers are derived in detail and the two different current controllers are compared. A double-closed-loop control method is then employed, which could directly control the high-frequency inductor current and eliminate the transient DC current bias of the transformer. Furthermore, load feedforward was introduced to further enhance the dynamic of the converter. With the proposed control method, the settling time could be reduced within several PWM cycles during load disturbance without transient DC current bias. A 5 kW IBDC converter prototype was built and the settling time of 6 PWM cycles during load change with voltage regulation mode was achieved, which verifies the superior dynamic performance of the control method.

Keywords: dual active bridge; deadbeat controller; load feedforward

Citation: Tian, C.; Wei, S.; Xie, J.; Bai, T. Dynamic Enhancement for Dual Active Bridge Converter with a Deadbeat Current Controller. *Micromachines* **2022**, *13*, 2048. https://doi.org/10.3390/mi13122048

Academic Editor: Wensheng Zhao

Received: 21 October 2022
Accepted: 18 November 2022
Published: 23 November 2022

Publisher's Note: MDPI stays neutral with regard to jurisdictional claims in published maps and institutional affiliations.

Copyright: © 2022 by the authors. Licensee MDPI, Basel, Switzerland. This article is an open access article distributed under the terms and conditions of the Creative Commons Attribution (CC BY) license (https://creativecommons.org/licenses/by/4.0/).

1. Introduction

The isolated bi-directional dual-active-bridge dc-dc converter (IBDC) has been a hot topic in recent years due to its simple structure, high efficiency and ultrafast response [1]. The transient DC current offset of the transformer and the inductor, which might saturate the transformer and increase the system's current stress during the abrupt load change, has attracted people's attention. Different dynamic modulation methods have been proposed to solve the problem [2–5]. Additionally, to increase the dynamics of IBDCs, the current mode controller could be a competitive alternative. It also has other inherent benefits including over-current protection, elimination of transient DC current offset and easy implementation of current sharing between multiple IBDCs [6].

Digital predictive current controllers based on conventional single phase shift (CSPS) modulation was proposed in [7,8], where the phase shift ratio was used to control the transformer current. In [7], the average current calculated by an analog integrator of the DC bus current was used as the feedback signal, which can achieve fast dynamic performance. However, transient DC current offset occurs during the sudden change in phase shift ratio for CSPS modulation.

The predictive duty cycle mode (PDCM) controller, shown in Figure 1b, was proposed in [6] to eliminate the transient DC current offset, which was applied in [8]. The drive signals of the primary side are fixed. The transformer current needs to be oversampled, and duty cycles d1 for S2,3 and d2 for S1,4 are calculated in turn in every half cycle. Another limitation of this method is that the controller only works in the ZVS range (IP1 > 0) shown in Figure 1b and may lose effectiveness when IP1 < 0.

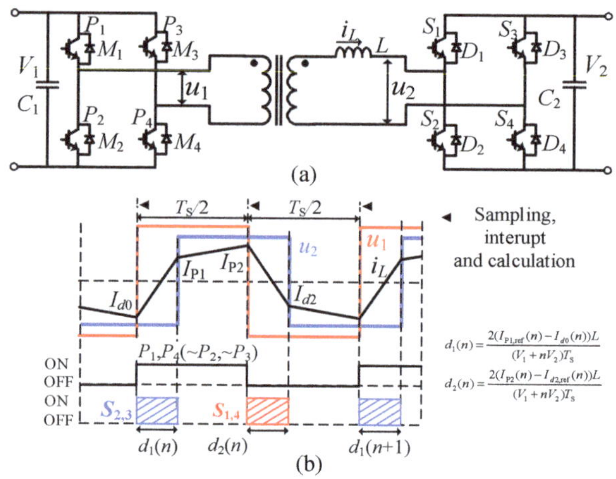

Figure 1. Principle of the PDCM controller. (**a**) IBDC, (**b**) waveforms.

A deadbeat current controller based on the middle current and enhanced PWM modulation was proposed in [9]. However, regulation of the output voltage and current was not introduced, which is more important in real application. To overcome the drawbacks, this paper investigated the deadbeat current controllers, including the peak current mode (PCM) and middle current mode (MCM). Based on the controllers, a double-closed-loop control method with load feedforward was introduced. Furthermore, a 5 kW IBDC converter prototype was built, and the settling time of 6 PWM cycles during load change could be achieved, which validates its superior dynamic performance.

2. Deadbeat Peak Current Mode Controller

2.1. Basic Model of IBDC for SPS Modulation

The basic model of SPS modulation-based IBDC is presented prior to introducing the proposed current controller. Figure 2 illustrates the theoretical waveforms of the IBDC using the SPS modulation method when converter voltage gain $k \geq 1$, where $k = V_1/(\mathrm{n}V_2)$ and n is the turn ratio of the transformer. The waveforms are symmetrical for the same transmission power of two opposite directions.

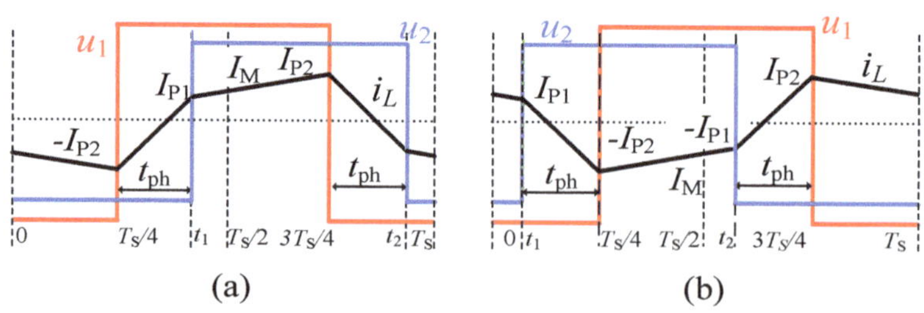

Figure 2. Waveforms of IBDC at steady state for CSPS modulation: (**a**) forward power transmission ($P > 0$), (**b**) reverse power transmission ($P < 0$).

The symbols in Figure 2 are defined as follows: TS is the switching cycle, f is the switching frequency, D is the phase shift ratio and tph is the shifted time. $D \geq 0$ ($t_{\mathrm{ph}} \geq 0$) stands for $P \geq 0$ and $D < 0$ ($t_{\mathrm{ph}} < 0$) for $P < 0$. I_{P1} and I_{P2} are the two "switching currents" and I_{P2} is the peak current when $k > 1$. The middle current IM, defined as the instantaneous current at $T_S/2$, is taken into consideration instead of the average current, which equals zero in one cycle. The basic equations for the IBDC are derived as follows:

$$\begin{cases} P = \frac{V_1^2}{2fLk}D(1-|D|)\,,\, t_{\mathrm{ph}} = \frac{DT_S}{2} \\ I_{\mathrm{P1}} = \frac{V_1(2k|D|-k+1)}{4fLk},\, I_{\mathrm{P2}} = \frac{V_1(2|D|+k-1)}{4fLk},\, I_{\mathrm{M}} = \frac{V_1 D}{2fLk} \end{cases} \tag{1}$$

The relationships among the variables P, I_{P1}, I_{P2}, I_{M}, D and t_{ph} at steady state can then be derived. Therefore, for a given value of one variable, other variables can be calculated.

2.2. Peak Current Mode Controller

Peak current mode (PCM) controllers are introduced in this section. Figure 3a,b show the transient waveforms of a PCM current controller in one cycle for forward and reverse power transmission, respectively. P_1, P_2, P_3 and P_4 are the drive signals for the primary side and S_1, S_2, S_3 and S_4 are the drive signals for the secondary side. The variable $t_{\mathrm{ph,ref}}$ is shifted-time at the steady state for the given $I_{\mathrm{P2,ref}}$ which can be derived from (1). A sawtooth carrier with the same frequency of the converter was utilized to generate the reference signals. The "switching on" moment t_1 and "switching off" moment t_2 should meet the constrains as: $0 < t_1 < T_S/2$ and $3T_S/4 < t_2 < T_S$. The variables t_{D} and t_{W} are defined as "delay time" and "width time", respectively. P_{ref} is the power reference, and $I_{\mathrm{P2,ref}}$ and $I_{\mathrm{P1,ref}}$ are the references for the corresponding "switching currents", respectively.

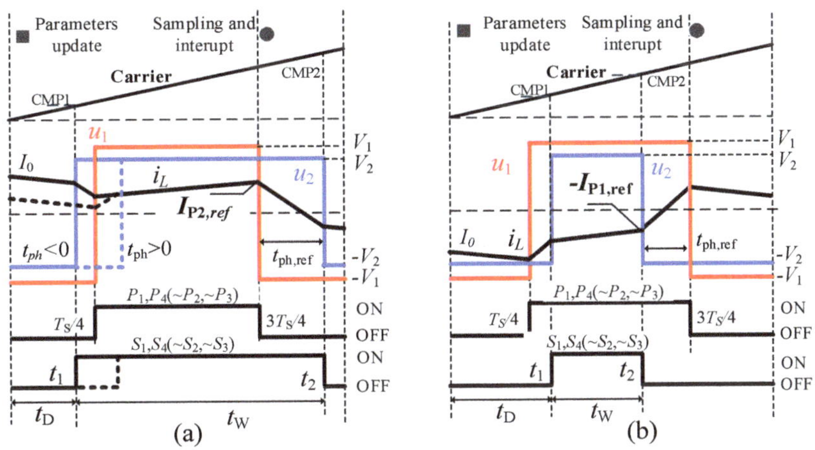

Figure 3. Transient waveforms of PCM controller: (**a**) forward power transmission ($P_{\mathrm{ref}} > 0$), (**b**) reverse power transmission ($P_{\mathrm{ref}} < 0$).

As shown in Figure 3a, there are two cases according to the initial current I_0 and reference current $I_{\mathrm{p2,ref}}$: u_2 leads u_1 ($t_{\mathrm{ph}} < 0$) for the solid line waveforms and u_2 lags u_1 ($t_{\mathrm{ph}} > 0$) for dotted line waveforms. For the sake of brevity, the superposition principle was used to derivate the inductor current when calculating t_{D} and t_{W}.

For forward power transmission, the requirement was imposed that $I_{\mathrm{P2}} = I_{\mathrm{P2,ref}}$ and $t_{\mathrm{ph,ref}} > 0$. According to the superposition principle, the inductor current ripple ΔI_L during 0 and $3T_S/4$ can be calculated by adding up the two ripple currents as follows:

$$\begin{cases} \Delta I_L = \Delta I_{L,u_1} + \Delta I_{L,u_2} \\ \Delta I_{L,u_1} = \frac{-V_1}{L}\cdot\frac{T_S}{4} + \frac{-V_1}{L}\cdot\frac{T_S}{2}\,,\, \Delta I_{L,u_2} = \frac{V_2}{L}\cdot t_{\mathrm{D}} + \frac{-V_2}{L}\cdot(\frac{3T_S}{4} - t_{\mathrm{D}}) \\ I_{\mathrm{P2}} = I_0 + \Delta I_L = I_{\mathrm{P2,ref}} \end{cases} \tag{2}$$

where $\Delta I_{L,u_1}$ and $\Delta I_{L,u_2}$ are current ripples generated by the two dependent voltage source u_1 and u_2, respectively. t_D is then derived as:

$$t_D = \frac{(I_{P2,ref} - I_0)Lk}{2V_1} + \frac{3-k}{8f} \tag{3}$$

Furthermore, t_W is derived as:

$$t_W = \frac{3T_S}{4} - t_D + t_{ph,ref} \tag{4}$$

With regard to reverse power transmission, the switching current at t_2 is set to be—$I_{P1,ref}$ and $t_{ph,ref} < 0$ as shown in Figure 4. Similar to forward power transmission, t_D and t_W can be obtained. Thus, the equations for the PCM controller are written as:

$$\begin{cases} t_D = \frac{kL(I_{P2,ref}-I_0)}{2V_1} + \frac{3-k}{8f};\ t_W = \frac{kL(I_{P2,ref}+I_0)}{2V_1} - \frac{k-5}{8f},\ P_{ref} \geq 0 \\ t_D = -\frac{kL(I_{P2,ref}+I_0)}{2V_1} + \frac{k+1}{8f};\ t_W = -\frac{kL(I_{P2,ref}-I_0)}{2V_1} + \frac{k+3}{8f},\ P_{ref} < 0 \end{cases} \tag{5}$$

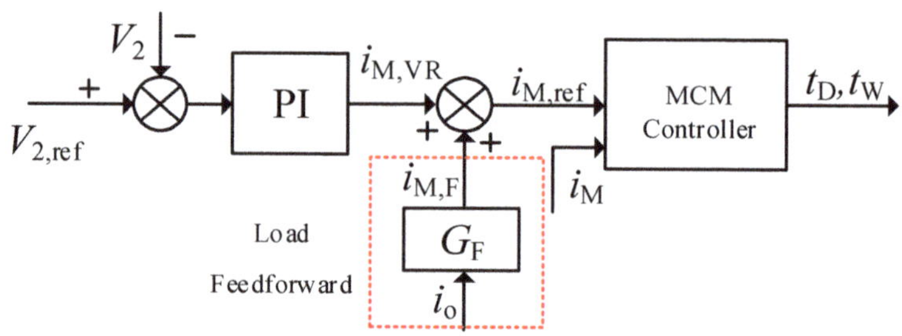

Figure 4. Voltage control scheme based on the MCM-ESPS controller.

As in the aforementioned Equations (6) and (7), initial current I_0 is sampled to calculate the t_D and t_W. However, a one-cycle delay exists between the sampling instant and control update due to the algorithm implementation of the digital processor. I_{P2} for the PCM controller is sampled at $3T_S/4$, and DSP interrupt occurs to calculate the new parameters shown in Figure 3. t_D and t_W update at the beginning of the next cycle. Assuming DC bus voltage V_1 and V_2 are constant in two adjacent periods, the relationships between the $I_{P2}(n-1)$, $I_M(n-1)$ in the $(n-1)$th cycle and the initial current in the nth switching cycle $I_0(n)$ could be derived as:

$$I_0(n) = \begin{cases} I_{P2}(n-1) - \frac{2V_1(k-1)}{Lk}(t_D(n-1) + t_W(n-1)) + \frac{V_1(7-k)}{4fLk},\ P_{ref}(n-1) \geq 0 \\ I_{P2}(n-1) - \frac{V_1(k-1)}{4fLk},\ P_{ref}(n-1) < 0 \end{cases} \tag{6}$$

According to the power transmission directions in two adjacent cycles, four situations are considered for the PCM controller: case 1 when $P_{ref}(n-1) \geq 0$ and $P_{ref}(n) \geq 0$; case 2 when $P_{ref}(n-1) < 0$ and $P_{ref}(n) \geq 0$; case 3 when $P_{ref}(n-1) \geq 0$ and $P_{ref}(n) < 0$; and case 4 when $P_{ref}(n-1) < 0$ and $P_{ref}(n) < 0$. Combining (6)–(9), the control variables t_D and t_W with delay compensation can be derived as shown in Table 1. With the control variables in Table 1, the inductor peak current could be tracked to the reference in two cycles, which is consistent with the idea of the deadbeat control in ref [10].

Table 1. Control variable calculation with one-cycle delay compensation.

Mode		$t_D(n)$	$t_W(n)$
CSPS ($P_{ref} \geq 0$)		$\frac{kL(I_{M,ref}(n)-I_M(n-1))}{2V_1} + t_D(n-1)$	$\frac{1}{2f}$
ESPS-PCM	Case 1	$\frac{kL(I_{M,ref}(n)-I_M(n-1))}{2V_1} + t_D(n-1) + t_W(n-1) - \frac{1}{2f}$	$\frac{kL(I_{P2,ref}(n)+I_{P2}(n-1))}{2V_1} - (k-1)(t_D(n-1) + t_W(n-1)) + \frac{6-k}{4f}$
	Case 2	$\frac{kL(I_{P2,ref}(n)-I_{P2}(n-1))}{2V_1} + \frac{1}{4f}$	$\frac{kL(I_{P2,ref}(n)+I_{P2}(n-1))}{2V_1} - \frac{k-3}{4f}$
	Case 3	$-\frac{kL(I_{P2,ref}(n)+I_{P2}(n-1))}{2V_1} + (k-1)(t_D(n-1)$ $+t_W(n-1)) + \frac{k-3}{4f}$	$-\frac{kL(I_{P2,ref}(n)-I_{P2}(n-1))}{2V_1} - (k-1)(t_D(n-1)$ $+t_W(n-1)) + \frac{5}{4f}$
	Case 4	$-\frac{kL(I_{P2,ref}(n)+I_{P2}(n-1))}{2V_1} + \frac{k}{4f}$	$-\frac{kL(I_{P2,ref}(n)-I_{P2}(n-1))}{2V_1} + \frac{1}{4f}$
ESPS-MCM		$\frac{kL(I_{M,ref}(n)-I_M(n-1))}{2V_1} + t_D(n-1) + t_W(n-1) - \frac{1}{2f}$	$\frac{kL(I_{M,ref}(n)+I_M(n-1))}{2V_1} - t_D(n-1) - t_W(n-1) + \frac{5}{4f}$

3. Double-Closed-Loop Control with Load Feedforward

In practice, instead of the high-frequency inductor current, the DC voltage, current or power should always be regulated. In this section, the voltage mode control strategy based on the MCM-ESPS controller is introduced.

Figure 4 shows the output voltage control scheme based on the deadbeat current controller, where two control loops are involved. The load feedforward control could substantially increase the system dynamic [11,12]. In order to improve the stability of output voltage under load disturbance, load feedforward under double-closed-loop control is presented. As shown in Figure 4, the feedforward current $i_{M,F}$ corresponding to the load was superimposed on the current reference $i_{M,VR}$, which is the output of the outer voltage loop, to form the final current reference value $i_{M,ref}$.

The relationships of the middle current were derived as:

$$I_M = \begin{cases} \frac{V_1}{2fLk}\left(\frac{1}{2} - \sqrt{\frac{1}{4} - \frac{2fLkP}{V_1{}^2}}\right), P \geq 0 \\ -\frac{V_1}{2fLk}\left(\frac{1}{2} - \sqrt{\frac{1}{4} + \frac{2fLkP}{V_1{}^2}}\right), P < 0 \end{cases} \tag{7}$$

Without considering the power loss of the converter, we could obtain:

$$P = V_2 i_o \tag{8}$$

Thus, the relationship between the middle current I_M and the load current i_o could be expressed as (9) and (10):

$$I_M = \begin{cases} \frac{V_1}{2fLk}\left(\frac{1}{2} - \sqrt{\frac{1}{4} - \frac{2fLi_o}{V_1}}\right), i_o \geq 0 \\ -\frac{V_1}{2fLk}\left(\frac{1}{2} - \sqrt{\frac{1}{4} + \frac{2fLi_o}{V_1}}\right), i_o < 0 \end{cases} \tag{9}$$

$$i_o = NI_M\left(1 - \frac{2fLkI_M}{V_1}\right) \tag{10}$$

The small signal model of the system, as shown in Figure 5, can be obtained from the control block diagram in Figure 4, where i_S is the average output current of an H bridge in a single period and $G_o(s)$ is the transfer function of capacitance voltage and capacitance current, denoted as:

$$G_o(s) = 1/(C_2 s) \tag{11}$$

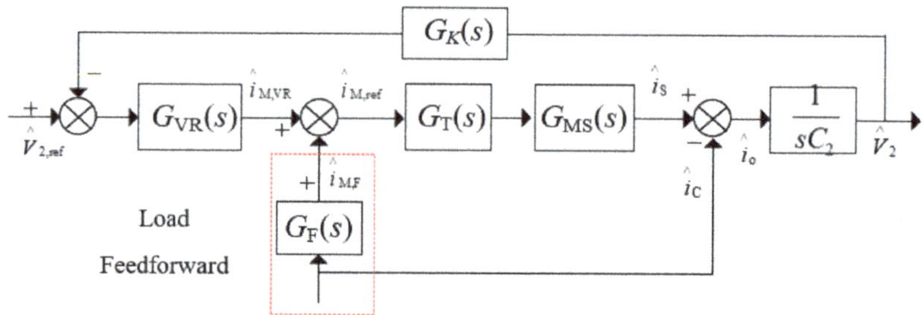

Figure 5. Small signal model of an IBDC with the double-closed-loop control strategy.

$G_{VR}(s)$ is the volatge regulation transfer function, where the conventional PI controller is always used. K_P and K_I are the proportional and integral coefficients of the PI regulator, respectively. Thus, we could obtain:

$$G_{VR}(s) = K_P + K_I s \tag{12}$$

$G_F(s)$ represents the transfer function of the load feedforward and $G_{MS}(s)$ is the relationship between i_M and i_s. $G_C(s)$ is the transfer function of the deadbeat current controller. Considering a one-cylce delay, it could be written as:

$$G_C(s) = \frac{1 - e^{-sT_s}}{s} \tag{13}$$

The feedforward transfer function $G_F(s)$ can be calculated using small-signal analysis based on Equation (10). To substitute $i_o = \bar{i}_o + \hat{i}_o$ and $I_M = \bar{I}_M + \hat{I}_M$ into (10), ignoring the higher-order terms, $G_F(s)$ be derived as:

$$G_F(s) = \hat{I}_{M,F}(s)/\hat{i}_o(s) = 1/(N(1 - 4fLkI_M/V_1)) \tag{14}$$

The average output current of H bridge in the secondary side is derived as:

$$i_s = \frac{V_1 N}{2fL} D(1 - D) \tag{15}$$

Combing (1) with (16), the $G_{MS}(s)$ could be derived as:

$$G_{MS}(s) = \hat{i}_S(s)/\hat{I}_M(s) = N(1 - 4fLkI_M/V_1) \tag{16}$$

According to the small signal model in Figure 5, the output impedance $R_{o1}(s)$ without and with feedforward could be calculated as (17) and (18), respectively.

$$R_{o1}(s) = \frac{\hat{V}_2(s)}{\hat{i}_o(s)} = -\frac{G_o(s)}{1 + G_{VR}(s)G_C(s)G_{MS}(s)G_o(s)} \tag{17}$$

$$R_{o2}(s) = \frac{\hat{V}_2(s)}{\hat{i}_o(s)} = \frac{(G_F(s)G_C(s)G_{MS}(s) - 1)G_o(s)}{1 + G_{VR}(s)G_C(s)G_{MS}(s)G_o(s)} \tag{18}$$

By substituting the circuit parameters and control parameters into Equations (17) and (18), baud diagrams of output impedance with different loads under the double-closed-loop control strategy can be drawn as shown in Figure 6. In this case, the inductance L = μH, the voltage V_1 = 300V and V_2 = 280V.The coefficients of the PI regulator are K_P = 2 and K_I = 4000.

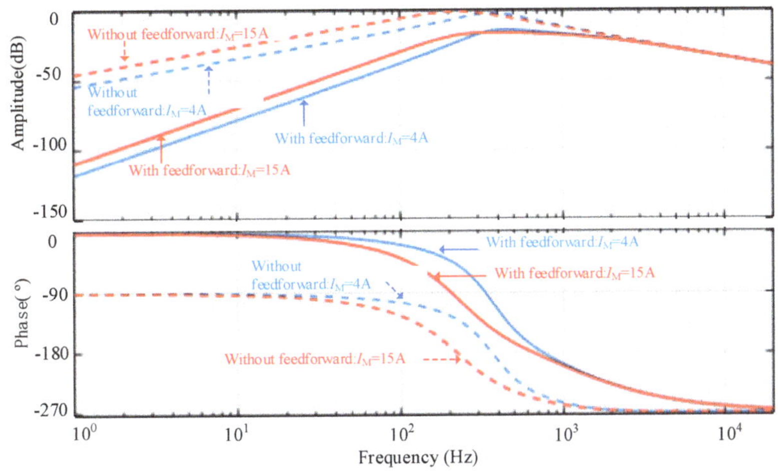

Figure 6. Output impedance of the converter with different loads.

As shown in Figure 6, the closed-loop output impedance at low frequency decreases significantly after the feedforward is added. When I_M = 4 A, the output impedance at a frequency of 100 Hz decreases from −15 dB to −35 dB, whereas when I_M = 15 A, the output impedance decreases from −8 dB to −30 dB at 100 Hz. If the frequency is further reduced, the amplitude attenuation of the closed-loop output impedance brought by the feedforward control become more obvious, which indicates a more robust output voltage under the load disturbance.

4. Experimental Verification

4.1. Experimental Platform

The laboratory IBDC experimental platform shown in Figure 7 was used to verify the proposed control method. The main circuit parameters are listed in Table 2. The current sensor LA55-P had a 200 kHz bandwidth from LEM. PE-Expert4 from Myway was utilized as the digital controller including DSP and FPGA cores. FPGA XC6SLX45 was used to generate PWM signals. The control variables were calculated in each cycle in DSP, and the corresponding PWM compare values CMP1 and CMP2 were updated at the beginning of the next cycle.

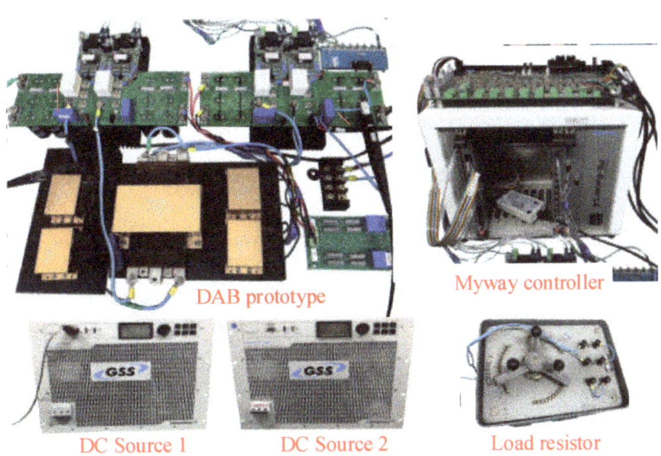

Figure 7. Experimental platform.

Table 2. System Parameters.

Input voltage V_1	300 V	Output voltage V_1	280 V
Turns ratio n	1:1	Switching frequency f	10 kHz
Primary capacitor C_1	2460 μF	Secondary capacitor C_2	2460 μF
Inductor L	652 μH	Equivalent resistor Rs	80 mΩ

4.2. Comparisons of Different Current Controllers for Forward Power Transmission

The performance of the CSPS modulation-based current controller in [5] and the proposed ESPS-PCM were compared for the forward power transmission mode.

Figure 8 shows the experimental waveforms of current i_L when the references have step changes. The references could be tracked for all the controllers when the current reference steps up from 3 A to 8 A and steps down from 8 A to 3 A. The settling time t_{set} in Figure 8a is obvious, while the settling time in Figure 8b is negligible. Additionally, the ESPS-PCM controller eliminates the transient DC current offset that exists in CSPS modulation-based controller, as shown in the dashed circle of Figure 8a.

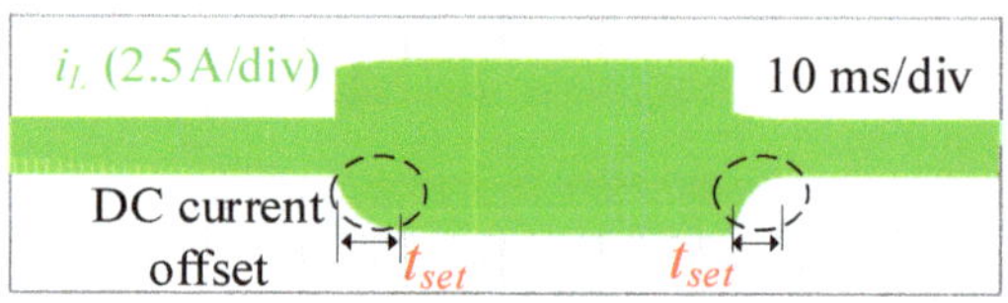

(**a**)

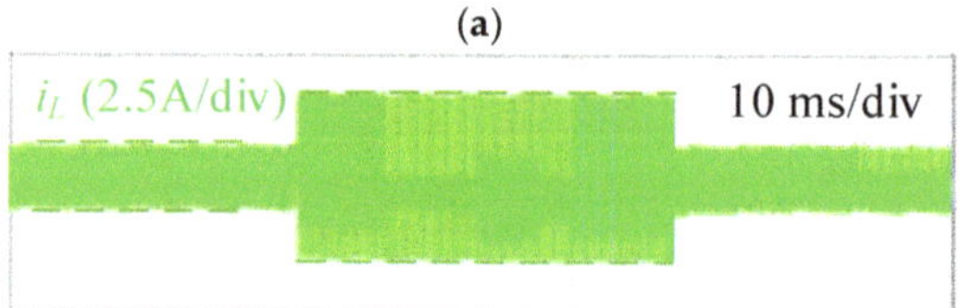

(**b**)

Figure 8. Zoomed-out waveforms during the step change of the current reference for forward power transmission. (**a**) CSPS modulation-based current controller. (**b**) ESPS-PCM controller.

To verify the performance of the current controllers during bidirectional power transmission, a sequence of current references was set to investigate the response. Figure 9 shows the waveforms of the ESPS-PCM controller. The transmission power P stepped up from 600 W to 1450 W at t_1, changed direction to -1450 W at t_3, reversed direction to 1450 W at t_5 and then stepped down to 600 W at t_7. The current references were smoothly reached with a one-cycle delay during the whole transient process, including the transition between two opposite power transmissions.

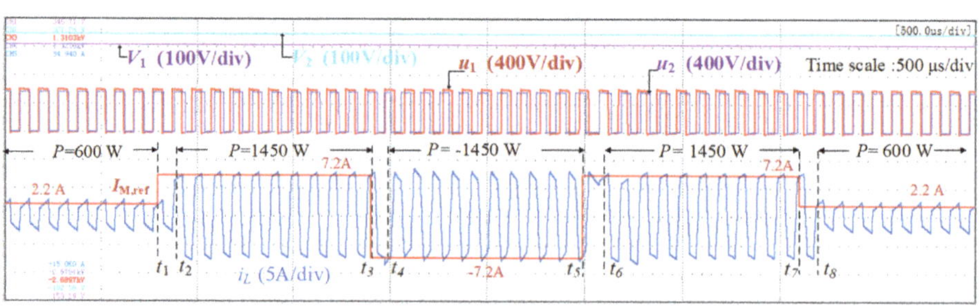

Figure 9. Transient waveforms of bidirectional power transmission.

4.3. Dynamic Performance Comparison between Different Control Methods

The experimetal results with traditional single-voltage loop control, double-closed-loop control and double-closed-loop control with feedforward under load disturbance are shown in Figures 10–12. The output voltage V_2 was 280 V, and the load increased abruptly at t_1 with load resistance decreases from 75 Ω to 25 Ω and decreased sharply at t_2 with load resistance increases from 25 Ω to 75 Ω. When the load increases, the DC voltage will fall, and the control loop will increase the phase shift angle to transfer more power to maintain the DC load. Simillarly, when the DC load decreases, the DC voltage will rise and the control loop will decrease the phase shift angle to reduce the transmitted power.

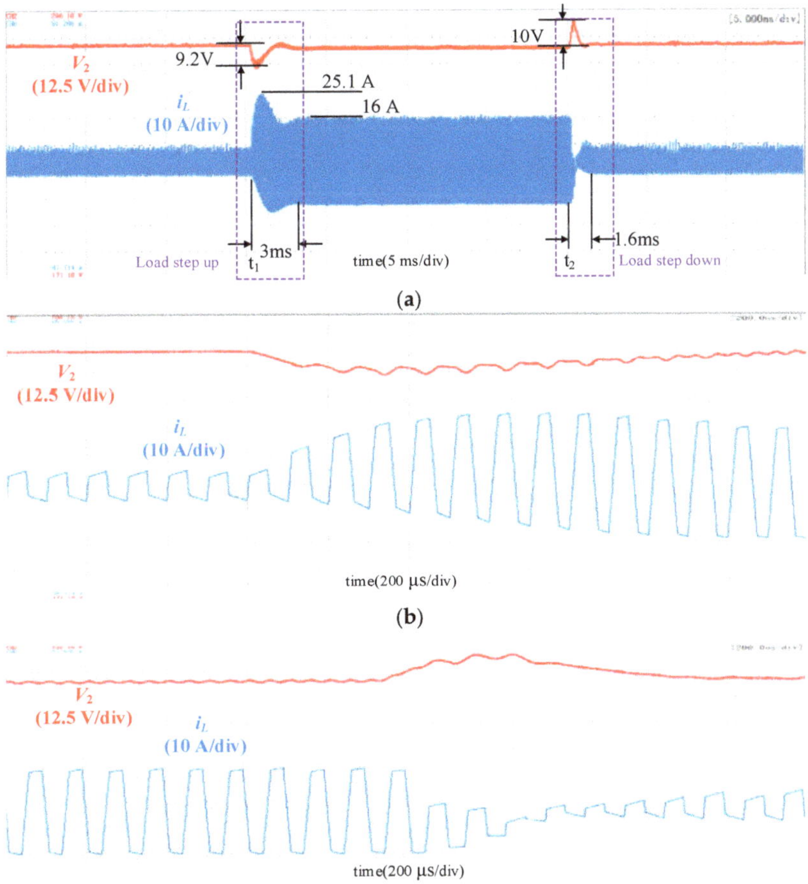

Figure 10. Waveforms using the conventional single-loop control. (**a**) Overall waveforms; (**b**) Zoomed-in waveforms during the load increase; (**c**) Zoomed-in waveforms during the load decrease.

Figure 10 shows the waveform using the conventional single-loop control. The inductor current shows obvious transient DC bias during abrupt load change. When the load increased, the peak current reached 25.1 A, and the current overshoot was 9.1 A. In the experiment, both the static and dynamic performances under load increase and decrease were considered when tuning the PI parameters. The waveform showed that the voltage sag was 9.2 V, and the settling time was 3 ms when the load increased. The voltage overshoot was 10 V, and the settling time was 1.6 ms when the load decreased. Figure 11 shows the waveform using the double-closed-loop control. The inductor current was symmetrical, and the transient DC bias was eliminated. When the load increased, the voltage sag was 8.2 V, and the settling time was 1.9 ms. The voltage overshoot was 9 V, and the settling time was 1.3 ms when the load decreased.

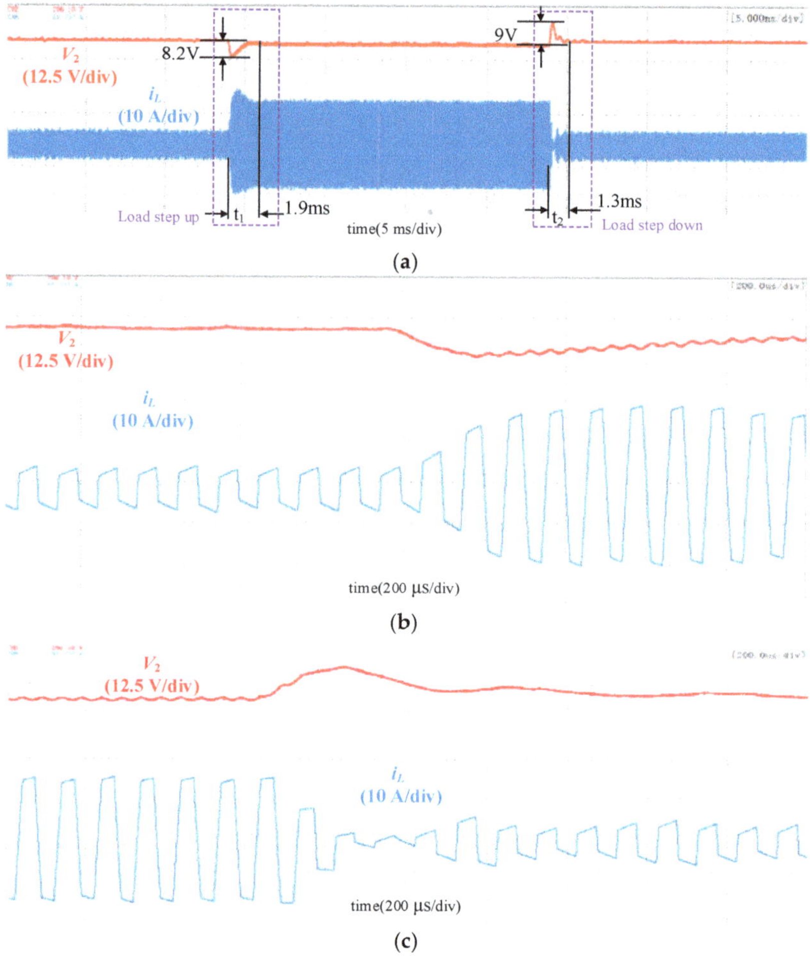

Figure 11. Waveforms using the double-closed-loop control. (**a**) Overall waveforms; (**b**) Zoomed-in waveforms during the load increase; (**c**) Zoomed-in waveforms during the load decrease.

Figure 12 shows the waveforms using the double-closed-loop control with load feedforward. The transient DC bias was eliminated. During the transient process, the DC voltage variation was significantly reduced, and the settling time was obviously shortened compared with the double-closed-loop control without load feedforward. The voltage sag was 3 V when the load increased, and the voltage overshoot was 4.4 V when the load decreased. In the process of load surge, the recovery time for DC voltage was 0.5 ms, which is five switching cycles. The recovery time was 0.6 ms, which is six switching cycles, in the process of load decreases.

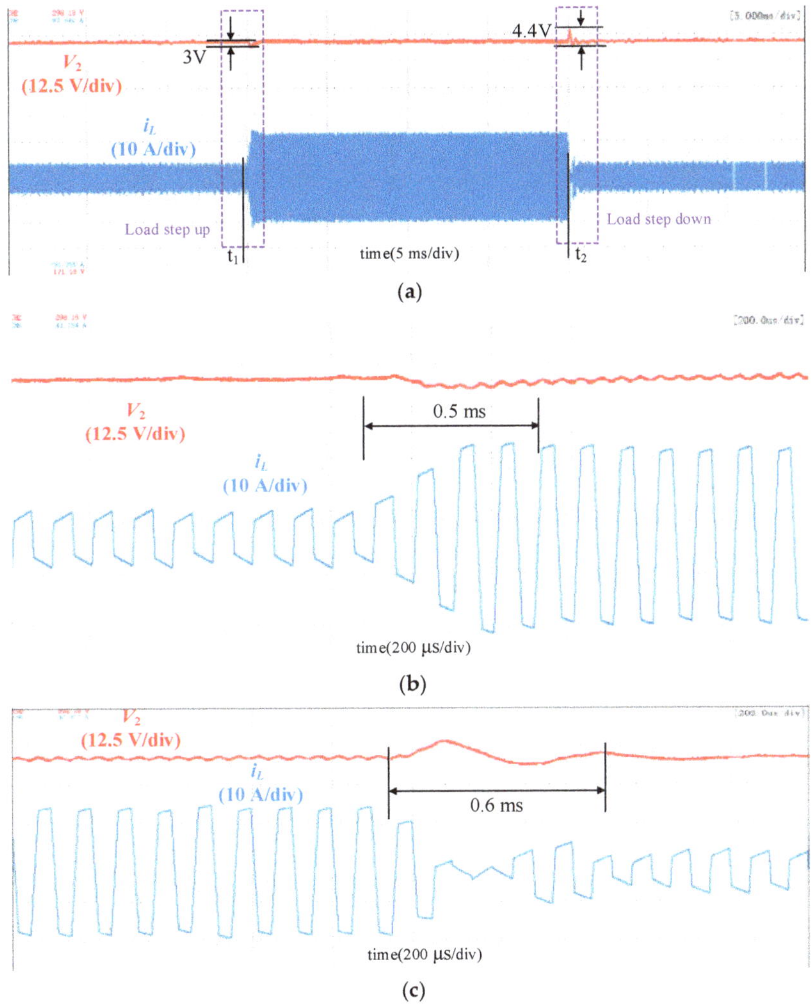

Figure 12. Waveforms using the double-closed-loop control with load feedforward. (**a**) Overall waveforms; (**b**) Zoomed-in waveforms during the load increase; (**c**) Zoomed-in waveforms during the load decrease.

Compared with the traditional single-voltage loop control, the double-closed-loop control utilizes the deadbeat current controller as the inner loop, which directly regulates the high-frequency AC current of the transformer. Thus, the transient DC current bias could be eliminated. Meanwhile, the dynamic performance the of the IBDC with voltage mode mainly depends on the bandwidth of the feedback signal [12]. With the feedforward control samples, the load changes directly, which could significantly increase the robutness of the DC voltage under the load disturblances. The dynamic performance enhancement of the proposed control can be seen by comparison of Figures 11 and 12.

5. Conclusions

A double-closed-loop control strategy based on the deadbeat current controller was proposed in this paper, which directly regulates the high-frequency inductor current to the reference and eliminates the transient DC current bias during the transient process.

Furthermore, load feedforward was introduced to enhance the dynamic of the converter. The proposed control method shows potential in the application of IBDC under voltage mode. With the proposed method, the settling time could be reduced to within several PWM cycles during load disturbance. A 5 kW IBDC converter prototype was built, and the superior dynamic performance of the proposed control strategy was verified by the experimental results.

Author Contributions: Conceptualization, C.T. and S.W.; methodology, S.W.; software, S.W.; validation, S.W.; formal analysis, J.X. and T.B.; investigation, C.T.; resources, S.W.; data curation, J.X.; writing—original draft preparation, T.B.; writing—review and editing, S.W. and J.X.; visualization, C.T.; supervision, C.T.; project administration, C.T.; funding acquisition, C.T. All authors have read and agreed to the published version of the manuscript.

Funding: This research received no external funding.

Data Availability Statement: Not applicable.

Conflicts of Interest: The authors declare no conflict of interest.

References

1. Hou, N.; Li, Y. Overview and comparison of modulation and control strategies for non-resonant single-phase dual-active-bridge dc-dc converter. *IEEE Trans. Power Electron.* **2020**, *35*, 3148–3172. [CrossRef]
2. Wei, S.; Di, M.; Wen, W.; Zhao, Z.; Li, K. Transient DC Bias and Universal Dynamic Modulation of Multiactive Bridge Converters. *IEEE Trans. Power Electron.* **2021**, *37*, 11516–11522. [CrossRef]
3. Hu, J.; Cui, S.; von den Hoff, D.; De Doncker, W.R. Generic dynamic phase-shift control for bidirectional dual-active bridge converters. *IEEE Trans. Power Electron.* **2021**, *36*, 6197–6202. [CrossRef]
4. Yang, C.; Wang, J.; Wang, C.; You, X.; Xu, L. Transient DC Bias Current Reducing for Bidirectional Dual-Active-Bridge DC–DC Converter by Modifying Modulation. *IEEE Trans. Power Electron.* **2021**, *36*, 13149–13161. [CrossRef]
5. Bu, Q.; Wen, H.; Shi, H.; Hu, Y.; Yang, Y. Universal Transient DC-Bias Current Suppression Strategy in Dual-Active-Bridge Converters for Energy Storage Systems. *IEEE Trans. Transp. Electrif.* **2021**, *7*, 509–526. [CrossRef]
6. Dutta, S.; Hazra, S.; Bhattacharya, S. A digital predictive current mode controller for a single-phase high-frequency transformer-isolated dual-active bridge dc-to-dc converter. *IEEE Trans. Ind. Electron.* **2016**, *63*, 5943–5952. [CrossRef]
7. Nasr, M.; Poshtkouhi, S.; Radimov, N.; Cojocaru, C.; Trescases, O. Fast average current mode control of dual-active-bridge dc–dc converter using cycle-by-cycle sensing and self-calibrated digital feedforward. In Proceedings of the Annual IEEE Conference on Applied Power Electronics Conference and Exposition (APEC), Tampa, FL, USA, 26–30 March 2017; pp. 1129–1133.
8. Shan, Z.; Jatskevich, J.; Iu, H.H.; Fernando, T. Simplified load -feedforward control design for dual-active-bridge converters with current-mode modulation. *IEEE J. Emerg. Sel. Top. Power Electron.* **2018**, *6*, 2073–2085. [CrossRef]
9. Wei, S.; Zhao, Z.; Li, K.; Yuan, L.; Wen, W. Deadbeat Current Controller for Bidirectional Dual-Active-Bridge Converter Using an Enhanced SPS Modulation Method. *IEEE Trans. Power Electron.* **2021**, *36*, 1274–1279. [CrossRef]
10. Saggini, S.; Stefanutti, W.; Tedeschi, E.; Mattavelli, P. Digital Deadbeat Control Tuning for dc-dc Converters Using Error Correlation. *IEEE Trans. Power Electron.* **2007**, *22*, 1566–1570. [CrossRef]
11. Bai, H.; Nie, Z.; Mi, C.C. Experimental Comparison of Traditional Phase-Shift, Dual-Phase-Shift, and Model-Based Control of Isolated Bidirectional DC–DC Converters. *IEEE Trans. Power Electron.* **2010**, *25*, 1444–1449. [CrossRef]
12. Segaran, D.; Holmes, G.D.; McGrath, P.B. Enhanced load step response for a bidirectional DC–DC converter. *IEEE Trans. Power Electron.* **2013**, *28*, 371–379. [CrossRef]

micromachines

MDPI

Article

Design and Application of a Flexible Blood Oxygen Sensing Array for Wearable Devices

Wen-Cheng Kuo *, Tzu-Chien Wu and Jun-Sheng Wang

Department of Mechatronics Engineering, National Kaohsiung University of Science and Technology, No. 2 Jhuoyue Rd., Nanzih, Kaohsiung 811532, Taiwan
* Correspondence: rkuo@nkust.edu.tw

Abstract: The performance of portable or wearable oximeters is affected by improper movement or wear, which causes an error in the blood oxygen concentration calculation. The error comes from external incident stray light or light leakage caused by the improper fit of the sensor to the skin. This study aimed to develop a flexible blood oxygen sensing system with a 3×3 array that uses a reflective-type blood oxygen sensing chip to sequentially measure the blood oxygen levels at nine locations through a time division pulse modulation method. Each sensing chip has light transmission and receiving parts. A flip chip package was used to integrate the sensing chip, and a flexible parylene substrate that could fit the curvature of the wrist and locate the array of photo diodes around the radial artery of the wrist was used. By scanning the sensor array in dynamic behavior, the correct light intensity can be extracted to obtain the blood oxygen concentration and prevent errors due to improper fit or sensor movement during exercise.

Keywords: parylene; blood oxygen; sensing array; wearable device

Citation: Kuo, W.-C.; Wu, T.-C.; Wang, J.-S. Design and Application of a Flexible Blood Oxygen Sensing Array for Wearable Devices. *Micromachines* **2022**, *13*, 1742. https://doi.org/10.3390/mi13101742

Academic Editor: Wensheng Zhao

Received: 15 September 2022
Accepted: 7 October 2022
Published: 14 October 2022

Publisher's Note: MDPI stays neutral with regard to jurisdictional claims in published maps and institutional affiliations.

Copyright: © 2022 by the authors. Licensee MDPI, Basel, Switzerland. This article is an open access article distributed under the terms and conditions of the Creative Commons Attribution (CC BY) license (https://creativecommons.org/licenses/by/4.0/).

1. Introduction

Blood oxygen sensing is currently widely used in medical applications for the immediate monitoring of human blood oxygen levels. Oxygen saturation directly reflects the oxygen supply capacity of blood, and oxygen is one of the primary factors for the aerobic metabolism of cell tissue. Hypoxia is harmful to human physiology and can be fatal.

An oxygen saturation percentage higher than 90% is considered normal. A score of 90% or less [1] indicates an aberration that must be further investigated to determine the presence of any undetected health concerns.

The early medical equipment was very large and impractical. In recent years, there have been some analysis methods of physiological signals such as the use of ECG [2] and photoplethysmography (PPG) [3,4] waveforms. The principle of biological radar detection is commonly used to analyze the ECG waveform [2], and the Beer–Lambert law [5,6] is commonly used to analyze the PPG waveform. PPG has been widely implemented in blood oxygen sensors. The concept of PPG is to measure the characteristics of the light absorption intensity when the light penetrates the skin and tissue, calculating the blood oxygen value through the use of red and infrared light that penetrates the biological tissue. When light travels through biological tissues, it is absorbed by various tissue absorption substances such as skin colors, bones and muscles, venous blood, and arterial blood and the arterial blood vessel diameter changes with heart contraction and relaxation. As depicted in Figure 1, the Beer–Lambert law states that the components that cause light attenuation are divided into two categories: alternating signal (AC) and direct current signal (DC). The composition changes from I_H (maximum value) to I_L (minimum value). AC refers to the heart pulse and changes, whereas DC refers to the components that do not change with heart pulsation. The amount of light absorbed by DC is fixed. Nowadays, numerous people are using these wearable devices to determine their current physical condition and monitor their health. Blood oxygen measuring instruments have gradually developed from large-scale device systems to small wearable blood oxygen measuring devices.

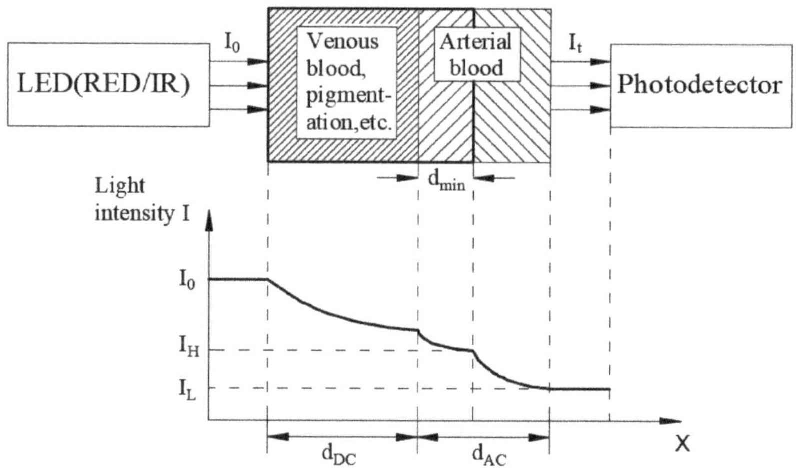

Figure 1. Illustration of the Beer–Lambert law.

In optical blood oxygen measurement methods, PPG has numerous advantages such as a small size, low power consumption, less dependency on a power supply, less electromagnetic interference, and a rapid response. Therefore, it has been gradually introduced into various applications including fitness trackers and homecare equipment.

PPG oximeters can be divided into two categories: transmitting and reflective oximeters, depending on the location of the light emitting and received signals. The light-emitting components (light-emitting diode (LED)) and -receiving components (photodiodes) of the transmitting oximeter are located on both sides of the measuring object. Furthermore, the light-emitting and -receiving parts of the reflective oximeter are located on the same side of the measuring object, there is a greater number of measuring positions, and the device is convenient to wear. The key advantage of the reflective blood oxygen sensor is convenience, which enables the device to be applied in the growing homecare market, and thus many of these wearable devices are available in the market. Depending on the measurement position, these can be divided as follows: ear-clip-type oxygen measurement [7,8], forehead oxygen measurement [7,9], and watch pulse oximetry [10].

Ear-clip-type blood oxygen sensors are easy to manufacture and are widely used in pulse oximetry. Spring-type ear clips were used in early designs. Although it is feasible to use them for measurement, its clamping pressure hinders operations during long-term measurement. A new type of microelectromechanical system (MEMS) technology can be used to create a lighter and more comfortable earphone oximetry device, which enables the monitoring of the ear oximetry sensor during physical activities [7].

The advantages of forehead oxygen measurement compared with other measurements in the body are that the thin cortex of the forehead and prominent skeletal structure help to redirect the light to the photodiode, thus providing a higher sensitivity and accuracy to changes in arterial signals [7] and reducing errors caused by physical activity. A forehead measurement device, which is generally installed on a helmet or sports headscarf, can be used during physical activity or to inform other soldiers of their current physiological condition on the battlefield [9].

In watch-type pulse oximetry, most wearable sports watch devices are used to measure the pulse and heart rate. However, because of the problem of light leakage, they are rarely used to measure the human blood oxygen concentration in dynamic behavior. A new PPG oximetry module (e.g., those of Garmin and Apple's iWatch) with a watch-type design has been developed as a wearable device to measure the blood oxygenation of the ulnar artery of the radial artery and wrist. Although these devices are easy to wear, they are not conventionally used for clinical purposes [10].

Improper site selection or sensor movement causes inaccurate light intensity reception by the sensor, and this generates the blood oxygen saturation concentration (S_pO_2) error. Limited by the width and curvature of the wrist, it is not possible to install too many arrays. Too many arrays would increase the size of the devices, and the outer sensor chips would be far away from the radial artery of the wrist, causing measurement errors. This research developed a 3 $\times$ 3 reflective sensing array for flexible wrist pulse oximetry to obtain S_pO_2. Each sensing chip has individual transmission (LED) and receiving (photodiode) components. The flip chip package was used to integrate the sensing chips, and a flexible parylene substrate that could fit the curvature of the wrist and locate the array of LEDs and photodiodes around the radial artery of the wrist was used. By scanning the sensor array, the correct light intensity can be extracted to obtain SpO_2 and prevent errors due to improper fitting or sensor movement during exercise.

2. Research Method and Design

To evaluate S_pO_2 in human veins or arteries, the blood oxygen saturation concentration was determined by measuring the difference in the ratio of oxygen in the light absorption spectrum to the difference in the absorption spectrum of oxygenated hemoglobin (H_bO_2) and deoxygenated hemoglobin (H_b). Figure 2 presents the light absorbance spectra for oxygenated hemoglobin and deoxygenated hemoglobin [11].

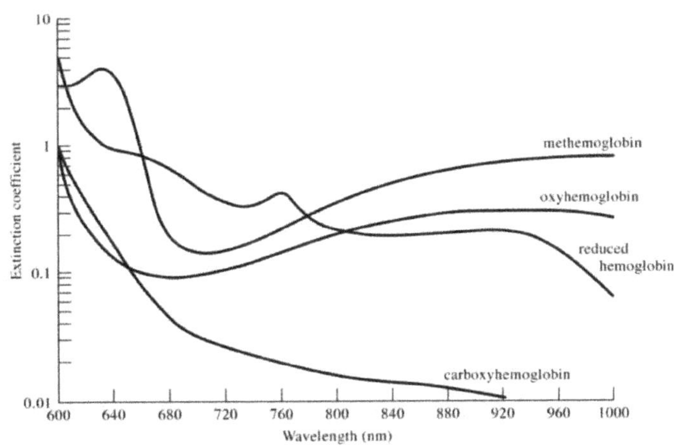

Figure 2. The light absorbance spectra for methemoglobin, oxyhemoglobin, reduced hemoglobin, and carboxyhemoglobin.

The Beer–Lambert law concerns the attenuation of light and the properties of the material through which the light travels [5]. As a solution, for a certain wavelength of light absorbed by the medium in the solution, the transmitted light penetrating the solution follows the medium in the solution. The intensity of light transmitted through a tissue that includes vessels with whole blood is represented as follows [12]:

$$I_t = I_0 \exp(-\varepsilon\, Cl),\ In(I_0/I_t) = \varepsilon\, Cl \quad (1)$$

where I_t, I_0, ε, C, and l represent the transmitted light intensity, incident light intensity, an extinction coefficient, the concentration of hemoglobin in the tissue, and the path of light, respectively.

The incident light intensity I_0 of the light that passes through the tissue to be measured is affected by components such as venous blood, arterial blood, skin, bones, and other tissues (DC component), which attenuate light intensity and are pulsed in the blood. I_L is lower light transmission through the tissue at a higher tissue blood volume, and I_H is higher transmitted light at a lower tissue blood volume. Therefore, Equation (1) can be rewritten as follows:

$$I_L = I_H \exp(-\varepsilon \Delta C l),\ \ln(I_H/I_L) = \varepsilon \Delta C l \tag{2}$$

Minor changes to blood vessels are expressed as ($I_H - I_L << I_L$) and $ln(I_H/I_L)$ is approximate to $(I_H - I_L)/I_L$, and thus Equation (2) can be expressed as follows:

$$(I_H - I_L)/I_L = \varepsilon \Delta C l \tag{3}$$

For the blood oxygen concentration measurement (Figure 3), this study was conducted using a red light source (660 nm) and infrared (905 nm) dual-wavelength light source, and light from these sources was transmitted into the human tissue. In Equation (3), I_d and I_s are substituted for I_H and I_L, respectively. The light absorption ratio (R) at the two wavelengths can be defined as follows:

$$R = \frac{[(I_H - I_L)/I_L]_{(660)}}{[(I_H - I_L)/I_L]_{(905)}} = \frac{[(I_D - I_S)/I_S]_{(660)}}{[(I_D - I_S) - I_S]_{(905)}} \tag{4}$$

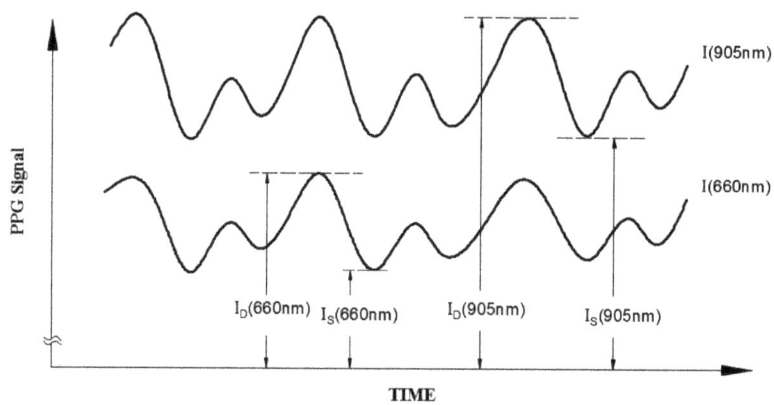

Figure 3. Illustration of the separated red (660 nm) and infrared (905 nm) patient signals with their I_D and I_S values caused by arterial pulsation.

The blood oxygen saturation concentration (S_pO_2) equation [13], where H_bO_2 and H_b represent the concentrations of oxygenated hemoglobin in the blood and no hemoglobin in the blood, is defined as follows [14]:

$$S_PO_2 = \frac{H_bO_2}{H_bO_2 + H_b} \times 100\% \tag{5}$$

The extinction coefficient ε of the hemoglobin is defined as the absorption constant of the blood sample divided by the concentration of the hemoglobin in the sample. The total extinction coefficient ε of the hemoglobin in the blood sample is related to its oxygen saturation because the hemoglobin in blood is a mixture of HbO_2 and Hb with the extinction coefficients ε_0 and ε_D, respectively, and can be represented as follows:

$$\varepsilon = \varepsilon_0 S_PO_2 + \varepsilon_D(1 - S_PO_2) = \varepsilon_D + S_PO_2(\varepsilon_0 - \varepsilon_D) \tag{6}$$

If the difference between the blood concentration change ΔC in the two wavelengths is neglected, this satisfies Equation (6), and R can be approximately expressed as follows [12]:

$$R \approx \varepsilon_{(660)} / \varepsilon_{(905)} \tag{7}$$

The relationship between the measured parameter R and SpO_2 in the arterial blood can be derived using Equations (6) and (7) through a simple manipulation as follows:

$$S_PO_2 = \frac{\epsilon_{D(660)} - R \cdot \epsilon_{D(905)}}{R\left(\epsilon_{0(905)} - \epsilon_{D(905)}\right) + \left(\epsilon_{D(660)} - \epsilon_{D(660)}\right)} \tag{8}$$

Equation (9) was calculated using the values of ε_D and ε_0 in Table 1 [11,15], presenting the relationship between the oxygen concentration value and R value.

$$S_PO_2 = \frac{0.81 - 0.21 \times R}{0.09 \times R + 0.73} \tag{9}$$

Table 1. The extinction coefficients ($\mathrm{Lmmol^{-1}cm^{-1}}$) of the reduced ($\varepsilon H_b$) and ($\varepsilon H_bO_2$) oxygenated hemoglobin at the wavelengths of 660 nm and 905 nm.

Wavelength (nm)	εH_b	εH_bO_2
660	0.81	0.08
905	0.21	0.30

However, Equation (9) is the theoretical SpO_2 value calculated using the R value. In practical applications, the performance of the LED emitter and photodiode needs to be calibrated to obtain the correct value. In this study, the FLUKE Index 2 Pulse Oximeter Simulator was used as the calibration instrument [16]. The test probe of the FLUKE Index 2 Pulse Oximeter Simulator detects the red light and infrared light emitted by the LED emitters, and then uses the LED emitter of the test probe to emit the red and infrared light by simulating 35–100% of the SpO_2 value to correct the photodiode used in this study and control the error rate within 1%. The calibration method is shown in Figure 4. The LED emitters used in this study emit red light (Red) and infrared light (IR), which are received by the photodiode on the test probe. The programmed calibrated default SpO_2 of the PPG waveform is sent to the LED emitter on the test probe to emit, and is received by the photodiode used in this study. By calculating the peak and trough values of the red light and infrared light waveforms, the corrected R value under the corresponding SpO_2 value can be obtained. Figure 5 shows the relationship between the theoretical and corrected blood oxygen saturation concentration values in relation to the R value.

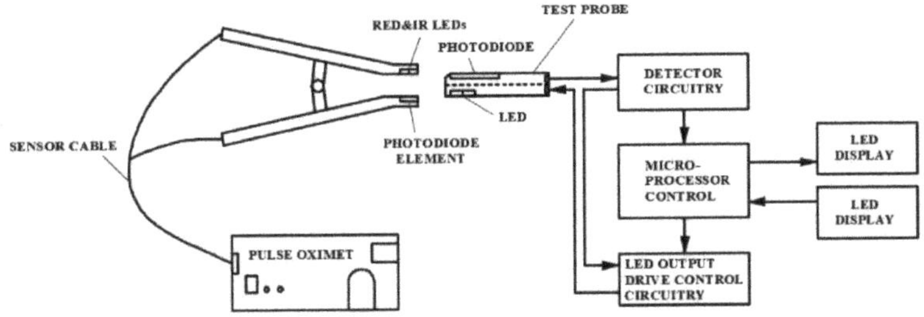

Figure 4. Schematic diagram of the LED emitter and photodiode calibrated by the pulse oximeter simulator.

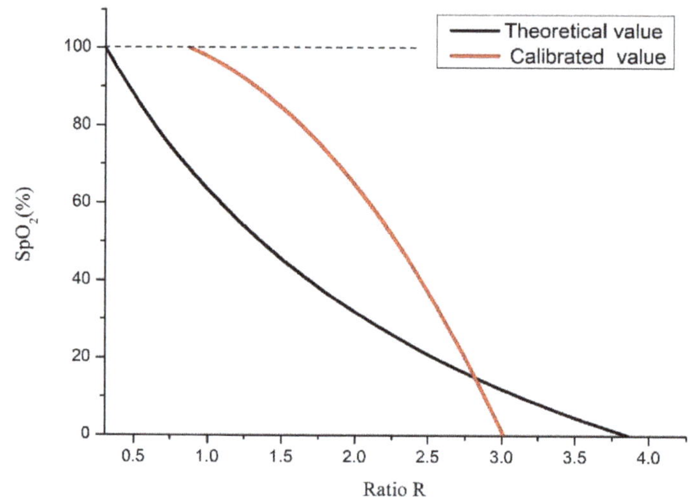

Figure 5. Relationship between the oxygen concentration value and R value.

In addition to the fact that the LED and photodiode must be calibrated, the emitter and receiver chips must be properly attached to the skin to avoid receiving external incident light and signaled light leakage, resulting in errors in SpO_2 measurement. This study developed a sensing system with a 3 × 3 array to sequentially measure the blood oxygen levels at nine locations through a time division pulse modulation method (TDPM).

If the LED emitter and receiver are not properly fitted with the skin, the external incident stray light will be received by the receiver, and some of the light emitted by the LED emitter cannot be received since it is too large (external incident stray light) or too small (signal light), and these wrong PPG waveforms are easy to identify. In real cases, it is not easy to generate a complete PPG waveform. In addition, compared to the normal received light intensity, the light intensity is higher than expected. These factors make it easy to identify incorrect R values and blood oxygen values. When the tester exercises a dynamic behavior, the sensing system can receive the PPG waveforms received by all receivers for calculation and extract the correct blood oxygen value; if one receiver can receive the signal correctly, it can provide the correct SpO_2 value.

3. Fabrication

A 3 × 3 array blood oxygen sensing system based on the PPG method and Beer–Lambert law was proposed that uses a reflective blood oxygen sensing module to sequentially measure the blood oxygen levels at nine locations.

The measurement position was on the radial artery of the left hand [17] (Figure 6). To increase the closeness between the sensor and skin on the curvature of the wrist, a biocompatible material (parylene) was used as the flexible substrate on which to fabricate the flexible circuit board that was packaged with nine commercial oxygen concentration sensing chips (DCM-05, AMPKorea; specifications shown in Table 2) using the flip chip technique. The sensing module chip can be attached to the skin for blood oxygen concentration measurement.

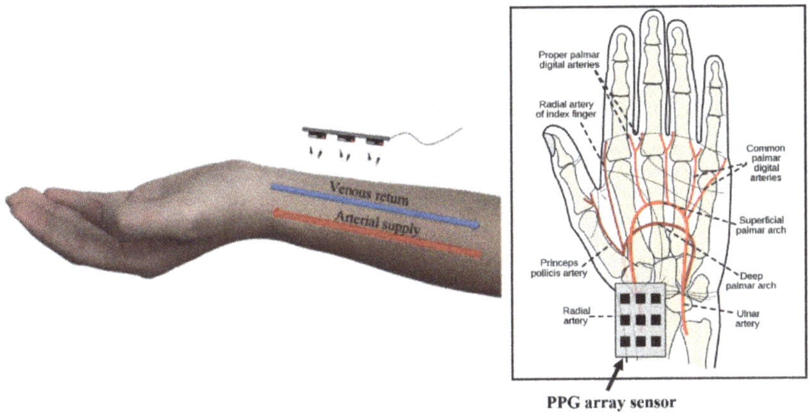

Figure 6. Illustration of the radial artery and sensing array location.

Table 2. The datasheet of DCM05.

SYMBOL	CHARACTERISCTIC	COMPONENTS	TEST CONDITION	MIN	TYP	MAX	UNITS
VF	Forward Voltage	LED1		-	1.9	2.2	V
		LED2	IF = 20 mA	-	1.3	1.5	
		PD	IF = 10 mA	0.5	-	1.3	
IR	Revers Breakdown Current	LED1, LED2	VR = 5 V	-	-	10	uA
PO	Output Power	LED1	IF = 20 mA	-	2	-	mW
		LED2		-	2	-	
λ PEAK	Peak Wavelength	LED1	IF = 20 mA	-	660	-	nm
		LED2		-	905	-	
		PD		-	940	-	
Δλ	Half Wave Width	LED1	IF = 20 mA	-	30	-	nm
Δλ	Half Wave Width	LED2	IF = 20 mA	-	60	-	nm
V BR	Reverse Breakdown Voltage	PD	IR = 100 Ua	35	-	-	V
I D	Reverse Dark Current	PD	V R = 10 V	-	-	20	nA
I L	Light Current		1mW@940 nm	-	21	-	uA
S	Spectral Response Range		-	400	-	1050	nm
CJ	Junction Capacitance		VR = 3 V, f = 1 MHz	-	20	-	pF

The proposed flexible blood oxygen sensing array system developed in this research can be divided into five categories: a blood oxygen sensing array, a modulation module, a blood oxygen data processing module, a zero-insertion-force (ZIF) [18] connector, and a biocompatible flexible circuit board (Figure 7).

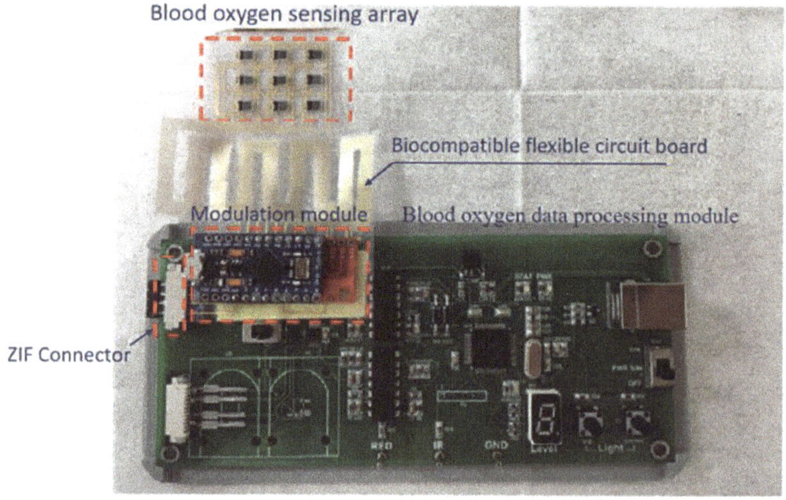

Figure 7. The actual blood oxygen sensing system.

3.1. Blood Oxygen Sensing Array

In this study, sensing chips were placed in a 3 × 3 array. The length and width of the array area were 25 and 18 mm, respectively. The distances between the sensing chips were 3 mm, 5 mm, and 7 mm (Figure 8). When the device was smoothly attached to the skin, the accurate PPG waveform could be measured, and SpO_2 could be obtained.

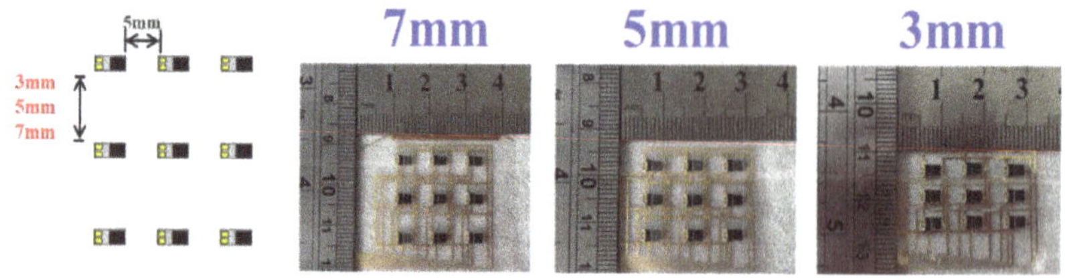

Figure 8. Three sensing element spacing designs: 3 mm, 5 mm, and 7 mm.

3.2. Modulation Module

The processing method for the data from the 3 × 3 sensing module (Ch0~Ch8) was based on the time division pulse modulation (TDPM) method (Figure 9), where the red and infrared light emitted by each channel were sequentially received by the photodetector. In this study, the TDPM was used to perform the sweep measurement of nine sensing chips, the sensing time of each channel was 5 s, and the switching time was 0.1 s. The modulation module comprised an Arduino Micro Development Board (Arduino Micro; specifications shown in Table 3) and a multiplexer CD74HC4067 (specifications shown in Table 4) that were used to perform and execute the TDPM. The Arduino Micro sent digital commands to control the CD74HC4067, causing it to perform a sweep, guiding the analog signals from nine channels to the blood oxygen data processing module.

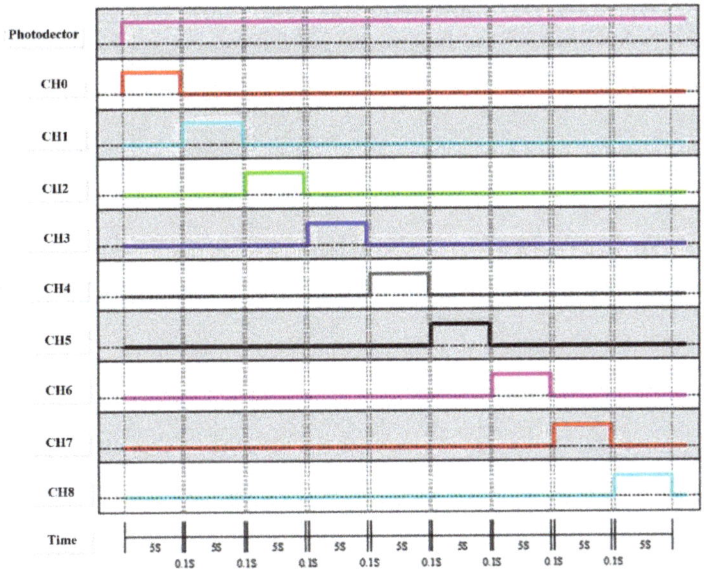

Figure 9. Time table of the photo detector and sensing element actuation.

Table 3. Datasheet of the Arduino Micro Development Board.

Dimension	35 mm × 18.5 mm × 4.5 mm
Microcontroller	ATmega32u4
Operating Voltage	5 V
Input Voltage (recommended)	7–12 V
Digital I/O Pins	20
PWM Channels	7
Analog Input Channels	12
DC Current for 5V Pin	40 mA
DC Current for 3.3V Pin	50 mA
Flash Memory	32 KB (ATmega32u4) of which 4 KB used by bootloader
SRAM	2.5 KB
EEPROM	1 KB
Clock Speed	16 MHz

3.3. Blood Oxygen Data Processing Module

The blood oxygen data processing module (SDPPG Kit, AMPKorea Inc.) received the analog PPG signals from nine channels to record and change the analog signals into digital signal data points. To integrate the modulation module and blood oxygen data processing module to yield a portable device, the Solidworks software was used to design the device case (Figure 10), and 3D printing technology was used to create it.

Table 4. Datasheet of the high-speed CMOS 16-channel analog multiplexer/demultiplexer CD74HC4067.

PARAMETER	SYMBOL	TEST CONDITIONS		VCC (V)	25 °C			−40 °C TO 85 °C		−55 °C TO 125 °C		UNITS
		VI (V)	VIS (V)		MIN	TYP	MAX	MIN	MAX	MIN	MAX	
HC TYPES												
High Level Input Voltage	V_{IH}	-	-	2	1.5	-	-	1.5	-	1.5	-	V
				4.5	3.15	-	-	3.15	-	3.15	-	V
				6	4.2	-	-	4.2	-	4.2	-	V
Low Level Input Voltage	V_{IL}	-	-	2	-	-	0.5	-	0.5	-	0.5	V
				4.5	-	-	1.35	-	1.35	-	1.35	V
				6	-	-	1.8	-	1.8	-	1.8	V
Maximum "ON" Resistance I_O = 1 mA	R_{ON}	VCC or GND	VCC or GND	4.5	-	70	160	-	200	-	240	Ω
				6	-	60	140	-	175	-	210	Ω
		VCC to GND	VCC to GND	4.5	-	90	180	-	225	-	170	Ω
				6	-	80	160	-	200	-	240	Ω
Maximum "ON" Resistance Between Any Two Switches	ΔR_{ON}	-	-	4.5	-	10	-	-	-	-	-	Ω
				6	-	8.5	-	-	-	-	-	Ω
Switch "Off" Leakage Current 16 Channels	I_{IZ}	$\bar{E}$ = VCC	VCC or GND	6	-	-	±0.8	-	±8	-	±8	μA
Logic Input Leakage Current	I_I	VCC or GND	-	6	-	-	±0.1	-	±1	-	±1	μA
Quiescent Device Current I_O = 0 mA	I_{CC}	VCC or GND	-	6	-	-	8	-	80	-	160	μA

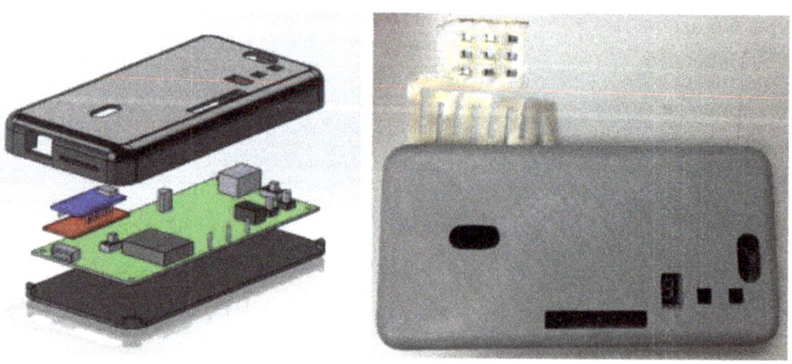

Figure 10. Illustration of the real 3D printing case.

3.4. ZIF Connector

Wire bonding is a conventional method for connecting the sensing chip and circuit board; however, the parylene substrate in this study could not withstand high-temperature soldering. To ensure the smooth connection between the sensor chip and parylene flexible circuit board, the ZIF connector [18] was used to connect the parylene flexible substrate and printed circuit board.

3.5. Biocompatible Flexible Circuit Board

This study used a biocompatible grade of parylene as the flexible substrate, and gold was then deposited as conduction wires and pads to connect the sensing chip. The flip chip packaging technology was used to fit the sensing chips on the pads. The biocompatible flexible circuit board enabled the attachment of sensing chips and the skin to measure the blood oxygen concentration. Figure 11 shows the fabrication process of the biocompatible flexible circuit board, in which there are two cross-sections, A-A′ and B-B′. A-A′ represents the sensor and ZIF contacts, respectively, and B-B′ represents the interconnection/jumper wire. The process steps are described stepwise as follows:

Step (a): PR/parylene coating

The photoresist (PR) was coated on the substrate using a spin coater, and the biocompatible parylene film was then deposited on the glass substrate through chemical vapor deposition. As a flexible substrate, its thickness was 30 μm.

Step (b): First Cr/Au deposition and PR patterning

Cr and Au were deposited through sputtering with a thickness of 20/200 nm to create a seed layer and the conduction wires, respectively. The unetched area was defined by photoresist AZ-4620 with a thickness of 5 μm.

Step (c): First Cr/Au etching and PR removal

Unprotected Au and Cr were etched using Au etchant and Cr-7, respectively. PR was removed using acetone.

Step (d): Second parylene deposition and PR patterning

A parylene film was deposited through chemical vapor deposition to protect and insulate the inner gold wire with a thickness of 5 μm. The etched pattern was defined by the photoresist AZ-4620 for the purpose of via opening.

Step (e): Parylene etched using reactive ion etching and PR removal

The unprotected parylene by the photoresist above the via opening was etched using reactive ion etching (RIE) and oxygen plasma to expose the via, and the photoresist was removed using acetone.

Step (f): Second Cr/Au deposition and PR patterning

Using sputtering, Cr and Au were deposited with a thickness of 20/200 nm to create the seed layer and second conduction wires, respectively. The unetched area was defined through photoresist AZ-4620 with a thickness of 5 μm.

Step (g): Second Cr/Au etching and PR removal

Unprotected Au and Cr were etched using the Au etchant and Cr-7, respectively. The photoresist was removed using acetone.

Step (h): Silver epoxy pasting

A biocompatible grade of silver was pasted on the gold connection pads for the flip chip mounting of sensing chips.

Step (i): Sensing chip mounting and ZIF connector mounting

After the silver glue was coated on the gold connection pads, nine sensing chips and ZIF connectors were attached to the gold connection pads, and were then baked in an oven at 80 °C for 3 h.

Step (j): Third parylene deposition

The third parylene deposition was applied to create the upper protection layer to protect and insulate the inner parts.

Step (k): Delamination form substrate

The biocompatible flexible circuit board was released from the substrate by peeling.

Three designs were proposed (Figure 8) for the sensing chip array location. The spacing in the x-direction was fixed at 5 mm, and the spacings in the y-direction were 3 mm, 5 mm, and 7 mm, respectively. Figure 12 depicts photos of the sensing chip number and location of three different pitch sensing arrangements. The red-dotted line indicates the radial artery position and crosses the center of the sensing chips (Nos. 1, 4, and 7).

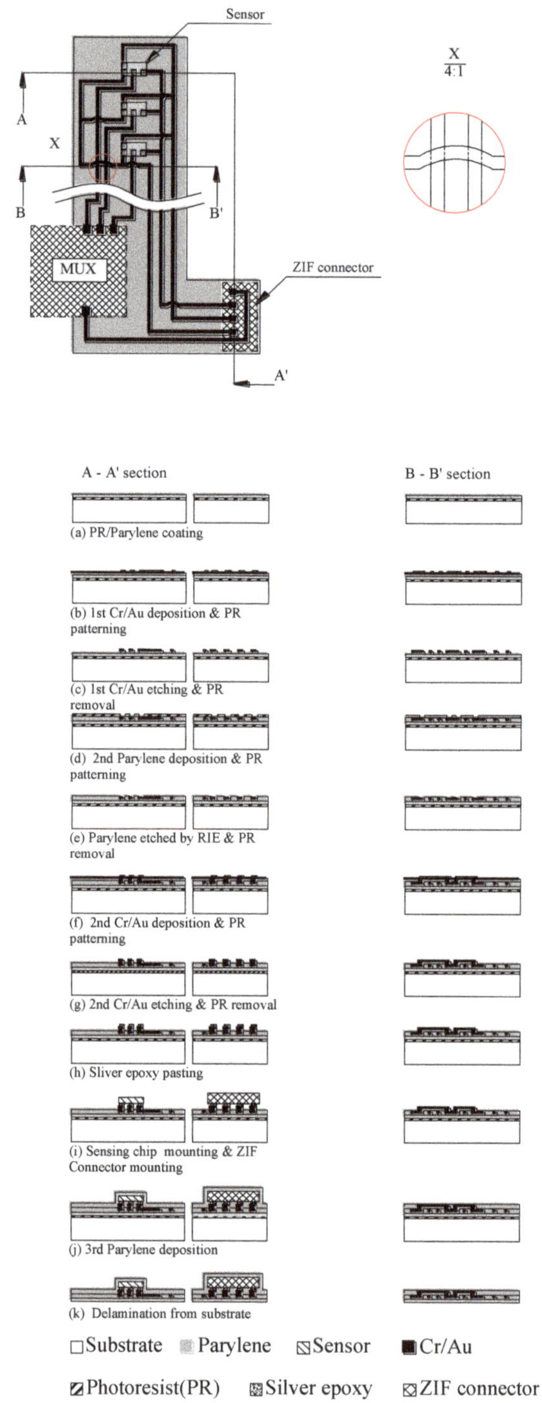

Figure 11. The fabrication process of the biocompatible flexible blood circuit board.

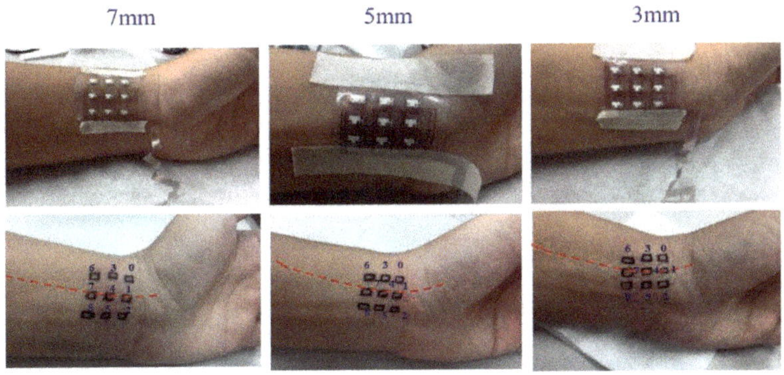

Figure 12. Photos of the sensing chip number and placement of three different pitch sensing chips. The red dotted line indicates the radial artery position and crosses the center of the sensing chips.

4. Measurements

In this study, the blood oxygen sensing array was used for dynamic measurements to examine the influence of the light leakage phenomenon caused by the improper fit between the sensor and the skin in the dynamic behavior of blood oxygen detection. To take a dynamic measurement, the arm was slightly swung to simulate the state of people walking during an activity. The arm swing frequency was approximately 70 times per minute. The arm was swung once to represent one step. A person walked one step and moved a distance of approximately 70–80 cm. A simulated speed of 3.15 km/h or 19.04 min/km was used to evaluate the feasibility of the device in the study. The measurement results indicate the sensor movement caused by the swing of the arm.

Figure 13, Figure 14, and Figure 15 show the PPG waveforms of the blood oxygen array design with spacings of 3, 5, and 7mm, respectively. It can be found in these PPG waveforms that due to the influence of improper fitting, the intensity of the external incident stray light is greater than the intensity of the light received by the emitter, and it is difficult to form a complete PPG waveform. It can be determined that it is not a correct PPG waveform. Figure 13 shows that only Ch1, 6, and 8 presented complete PPG waveforms under the design with a sensor spacing of 3 mm. Figure 14 shows that only Ch1, 2, and 4 presented complete PPG waveforms under the design with a sensor spacing of 5 mm. Figure 15 shows that only Ch0 presented complete PPG waveforms under the design with a sensor spacing of 5mm. Comparing the complete PPG waveforms in Figure 13, Figure 14, and Figure 15, the received light intensity of the 7 mm spacing design was smaller than that of the 3 and 5 mm designs because the sensor was far from the radial artery, so the received signal of the radial artery was relatively small.

In the calculation of the blood oxygen concentration, if there are multiple sensors that receive the correct PPG waveform, and because the abnormal reception will reduce the light intensity, the R value with the largest light intensity was selected for the calculation of the blood oxygen concentration. Tables 5–7 show the calculated blood oxygen values of the array sensors under the design with sensor spacings of 3, 5, and 7 mm, respectively. From these tables, it can be seen that the values of the blood oxygen values calculated from the complete PPG waveform were very close. The reason for this is that the R value is the ratio of the intensity of red light to infrared light. If the received intensities of red light and infrared light are both attenuated to the same degree, the difference in the ratio is not significant.

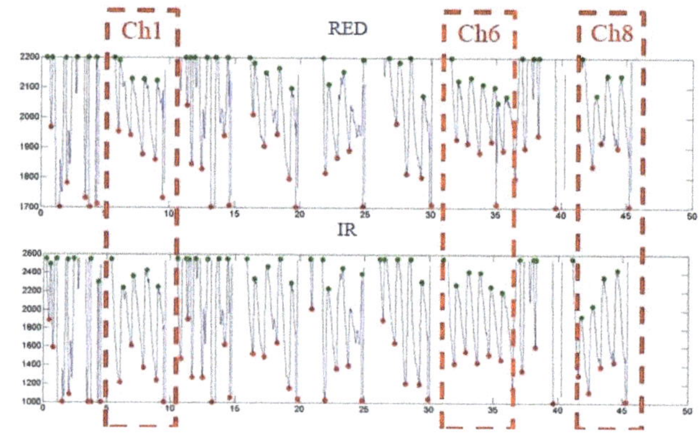

Figure 13. Dynamic state of the blood oxygen concentration value measurement results of a flexible blood oxygen sensing array with a pitch of 3 mm.

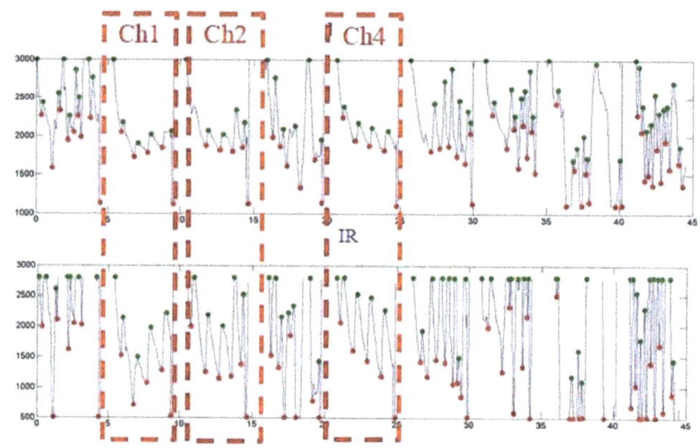

Figure 14. Dynamic state of the blood oxygen concentration value measurement results of a flexible blood oxygen sensing array with a pitch of 5 mm.

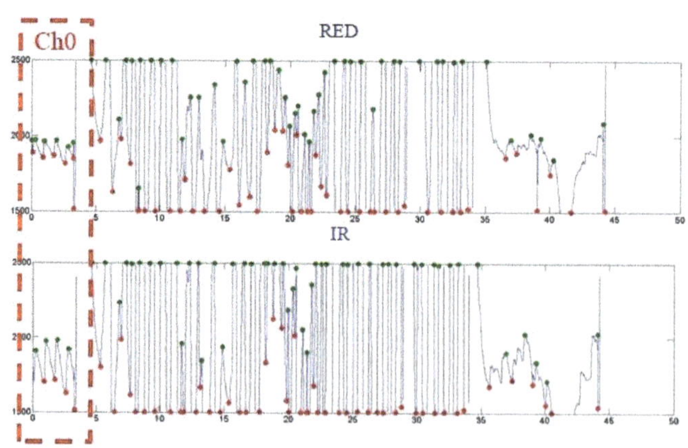

Figure 15. Dynamic state of the blood oxygen concentration value measurement results of a flexible blood oxygen sensing array with a pitch of 7 mm.

Table 5. The calculated blood oxygen concentration of the array sensors using the design with a sensor spacing of 3 mm.

Ch0	Ch3	Ch6
Ave. S_pO_2 N.A	Ave. S_pO_2 N.A	**Ave. S_pO_2** **~99**
Ch1	Ch4	Ch7
Ave. S_pO_2 **~76**	Ave. S_pO_2 N.A	Ave. S_pO_2 N.A
Ch2	Ch5	**Ch8**
Ave. S_pO_2 N.A	Ave. S_pO_2 N.A	**Ave. S_pO_2** **~99**

Table 6. The calculated blood oxygen concentration of the array sensors using the design with a sensor spacing of 5 mm.

Ch0	Ch3	Ch6
Ave. S_pO_2 N.A	Ave. S_pO_2 N.A	Ave. S_pO_2 N.A
Ch1	**Ch4**	Ch7
Ave. S_pO_2 **~99**	**Ave. S_pO_2** **~98**	Ave. S_pO_2 N.A
Ch2	Ch5	Ch8
Ave. S_pO_2 **~99**	Ave. S_pO_2 N.A	Ave. S_pO_2 N.A

Table 7. The calculated blood oxygen concentration of the array sensors using the design with a sensor spacing of 7 mm.

Ch0	Ch3	Ch6
Ave. S_pO_2 **~99**	Ave. S_pO_2 N.A	Ave. S_pO_2 N.A
Ch1	Ch4	Ch7
Ave. S_pO_2 N.A	Ave. S_pO_2 N.A	Ave. S_pO_2 N.A
Ch2	Ch5	Ch8
Ave. S_pO_2 N.A	Ave. S_pO_2 N.A	Ave. S_pO_2 N.A

5. Conclusions

This study proposed a 3 × 3 array blood oxygen sensing system based on the PPG method and Beer–Lambert law that uses a reflective blood oxygen sensing module to sequentially measure the blood oxygen levels at nine locations. A flexible parylene based substrate to keep the sensor fit to the skin can reduce the blood oxygen concentration calculation error caused by light leakage. The dynamic behavior measurement results show that the smaller the spacing of sensors away from the radial artery, the more sensors that can receive the complete PPG waveform, the greater the received light intensity. The array type sensor design can detect the blood oxygen concentration value at multiple points. Under dynamic behavior, one sensor is correctly attached to the skin, and the blood oxygen concentration value can be obtained.

Author Contributions: Conceptualization, W.-C.K. and T.-C.W.; investigation, W.-C.K.; resources, W.-C.K., T.-C.W. and J.-S.W.; writing—original draft preparation, W.-C.K.; writing—review and editing, W.-C.K. All authors have read and agreed to the published version of the manuscript.

Funding: This research was funded by the Ministry of Science and Technology, Grant/Award Number: MOST 109-2221-E-992-033-MY3.

Data Availability Statement: The data presented in this study are available upon request from the corresponding author. The data are not publicly available due to funders, and so cannot be made freely available.

Conflicts of Interest: The authors declare no conflict of interest. The funders had no role in the design of the study; in the collection, analyses, or interpretation of data; in the writing of the manuscript; or in the decision to publish the results.

References

1. Sugathan, A.; Roy, G.G.; Kirthyvijay, G.J.; Thomson, J. Application of arduino based platform for wearable health monitoring system. In Proceedings of the 2013 IEEE 1st International Conference on Condition Assessment Techniques in Electrical Systems (CATCON), Kolkata, India, 6–8 December 2013; pp. 1–5.
2. Ji, S.; Zhang, Z.; Xia, Z.; Wen, H.; Zhu, J.; Zhao, K. RBHHM: A novel remote cardiac cycle detection model based on heartbeat harmonics. *Biomed. Signal Process. Control* **2022**, *78*, 103936. [CrossRef]
3. Kamal, A.A.R.; Harness, J.B.; Irving, G.; Mearns, A.J. Skin photoplethysmography—A review. *Comput. Methods Programs Biomed.* **1989**, *28*, 257–269. [CrossRef]
4. Alen, J. Photoplethysmography and its application in clinical physiological measurement. *Physiol. Meas.* **2007**, *28*, R1–R39. [CrossRef] [PubMed]
5. Hay, O.Y.; Cohen, M.; Nitzan, I.; Kasirer, Y.; Shahroor-karni, S.; Yitzhaky, Y.; Engelberg, S.; Nitzan, M. Pulse oximetry with Two infrared wavelengths without calibration in extracted arterial blood. *Sensors* **2018**, *18*, 3457. [CrossRef]
6. Jensen, L.A.; Onyskiw, J.E.; Prasad, N.G.N.N. Meta-analysis of arterial oxygen saturation monitoring by pulse oximetry in adults. *Heart Lung* **1998**, *27*, 387–408. [CrossRef]
7. Tamura, T.; Maeda, Y.; Sekine, M.; Yoshida, M. Wearable Photoplethysmographic Sensors—Past and Present. *Electronics* **2014**, *3*, 282–302. [CrossRef]
8. Budidha, K.; Kyriacou, P.A. The human ear canal: Investigation of its suitability for monitoring photoplethysmographs and arterial oxygen saturation. *Physiol. Meas.* **2014**, *35*, 111–128. [CrossRef] [PubMed]
9. Dresher, R. Wearable Forehead Pulse Oximetry: Minimization of Motion and Pressure Artifacts. Master's Thesis, Department of Biomedical Engineering, Worcester Polytechnic Institute, Worcester, MA, USA, 2006.
10. Lee, Y.; Shin, H.; Jo, J.; Lee, Y.K. Development of a wristwatch-type PPG array sensor module. In Proceedings of the 2011 IEEE International Conference on Consumer Electronic, Berlin Germany, 9–12 January 2011; pp. 168–171.
11. Zijlstra, W.G.; Buursma, A.; Meeuwsen-Van der Roest, W.P. Absorption spectra of human fetal and adult oxyhemoglobin, de-oxyhemoglobin, carboxyhemoglobin, and methemoglobin. *Clin. Chem.* **1991**, *37*, 1633–1638. [CrossRef] [PubMed]
12. Nitzan, M.; Engelberg, S. Three-wavelength technique for the measurement of oxygen saturation in arterial blood and in venous blood. *J. Biomed. Opt.* **2009**, *14*, 024046. [CrossRef] [PubMed]
13. Gao, F.; Peng, Q.; Feng, X.; Gao, B.; Zheng, Y. Single-wavelength blood oxygen saturation sensing with combined optical absorption and scattering. *IEEE Sens. J.* **2016**, *16*, 1943–1948. [CrossRef]
14. Reichelt, S.; Fila, J.; Werber, A.; Foerster, K.; Heilmann, C.; Klemm, R.; Zappe, H. Development of an implantable pulse oximeter. *IEEE Trans. Biomed. Eng.* **2008**, *55*, 581–588. [CrossRef] [PubMed]
15. Zwart, A.; van Kampen, E.J.; Zijlstra, W.G. Results of routine determination of clinically significant hemoglobin derivatives by multicomponent analysis. *Clin. Chem.* **1986**, *32*, 972–978. [CrossRef] [PubMed]
16. Index 2 Pulse Oximeter Simulator. Available online: https://www.qrs-solutions.com/products/fluke-index-2-pulse-oximeter-simulator/ (accessed on 1 January 2022).
17. Blitz, A.; Osterday, R.M.; Brodman, R.F. Harvesting the radial artery. *Ann. Cardiothorac. Surg.* **2013**, *2*, 533–542.
18. Gutierrez, C.A.; Lee, C.; Kim, B.; Meng, E. Epoxy-less packaging methods for electrical contact to parylene-based flat flexible cables. In Proceedings of the 16th International Solid-State Sensors, Actuators and Microsystems Conference (Transducers 2011), Beijing, China, 5–9 June 2011; pp. 2299–2302.

micromachines

MDPI

Review

Recent Progress and Challenges Regarding Carbon Nanotube On-Chip Interconnects

Baohui Xu [1], Rongmei Chen [2], Jiuren Zhou [3] and Jie Liang [1,*]

1 School of Microelectronics, Shanghai University, Shanghai 201800, China; xbh1@shu.edu.cn
2 Interuniversity Microelectronics Centre (IMEC), 3001 Leuven, Belgium; rongmei.chen@imec.be
3 Emerging Device and Chip Laboratory, Hangzhou Institute of Technology, Xidian University, Hangzhou 311200, China; zhoujiuren@163.com
* Correspondence: liangjieclair@shu.edu.cn

Abstract: Along with deep scaling transistors and complex electronics information exchange networks, very-large-scale-integrated (VLSI) circuits require high performance and ultra-low power consumption. In order to meet the demand of data-abundant workloads and their energy efficiency, improving only the transistor performance would not be sufficient. Super high-speed microprocessors are useless if the capacity of the data lines is not increased accordingly. Meanwhile, traditional on-chip copper interconnects reach their physical limitation of resistivity and reliability and may no longer be able to keep pace with a processor's data throughput. As one of the potential alternatives, carbon nanotubes (CNTs) have attracted important attention to become the future emerging on-chip interconnects with possible explorations of new development directions. In this paper, we focus on the electrical, thermal, and process compatibility issues of current on-chip interconnects. We review the advantages, recent developments, and dilemmas of CNT-based interconnects from the perspective of different interconnect lengths and through-silicon-via (TSV) applications.

Keywords: on-chip interconnect; carbon nanotube; through-silicon-via (TSV); Cu-CNT composite

Citation: Xu, B.; Chen, R.; Zhou, J.; Liang, J. Recent Progress and Challenges Regarding Carbon Nanotube On-Chip Interconnects. *Micromachines* **2022**, *13*, 1148. https://doi.org/10.3390/mi13071148

Academic Editor: Wensheng Zhao

Received: 18 June 2022
Accepted: 18 July 2022
Published: 20 July 2022

Publisher's Note: MDPI stays neutral with regard to jurisdictional claims in published maps and institutional affiliations.

Copyright: © 2022 by the authors. Licensee MDPI, Basel, Switzerland. This article is an open access article distributed under the terms and conditions of the Creative Commons Attribution (CC BY) license (https://creativecommons.org/licenses/by/4.0/).

1. Introduction

Transistors and interconnects are vital components of integrated circuits (IC). With the advancement of technology nodes and the reduction of critical dimensions, the performance of transistors has been greatly improved, while the interconnect has become the bottleneck limiting the development of IC.

Since the 1990s, transistors are no longer the only major factor affecting integrated circuits, and interconnects have gradually become another breakthrough for improving chip performance. With scaling down, in the technology node of 180 nm, the demands of high performance and reliability drove the transition from Aluminum (Al) interconnects to Dual Damascene Copper (Cu) interconnects due to the better conductivity and electromigration (EM) lifetime of copper [1]. TEM images of the EM effect are shown in Figure 1. A few years later, in 2003, low-K dielectric was introduced to separate copper lines, reducing parasitic capacitance, enabling faster switching speeds, and lowering heat dissipation [2]. Low-K dielectric is one of the strategies employed to carry out scaling followed by Moore's law. Since then, the composition of Cu and low-K material has become the basic structure of integrated circuits. Figure 2a show a simple conventional interconnect structure.

The main concerns with the continued scaling of copper-based interconnects are the significant increase in resistivity and electromigration issues resulting from the surface, grain boundary, and line edge roughness scattering [3]. In the back-end-of-line (BEOL) integration of copper-based interconnects, copper is surrounded by a barrier and liner. A barrier is necessary to maintain a certain thickness of about 2 nm to prevent the diffusion of copper through the interlayer dielectric (ILD); therefore, the effective conducting part of copper decreases with continuously scaling [4]. The decreasing dimension leads to higher resistivity due to electron scattering, as shown in Figure 2b, and reduced electromigration resistance. Copper-based interconnects are now facing similar challenges to Aluminum-based interconnects in terms of performance and reliability. In this case, recent research proposed by the IMEC (Interuniversity Microelectronics Centre) suggested the replacement of copper with Ruthenium (Ru) and Cobalt (Co) [3,5]. Although the resistivity of Ru and Co is higher than the resistivity of Copper, the interconnect lines developed with Ru and Co can conduct without barrier layers, which shows better overall conductance than that of copper lines. Furthermore, the fabrication of interconnect lines using Ru and Co is compatible with the process used currently in the back end of the line, which means lower replacement costs.

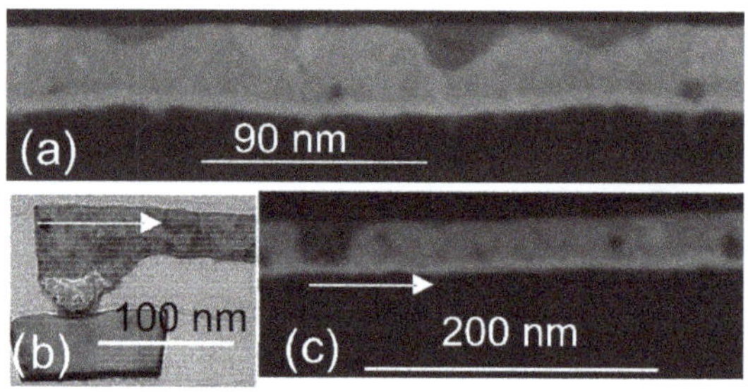

Figure 1. Transmission Electron Microscope (TEM) images of Cu lines and via, including the EM-induced via void and line voids. (**a**) is an unstressed sample of Cu line with Cu fill voids and surface gouges. (**b**) is the via after EM stress with EM-induced voids. (**c**) is the M1 (metal 1) line after stress with EM-induced voids. The arrows show directions of electron flows. Reprinted with permission from ref. [6]. Copyright 2018, IEEE.

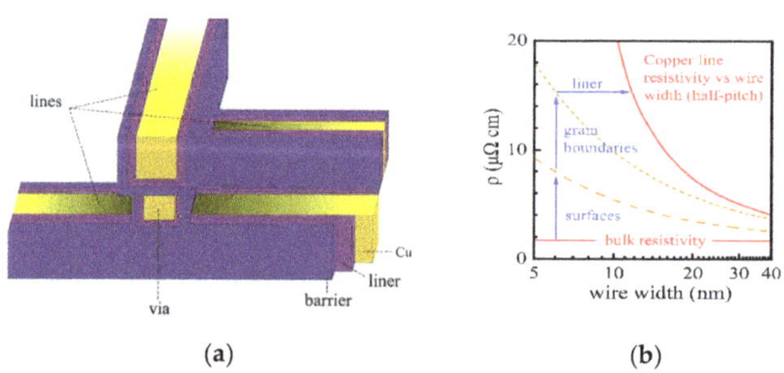

Figure 2. (**a**) Schematic of the interconnect structure. Two layers of horizontal metal lines are connected vertically by via. Barrier and liner are indicated. (**b**) The resistivity of Cu interconnects wire versus wire width. Reprinted with permission from ref. [7]. Copyright 2021, Springer Nature.

Recently, in advanced technology nodes, the semiconductor industry has focused on seeking a candidate to replace copper. We summarize the recent development of on-chip interconnect alternatives in Table 1. In the 14 nm node, Intel studied Copper and Tungsten (W) for new processes and new barrier materials [8,9]. In 10 nm and 7 nm nodes, Intel and Global Foundries explored the use of Co without barriers [5,10]. Regarding 5 nm and 3 nm nodes, IMEC expanded the possibility of employing Ru, as its resistivity is dependent on thickness [5,10–12]. When going beyond 3 nm nodes, it is necessary to find new materials, such as carbon-based materials, to meet the urgent demands of downscaling for interconnects in back-end-of-line fabrication. TSMC (Taiwan Manufacturing Company) has shown interest in multilayer graphene nanoribbon (GNR), which exhibits superior performance in intermediate and global interconnects [13,14]. Carbon nanotubes (CNTs) are the materials applied in interconnects with great potential due to their long mean free paths (MFP) [15], high thermal conductivity [16], large current carrying capacity [17], and excellent mechanical properties [18]. However, contact resistance and integration processes for CNT-based interconnects remain a great challenge for the application of CNT-based on-chip interconnects [19].

Table 1. The Development of On-chip Interconnects.

Technology Node	Material	Advantage	Limitation	Industry
14 nm	Cu/W [8,9]	Lower resistivity	Barrier effect	Intel
10/7 nm	Co [5,10]	Barrierless Thin liner	High resistivity	Intel/Global Foundries
5/3 nm	Ru [5,10–12]	Barrierless Thin liner	Surface scattering	IMEC
<3 nm	GNR, CNT [13–15]	Ballistic transport	Integration/Contact resistance	TSMC

In this paper, we mainly discuss recent developments and challenges in CNT-based interconnects. For local interconnects, size effects are the main issue due to the shrinking of critical dimensions. We discuss the conventional local interconnect physical resistance limitations and capacitance drawbacks and how single-walled (SWCNTs) and multi-walled CNTs (MWCNTs) could potentially solve these challenges by means of doping. Non-ideal factors in the practical CNT integration process are also covered. At the intermediate level of interconnects, performance is dominated by the product of resistance and capacitance due to larger transmission lines. Additionally, the ballistic transport properties of CNTs are highlighted gradually in intermediate levels. We analyze CNT-based interconnects with regard to their electrical properties. Among them, double-walled carbon nanotube (DWCNT) shows its particularity but still requires the support of specific processes. In the global interconnect, the reliability of the transmission line is especially important. We not only discuss the electrical and thermal properties of CNT-based interconnects but also give more attention to the analysis of reliability. The significant advantages of CNT-based interconnects in global interconnects are revealed. We introduce the process progress and high-frequency characteristics of CNT TSVs, and the feasibility of Cu-CNT TSVs is overviewed. Fault diagnosis techniques and the CNT-immune technique are also considered.

A summarized timetable is shown in Figure 3, which includes the main research on the electrical, thermal, and power analysis of CNT-based interconnects in recent decades. We notice that the electrical properties of CNT-based interconnects are the main point of interest, and there is a higher propensity for CNT-based interconnects to be applied on the global stage. Following CNT interconnect technology evolutions, the first compact model of standalone CNT was established in 2002 [20]. Based on that, complex electrical models considering more physical properties on CNT bundles and MWCNTs for interconnect applications were developed further in 2005 [21] and 2008 [22]. Regarding thermal analysis, one of the most important studies, conducted in 2005 [23], used three-dimensional finite element electrothermal simulations to conduct a comprehensive analysis of the thermal power of CNT-based interconnects. Based on this method, in 2009, the characteristics and requirements of the SWCNT bundle used as VLSI interconnects were discussed in depth [24]. Moreover, the first CNT interconnect integration process was introduced by Franz Kreupl in the year 2002 [25]. Decades after that, various integration processes were developed [26], and the most Silicon-compatible CVD (chemical vapor deposition) growth process was achieved at around 550 °C. TSVs using CNTs as materials were first discussed in 2007 [27]. The first generation mentioned CNT bundles can be uniformly grown to bond two wafers, namely through-wafer-interconnects at that time. In the following years, the technology of CNT TSV was continuously developed and advanced. In 2016, a unique process for fabricating Cu-CNT composite TSVs was proposed [28]. It demonstrated the advanced TSV process and included thermal analysis, which provided experimental foundations for subsequent development. Additionally, the Cu-CNT composite used for interconnect applications was revealed to be the potential solution for global interconnects in 2013 [29]; its electrical modeling investigations are presented in [30,31].

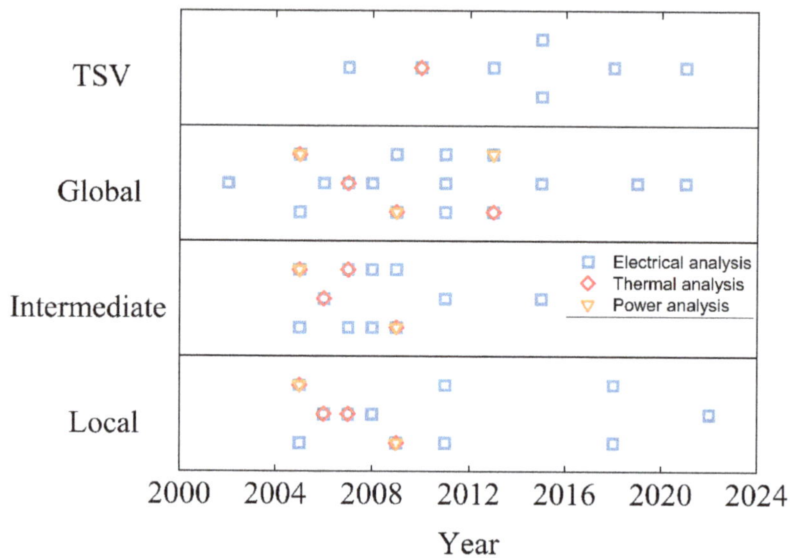

Figure 3. A timetable summarizing recent developments in CNT-based interconnects on local, intermediate, global, and TSV stages with their corresponding electrical, thermal, and power analysis. For TSV level, square patterns represent the related references [27,32–37], from left to right respectively. Rhombus patterns represent the related reference [32]. For global level, square patterns represent the related references [20–24,38–48], from left to right respectively. Rhombus patterns represent the related references [23,24,45,49], from left to right respectively. Triangle patterns represent the related references [23,24,44], from left to right respectively. For intermediate level, square patterns represent the related references [21–24,38–40,42,46,50,51], from left to right respectively. Rhombus patterns represent the related references [23,24,39,52], from left to right respectively. Triangle patterns represent the related references [23,24] from left to right respectively. For local level, square patterns represent the related references [21–24,38,39,41,53–56], from left to right respectively. Rhombus patterns represent the related references [23,24,39,52], from left to right respectively. Triangle patterns represent the related references [23,24], from left to right respectively.

The following content will be divided into four parts. Section 2 analyzes the advantages and disadvantages of CNTs in view of physical models according to different interconnect lengths (local, intermediate, and global). Section 3 presents the application of CNT and Cu-CNT composites on through-silicon-via (TSV). The discussion and perspectives will be driven in Section 4. Conclusions will be presented in Section 5.

2. On-Chip Interconnect

Interconnects can be classified as local, intermediate, and global, depending on their typical dimensions. We summarized the typical length values for the three different interconnect stages in Table 2 for easy distinction. However, it should be emphasized that the three interconnect stages are not divided by precise values but by a concept of scope. Local interconnects are the shortest and the narrowest, ranging from several nanometers to a few micrometers. They connect various transistors and logic blocks. The length of intermediate interconnects is located between the local and global interconnects, ranging from a few micrometers to tens of micrometers. Intermediate interconnects connect complex logic blocks and different cores. Global interconnects have a length greater than hundreds of microns. Global interconnects usually transmit power, ground, and clock signals to all the cores in circuits.

Table 2. Typical values for interconnect size.

Type of Interconnect	Dimensions
Local	<~2 μm [57,58]
Intermediate	2~100 μm [57,58]
Global	>~100 μm [57,58]

2.1. Local Interconnect

Local interconnects are on the lowest levels of the interconnect stack connecting nearby transistors and logic blocks, usually at the nanoscale. Logically, they are susceptible to the size effect, which leads to circuit performance degradation and tends to worsen with future device miniaturization. As shown in Figure 4, narrower wires have smaller grains and larger grain boundary densities, which increase electron scattering and thus increase resistance. Moreover, local interconnects occupy over 50% of power dissipation among interconnects [59]. The performance of local interconnect lines is vital to the chip.

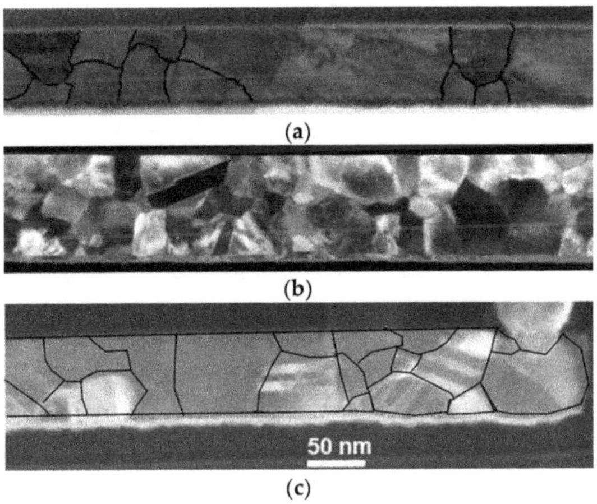

Figure 4. TEM images along Cu (**a**) 80 nm (**b**) 50 nm, and (**c**) 28 nm wide lines. Black lines represent grain boundary locations. Reprinted with permission from ref. [60]. Copyright 2013, Springer Nature.

In addition to physical resistance limitation, capacitance is also a critical issue that affects local interconnects. For nanoscale interconnect lines, their resistances are too small to be comparable to those of transistors and blocks, while their capacitances almost remain unchanged except for the effect of crosstalk [39]. The capacitance, in that case, which depends on the geometry and the dielectric constant of the surrounding insulating material, mainly contributes to the performance of transmission lines over short distances by affecting RC delay. Furthermore, downscaling leads to a higher area density, indicating that the distance between the adjacent wires has become smaller. The three primary capacitance components (the line-to-line capacitance, the line-to-ground capacitance, and the cross-over capacitance), as indicated in Figure 5, are all inversely proportional to the spacing between the wires [61]. Meanwhile, crosstalk, another critical issue, arises. In the empirical model and aggressor-victim model [61,62], the width, thickness, height, spacing between neighboring wires, the aspect ratio of wires, and switching signals are all influencing parameters of the crosstalk effect. In particular, the distance between adjacent wires decreases, leading to higher coupling capacitance. Coupling capacitance can delay the signal propagation in wires and result in logic errors and noises, so-called crosstalk-induced delay [62,63]. With further scaling, the negative impacts caused by capacitance and crosstalk greatly increase.

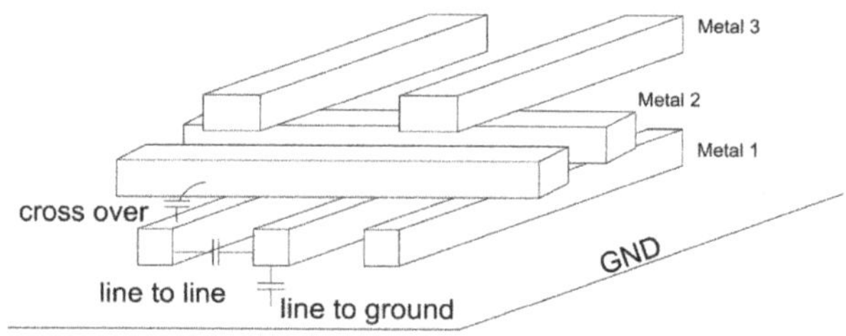

Figure 5. Schematic of interconnect capacitance model. Line to line, line to ground, and cross-over capacitance are shown accordingly.

Carbon nanotubes have shown great potential to solve these issues. According to the transmission line model, the capacitance of the SWCNT is composed of the quantum capacitance and the electrostatic capacitance in series, which is theoretically smaller than that of traditional Cu wires [20]. Compared to copper-based interconnects in 22 nm nodes, for local interconnects, several layers of SWCNT can offer up to 50% reduction in capacitance and power dissipation with up to 20% in latency if they are short enough (<20 μm) [39,64]. For the MWCNT, the shell-to-shell capacitance needs to be further considered. However, it is more complicated and needs to be analyzed according to the actual application. MWCNTs exhibit up to 10% improvement in capacitance for local interconnects compared to copper-based interconnects in 14 nm nodes [22]. Logically, reduced capacitance leads to better performance on crosstalk. Research shows that crosstalk-induced overshoot/undershoot remains nearly the same as technology node scaling and increases as the length of the copper-based interconnect increases. Whereas in CNT-based interconnects, the overshoot/undershoot remains the same and is almost unaffected by the scaling of interconnect length [43]. However, CNTs do not perform as well as expected in local interconnects. In short-distance interconnects, although the delay is mainly capacitance-dominated, fabrication process factors such as contact resistance, CNT chirality, and CNT density are also taken into account. These non-ideal factors are unavoidable in practical processes and degrade the electrical properties of CNTs. Especially in the local stage, CNTs cannot entirely take advantage of their high electron mobility, which makes the performance of CNT-based interconnects far worse than ideal. In addition, the advantage of MFP length cannot be fully benefited, and the advantage of capacitance is not significant. The actual delay of CNT is usually greater than that of Cu [22]. Theoretically, in the case of short local interconnects, CNT-based interconnects require harsh conditions to potentially outperform copper-based interconnects, such as in all-carbon circuits, and warrant a fairly small CNT pitch in CNTFETs (carbon nanotube field-effect transistors) (<9 nm) [65].

Generally, at nanoscale interconnect lengths, compared to Cu lines, SWCNT bundles and MWCNT have better performance in terms of delay, as well as multilayer SWCNT interconnects [22,24,39]. Recently, Chen et al. studied the impact of charge transfer doping on the performance and variability of MWCNT interconnects using enhanced compact models [53,54]. They further explored an all-carbon SRAM (ACS) (SRAM: Static Random-Access Memory) using a 5 nm technology node, which showed a great improvement in power efficiency with 72% lower energy-delay-product (EDP) and 48% lower static power, despite a slight speed reduction, compared to 7 nm FinFET (fin field-effect transistor) SRAM cells with copper-based interconnects [56,66].

From the point of view of CNT-based interconnect integration, both the capacitance and delay of CNT-based interconnects still have room to be improved compared to copper-based interconnects in the local interconnect region due to large CNT-metal contact resistance, the density of CNT bundles, and so on [39,43,44,67]. However, the study in [33] showed that MWCNTs can exhibit conductivity surpassing that of Cu lines in local interconnects by doping with certain concentrations of Pt salts [55,68].

For local interconnects, doping of CNT provides a potential solution to improve the problems caused by increased capacitance and gradually deteriorated crosstalk and has an advantage regarding power efficiency; however, CNT-based interconnects still require further study to overcome CNT–metal interface contact resistance with respect to the current transistor fabrication process in order to outperform Cu line local interconnects [22].

2.2. Intermediate Interconnect

Intermediate interconnects are usually at a micron-meter scale. They mainly serve to connect logic blocks and cores. At this level, interconnect delay depends mostly on the product of resistance and capacitance due to larger transmission lines. Along with advanced technology nodes, the shrinking of Cu lines leads to the reduction of effective conducting parts of interconnect lines, which further impact the performance of intermediate interconnects. Moreover, repeaters are used at the intermediate level to increase the drive capability and reduce the signal delay, but bring some negative effects regarding power dissipation, area resources, and design [69,70].

The long MFP and ballistic transport properties of CNTs are highlighted gradually in intermediate interconnects [15,71,72]. For densely packed bundles at the intermediate interconnect level, the bundle resistance is less than that of Cu for a wide range of interconnect lengths, even for low metallic CNT density (45%) [24]. Dense SWCNT bundle interconnects can easily surpass Cu with at least 30% less latency [24]. Similarly, the resistivity of MWCNTs could be several times lower than that of Cu wire and becomes increasingly comparable to that of SWCNT bundles for long lengths (>10 μm) [22,38]. CNTs not only improve latency but also advances the use of repeaters. With the same repeater size, carbon-based interconnects have significantly less delay than that of copper-based interconnects. The optimal number of repeaters could also be reduced [41,47].

Additionally, research shows that both DWCNT and SWCNT bundles provide a significant improvement in performance compared to Cu wires [40]. In particular, the performance of DWCNT is even better than that of SWCNT. In the 14 nm technology node, DWCNT has a 20% improvement in crosstalk-induced time delay over SWCNT at the intermediate level [42]. From a process point of view, DWCNT is easier to achieve with a high metallic chirality ratio than SWCNT, and it also has small capacitance [57]. Therefore, in the intermediate stage, high-density DWCNT bundles could be a better solution under the consideration of delay based on RC product, as mentioned above. However, in general, DWCNT is far from SWCNT and MWCNT in terms of application range and process compatibility, and there are few studies investigating the feasibility of DWCNT. Therefore, the potential application prospects of DWCNT and the targeted process issues remain to be further explored.

2.3. Global Interconnect

Global interconnects are usually more than 100 micronmeters. They carry power, ground, and clock signals, which indicate a high current carrying capacity. Therefore, for global interconnects, performance improvement should be considered, but more importantly, the optimization of interconnect reliability should be taken into account to meet the requirements in advanced nodes.

At the global level, a reduction in the size of Cu lines results in the increase of resistance and current density, which in turn causes critical issues on delay, IR drop, power dissipation, and electromigration effect. Moreover, the number of repeaters grows with interconnect length, increasing power consumption while reducing latency. Under certain conditions, it could cause a blockage problem, which reduces the transmission efficiency [73]. Power consumption is not only a problem for energy transmission efficiency but also a thermal problem caused by high current flow. Thermal management has naturally become one of the key optimization directions.

In the global interconnect with resistance as the main factor affecting delay, CNT takes full advantage of its high mobility and long MFP, surpassing Cu with lower resistance. For the global interconnect scale, CNT-based interconnect lines have at least 20% lower resistance compared to that of Cu lines and an optimized RC delay of more than 30%, which greatly enhances the performance of interconnecting transmission [22,39,74,75]. Additionally, because the transmission length is long enough, the negative effects of non-ideal factors such as contact and defects are relatively reduced [22,50,76]. Theoretically, the transport speed of CNT for long-distance transport is much better than that of Cu. It is worth mentioning that some studies have shown the mixed-CNT bundle has better performance than using SWCNT bundles [48,77]. However, it is limited by complex distribution requirements and process challenges; its practical application still needs further investigation. Figure 6a–g show the SEM images of CNT vias. They display the different steps in the CNT growth process. They demonstrate that the width, length, and area density of CNTs in the growth process are difficult to precisely control. Each step in the process causes changes in the CNTs, as the CNTs are highly variable. The idea of mixed CNT bundles is practical for interconnect processes. However, using the Gaussian distribution to simulate the change of CNT diameters at different positions still has a certain discrepancy with the actual situation, so further exploration is required. The challenge of dimension control is also applicable to SWCNT bundles and MWCNTs.

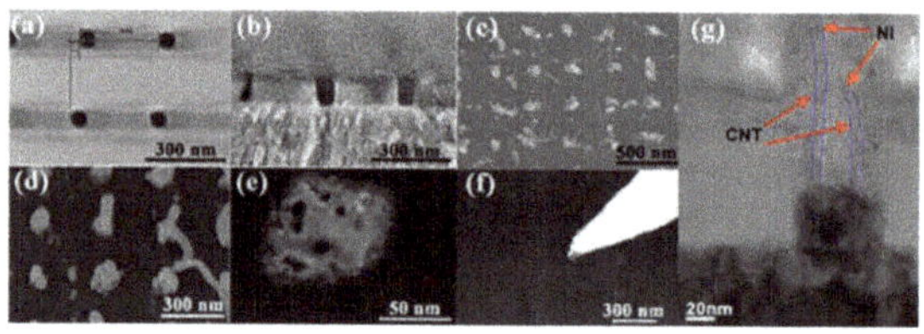

Figure 6. Scanning electron microscope (SEM) images of CNT vias. (**a**) Top-view image after patterning; (**b**) cross-section image after etching; (**c**) image of CNT vias; (**d**) image of CNT vias after dielectric filling; (**e**) top-view of a single CNT after ion milling; (**f**) nanoprobe landing on single CNT and making contact with CNT tips; (**g**) image of a 60 nm CNT via before ion milling, where red arrows indicate Ni catalyst particle at CNT tip and blue lines indicate the CNT sidewalls. Reprinted with permission from ref. [9]. Copyright 2015, IEEE.

A densely packed SWCNT bundle can reduce power consumption by up to eight times compared to Cu-based interconnects at the 14 nm node [24]. Theoretically, at a length of 100 microns, its EDP can be 12 times better than Cu, reducing the energy-per-bit by at least three times at the 11 nm node [45]. The optimization of power consumption is mainly attributed to the low resistance, low capacitance, and excellent thermal properties of CNTs. Power consumption issues and thermal issues are often side-by-side.

According to molecular dynamics simulations, the thermal conductivity of CNTs can reach 6600 $Wm^{-1}K^{-1}$ [78]. It depends on multiple factors such as CNT length, diameter, defects, bundle density, etc. [79]. Therefore, CNTs, under different conditions, may be excellent thermal conductors or thermal insulators [80]. Experiments show that the thermal conductivity of SWCNTs with a length of 2.6 microns and a diameter of 1.7 nm can reach 3500 $Wm^{-1}K^{-1}$ at room temperature, while the thermal conductivity of a single MWCNT with a diameter of 14 nm is more than 3000 $Wm^{-1}K^{-1}$ at room temperature [52,81]. In contrast, the thermal conductivity of copper is only about 400 $Wm^{-1}K^{-1}$. Moreover, CNTs can maintain the basic stability of the structure at a temperature of 1000 K, which is enough to cover the upper temperature limit of on-chip interconnects at present [46,82]. CNTs are excellent thermal conductive materials that can be used to enhance interconnect reliability. Table 3 summarize the electrical conductivity, thermal conductivity, electron mean free path, and dielectric constant of various interconnect-based materials.

Table 3. A summary of properties of SWCNT, MWCNT, Cu-CNT, Cu Co, and Ru.

	SWCNT	MWCNT	Cu-CNT	Cu	Co	Ru
Conductivity (S/cm)	7×10^5 [83,84]	2.7×10^5 [84]	2.3–4.7×10^5 [29,85]	5.8×10^5	1.6×10^5	1.4×10^5
Thermal Conductivity @300k (W/mK)	>3500 [52]	3000 [81]	637 [86]	385	100	117
Electron mean free path @300K (nm)	>1 μm [87]	>30 μm [88]	NA	39	19 [3]	6.7 [3]
Dielectric constant *k* *	graphene oxide-polyimide ($k = 2$) [89]	graphene oxide-polyimide ($k = 2$) [89]	NA	SiCOH (k = 2.4~2.55) [58]	SiCOH (k = 2.7~3.2) [58]	SiCOH (k = 2.4) [11]

* For air gap, $k = 1$ can theoretically be used as long as the process is compatible.

In CNT thermal models for circuit-level simulations, the temperature is usually introduced as an inverse function of MFP to show its effect on electron–phonon scattering [90,91]. When the interconnect length is longer than the MFP of CNT, the electron–phonon scattering in the length direction of CNT will cause self-heating and temperature increase. Increased temperature, in turn, enhances scattering and degrades the performance. On the other hand, for large-diameter CNTs, the band gap between sub-bands is smaller than that for small-diameter CNTs. The rise of temperature provides sufficient thermal energy for the carriers to cross the bandgap. That is, the increase in temperature can also increase the number of conducting channels and enhance the transport ability of CNTs [49]. According to the model results, at the global length, the temperature is proportional to the resistance of the CNT [92]. The TCR (temperature coefficient of resistance) value of MWCNT with a diameter of 50 nm at 100 μm length is 2.7, while the TCR value of Cu is 4 [49]. This behavior can be explained as follows: for large-diameter MWCNTs, the increase in the number of conducting channels caused by the temperature increase reduces the scattering resistance and contact resistance so that the resistance growth and self-heating effects are not significant compared to Cu. Therefore, CNTs have less resistance sensitivity to temperature. Combined with their lower resistance and inherently better thermal conductivity, CNT-based interconnects have lower power dissipation and better heat dissipation. For ideal SWCNT bundle interconnects at dimensions of 10–20 nm wide and 1 mm long, the delay is five times less than that of Cu interconnects under the same conditions over the temperature range from 300 K to 373 K [90]. As shown in the simulation in Figure 7, the maximum temperature of CNT-based via is almost two times better than that of Cu via [23,24]. In Figure 7, it can also be clearly observed that CNTs can reduce the maximum temperature regardless of the length of the interconnect used. It demonstrates that the heat dissipation capability of CNTs is superior to that of Cu.

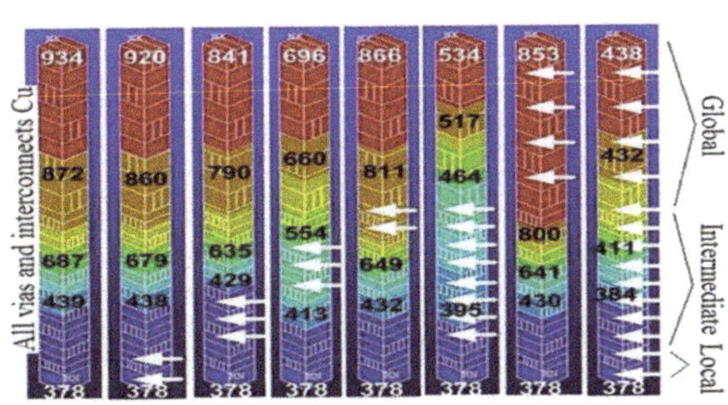

Figure 7. Interconnect temperature map of vias at different levels obtained from 3D finite-element electrothermal simulation at 22 nm node. The parts pointed by the white arrows are composed of CNT bundles, and the other parts are composed of Cu. Reprinted with permission from ref. [24]. Copyright 2009, IEEE.

The electromigration problem is especially significant in global interconnects, where interconnect lines carry large current flows that may lead to voids or even circuit failure, as shown previously in Figure 1. Electromigration is a mass transport process due to the self-diffusion of metallic ions in response to an electric field applied across interconnects [93]. With the scaling down of dimensions, Cu lines shorten the EM failure time because the maximum allowed current density decreases with dimension. Research shows that, for Cu interconnects, the normalized median EM failure time approximately degrades 100% scaling from the 180 nm node to the 14 nm node [94]. In contrast, CNT-based interconnect lines have a large current carrying capacity of 109 A/cm^2, at least two orders of magnitude higher than the maximum current density of Cu interconnect [95]. This shows that using CNT-based interconnects could be a potential solution to alleviate EM problems. In addition to electromigration, time-dependent-dielectric-breakdown (TDDB) is also an important reliability issue, which limits the spacing between adjacent wires for a specific low-k dielectric. After long-term use, the drift of copper ions can cause the deterioration and breakdown of the interconnect structure [96]. The TDDB-induced damage is shown in Figure 8. It indicates that after the TDDB test, the TaN/Ta barrier was damaged at the bottom corner, and more Cu atoms migrated into SiO_2 as the test time increased. This phenomenon seriously affects the performance and reliability of the interconnect. Therefore, both TDDB and EM are regarded as the key issues surrounding interconnect reliability. CNT-based interconnects are expected to use dielectrics with a lower dielectric constant ($k < 2.5$) [89], thus mitigating the TDDB problem.

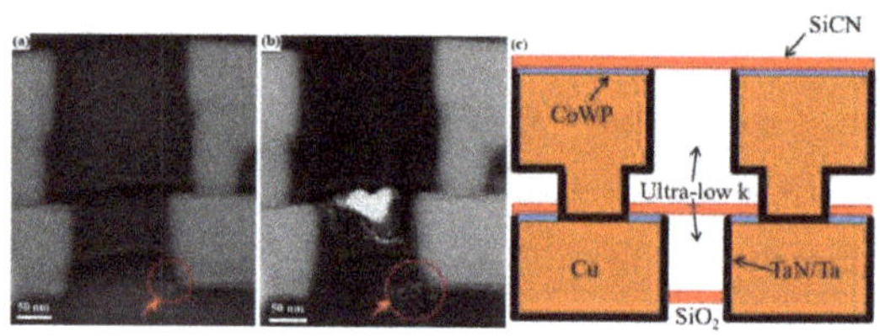

Figure 8. Electron spectroscopic imaging (ESI) analysis of the Cu distribution after the electrical test of (**a**) 20 V/106 min + 25 V/638 min and (**b**) 20 V/106 min + 25 V/872 min. (**c**) is the schematic structure of (**a**,**b**). The structure consists of Cu interconnects in M1 and M2 metal layers, which are encapsulated by a TaN/Ta barrier, SiCN capping layer, CoWP top coating, and insulated by ultralow dielectric permittivity material (porous organosilicate glass). Reprinted with permission from ref. [97]. Copyright 2015, Elsevier.

3. Through-Silicon-Via (TSV)

Through-Silicon-Via is a vital part of modern three-dimensional (3D) integration technology. In the case of high-density integration in the horizontal direction of a single chip, the integration direction is further expanded by stacking multiple layers vertically through TSVs. TSVs further expand the feasibility of continued size reduction and also shorten the interconnect path. TSV usually transmits periodic power and ground signals. It has been widely used in 3D integrated circuits and 3D packaging [98], which has great application prospects in the development of SoC (System on Chip) and heterogeneous integration.

3.1. Carbon Nanotube

A commonly used CNT line integration is vertical growth at high temperatures (>700 °C) by catalyst-enhanced chemical vapor deposition (CCVD). There are fewer cases of CNT growth directly in the horizontal direction [57]. Therefore, compared to the horizontal CNT interconnect line process, growing CNTs directly in Vias or TSVs could be a more friendly process and takes full advantage of vertical CNT bundles [9,27]. Using a flip and roll technique to form vertical–horizontal combined structures was investigated in [99].

Contact and CMOS process compatibility are the two major challenges of CNT growth, which affect the resistance and reliability of CNT-based interconnects. Studies have shown that aligned CNTs can be grown vertically as TSVs by different methods at a temperature below 550 °C that is compatible with the CMOS process and device fabrication [34,35]. Experiments show that the aspect ratio (AR) of CNT TSV with a length of 50 μm can achieve 5 or 10, while the resistance is only 69.7 Ω [33]. Moreover, CNTs have excellent thermal conductivity and thermal stability and are very suitable for long-distance transmission within large current-carrying TSV structures. Recently, a method to fabricate CNT-filled TSV using the vacuum-assisted spin coating of polyimide (PI) liners was reported in [37]. It is possible to manufacture a CNT TSV with a diameter of 15 microns and a depth of 200 microns (AR~13.3) at low cost and low temperature (<~240 °C). The temperature requirements for CNT TSVs growth have been successfully reduced to be compatible with CMOS processes. However, using CNTs as Via or in TSV still remains a challenge for mass production and industrial compatibility [100,101].

For high-frequency applications, with current studies, all metallic SWCNT bundles have lower resistance than Cu due to their long MFP, while MWCNTs have poor high-frequency performance due to their large inductance [32,57]. Moreover, the electrical properties of CNT TSVs are easily affected by changes in kinetic inductance. At high frequencies, the electrical properties of CNTs degrade with increasing kinetic inductance [31].

Therefore, CNT TSVs present great improvement in reliability at the expense of the loss of conductivity, showing excellent stability and scalability. However, although the process compatibility of CNTs used as TSVs has been investigated, more attention is required regarding research on interface contact resistance.

3.2. Cu-CNT Composite

Both TSVs and global interconnects need to transmit large currents as well as AC (alternative current) signals, which are prone to EM. Faced with this problem, on the basis of CNT-based interconnects, people thought to combine Cu and CNTs to meet the demand. Cu-CNT composite has been experimentally proved to be a material with high conductance (2.3–4.7 × 105 Scm^{-1}) and high current capacity (6 × 108 Acm^{-2}) [29,102].

In terms of electrical properties, it makes up for the low electrical conductivity of CNT TSV, reaching a level comparable to that of Cu (5.8 × 105 Scm^{-1}). At the same time, the ampacity is almost 100 times that of copper, which is more resistant to the EM effect [103]. The performance at high frequencies remains similar to Cu [31]. Compared with CNT, the Cu-CNT composite has better electrical conductivity and the ability to suppress kinetic inductance changes [104]. In addition, its stability and reliability are much better than Cu, which makes the Cu-CNT composite TSV a potential candidate for high-frequency 3D-IC applications. Moreover, studies have shown that surface modification based on Cu-CNT composite can continue to improve its electrical and mechanical properties [105,106]. Although the improvement effect of surface modification is not significant, it provides a direction for further development.

In terms of thermal properties, Cu-CNT composite has higher thermal conductivity and lower CTE (coefficient of thermal expansion) than copper. When the CNT concentration reaches 250 mg/L, the thermal conductivity of the Cu-CNT composite can reach 637 $Wm^{-1}K^{-1}$ [86], which is inferior to CNT (>2000 $Wm^{-1}K^{-1}$) but still better than Cu (385 $Wm^{-1}K^{-1}$). Compared with pure copper, the Cu-CNT composite with 50% copper can reduce the CTE by 90% [51]. At the circuit level, Cu-CNT TSV shows its excellent thermal properties. For Cu-CNT TSVs, the filling rate of CNTs is inversely proportional to the temperature increment and the maximum temperature [107]. Additionally, it exhibits less temperature sensitivity than Cu in both conductivity and high-frequency transmission performance [108].

From a process point of view, the use of Cu-CNT composite materials can better combine with the existing Cu process and have good interface compatibility. It is possible to achieve void-free filling, CMP (chemical mechanical polishing), or patterning integration and decrease variability [36]. Figure 9 show the fabrication process of Cu-CNT TSVs. Grid arrays of carbon nanotubes are grown by CVD (chemical vapor deposition). After a series of steps, including sputtering, transfer, etc., Cu is injected into the pre-reserved gaps during the electroplating process to realize the Cu-CNT composite, as shown in Figure 10. Unfortunately, the major challenge is the process still needs to be improved and explored to be suitable for VLSI manufacture [109,110]. Specific developments are required for Cu-CNT composites.

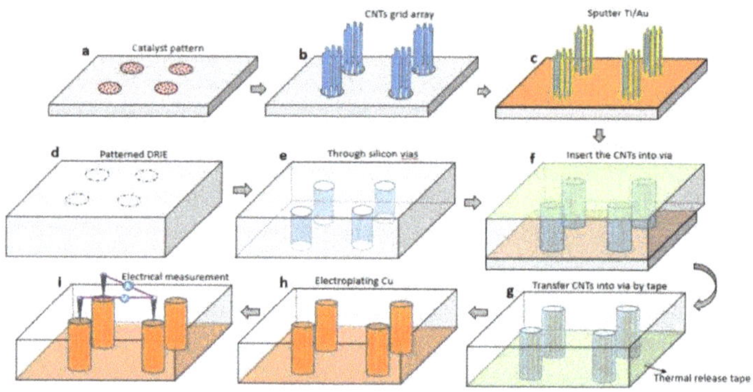

Figure 9. Process flow diagram of Cu-CNT TSV fabrication. (**a**) Patterned catalysts were deposited by E-beam evaporation. (**b**) A CNT grid array was synthesized by using CVD. (**c**) Sputtered 10 nm Ti and 20 nm Au onto the CNT grid array. (**d**,**e**) In parallel to these steps, the target Si wafer/chip with via was prepared by deep reactive ion etching (DRIE). (**f**) Thermal release tape was attached onto the front surface of the target wafer/chip and the CNT grid array were transferred into the via. (**g**) The donor wafer/chip was removed. (**h**) Cu was transferred into the vias by electroplating to form the composite CNT/Cu TSV and the adhesive tape was removed. (**i**) Electrical performance was characterized by the four probe method. Reprinted with permission from ref. [28]. Copyright 2016, IOP Publishing.

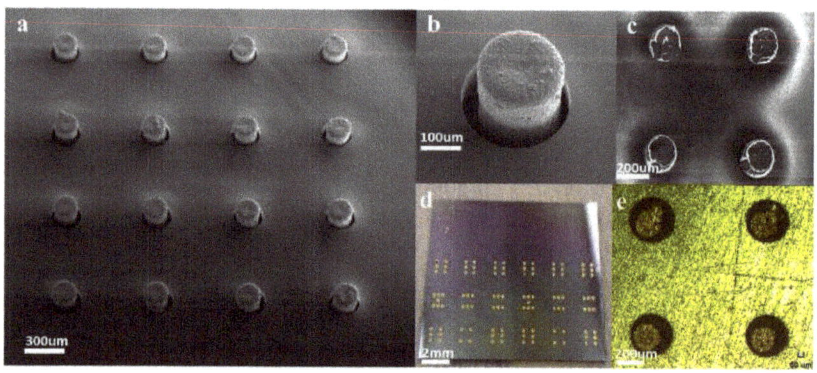

Figure 10. Images of Cu-CNT composite TSVs. (**a**) Cu-CNT composite structures inside the silicon vias; (**b**) the image of a Cu-CNT composite in the via; (**c**) the surface topography of Cu-CNT composite TSVs after polishing; (**d**) a test sample with Cu-CNT TSVs; (**e**) the image showed top surface topography of Cu-CNT composite TSVs after polishing. Reprinted with permission from ref. [28]. Copyright 2016, IOP Publishing.

In conclusion, although the manufacturing process has yet to be further developed toward industrial requirements, the use of Cu-CNT composite as a TSV or global interconnect material enhances both the performance and reliability of the integrated circuit, even better than CNT in the aspect of electrical performance. In addition, surface modification provides room for further development. Therefore, Cu-CNT composite material is one of the potential solutions for future global interconnects.

4. Discussion and Perspectives

With the continuous reduction of the overall size of integrated circuits, entering the nanometer and angstrom era, Cu/low-k interconnects have not been replaced as predicted by some articles [13,111] and have stood the test of time. For CNT-based interconnects, there have been many achievements in physical models, designs, and processes in the past few decades. Most of them indicate that CNT-based interconnects have three advantages, high performance, improved reliability, and better energy efficiency [84], which are suitable for on-chip interconnect applications. Moreover, many studies have shown that there is great potential for CNT-based interconnects to surpass Cu lines under certain conditions.

As far as physical models are concerned, efforts should be made to develop realistic, high-reliability physical models. It is not only necessary to consider the ideal situation but also to further consider various factors such as variability and defect-containing factors. Hierarchical physical models should be considered from material to device to circuit, considering the possible non-ideal effects in the actual fabrication process as comprehensively as possible. In that case, design technology co-optimization (DTCO) should be investigated to provide a theoretical basis and guidance for future CNT-based interconnect architectures. In terms of the CNTs integration process, it should focus on optimizing issues including contact, CNT density and distribution, chirality control, growth temperature, and cost to meet the requirements for VLSI circuit manufacture. In addition, doping of CNT and Cu-CNT composite material needs specific research to enrich the full schema of carbon-based BEOL integration.

Moreover, for CNT-based interconnect circuit-level simulations and behavior analysis, fault diagnosis should be properly incorporated. The purpose is to detect, identify, and isolate various anomalies and faults as early as possible in order to minimize their damage and facilitate the VLSI manufacturing process. However, there is little research in this area. Most of these studies focus on diagnosing and measuring delay faults. A ring oscillator-based testing technique [112] and a model with an inverter chain followed by a D flip-flop [113] are, respectively, proposed to diagnose and measure delay–fault caused by the variability of CNTs [53]. Such techniques help to judge whether the dimension of CNTs is as expected. However, for fault-tolerant technology, the traditional method is not applicable due to the large power consumption and area [84]. Hence, a defective CNT-immune design technique is proposed with a smaller power, area, and speed impact than traditional defect and fault tolerance techniques [114]. It solves the problems of misposition and misalignment of CNTs and has good compatibility with the current VLSI circuit design flows. Nevertheless, the above-mentioned techniques do not comprehensively cover the possible factors of CNT-based interconnect failure, such as the chirality of CNTs where all-metallic CNTs are favorable for interconnect applications. Therefore, the fault diagnosis and tolerance technology for CNT-based interconnects still need to be further explored to serve VLSI circuit manufacturing.

While continuing to delve into the physical nature of CNTs, novel application-oriented optimization designs should also be carried out based on the advantages of CNTs themselves. For example, the application of CNT in N3XT and N3XT 3D MOSAIC structures (N3XT: Nano-Engineered Computing Systems Technology; MOSAIC: MOnolithic/Stacked/Assembled IC) [115,116]. The N3XT structure contains energy-efficient field-effect transistors (FETs), high-density nonvolatile memories, fine-grained monolithic 3D integration, efficient heat removal, and computation immersed in memory. Moreover, CNTs hold the promise of optimizing thermal management in the N3XT 3D MOSAIC.

5. Conclusions

In this paper, we reviewed the state-of-the-art CNT-based interconnects from the perspective of different interconnect levels and TSV applications. Carbon nanotubes exhibit different properties at different length stages. Theoretically, although CNTs have a good crosstalk suppression in the local interconnect, their resistance and delay are still larger than Cu. CNT-based interconnects also have critical process requirements for defects, contacts, and variability in the local stage. Based on the reduced resistance caused by long MFP, CNTs with high purity and high metallic fraction begin to show performance beyond that of Cu in the intermediate interconnects. At the global level, not only the electrical performance surpasses Cu, but also the advantages in reliability are fully exerted, such as high thermal conductivity, large current carrying capacity, and large EM resistance. The integration process for achieving high density and large MFP would be the key solution to enhance the performance of CNT interconnects on the global stage. In TSV applications, the excellent reliability of CNTs is demonstrated again. Nevertheless, the electrical property of CNT TSV is still limited by the integration process. A low-temperature Si-compatible CNT growth process is always necessary for VLSI circuit manufacturing. The emergence of Cu-CNT composites makes up for the deficiency of CNT's electrical property and achieves both excellent electrical conductivity comparable to Cu and improved reliability. The performance, reliability, and process requirements of CNT-based interconnects at different lengths are briefly summarized in Table 4.

Table 4. A summary of the performance, reliability, and process requirements of CNTs at different length scales by comparing with Cu.

	Local	**Intermediate**	**Global**	**TSV**
Performance	Inferior to Cu (non-ideal) Exceeds Cu (ideal)	~30% lower delay	>30% lower delay 8 times less power consumption 7 times larger thermal conductivity	Inferior to Cu (non-ideal) Exceeds Cu at high frequency (ideal) Better thermal performance
Reliability	Exceeds Cu	Exceeds Cu	Exceeds Cu	Exceeds Cu
Process requirements	Defectless CNTs [50], Low variability [53], Low contact resistance [53], Low temperature (Si process compatible) [36]	Defectless CNTs [50], Dense bundle, High purity [50,51], High metallic fraction [22], Low temperature (Si process compatible) [51]	Dense bundle [39], Large electron MFP [39], High metallic fraction [22], Low temperature (Si process compatible) [117]	Dense bundle [118], Large electron MFP, High metallic fraction [32], Cu-CNT composite [28], Low temperature (Si process compatible) [37]

Integration processes and compatibility are always the key challenges. Contact, density, chirality, defects, and CNT growth temperature (process temperature) have always been obstacles that need to be overcome in the development of CNTs. In particular, some progress has been made in reducing the CNT growth temperature, which improves compatibility. Carbon nanotube technology provides opportunities to realize power-consumption-oriented, energy-efficient on-chip interconnect applications. Thus, CNT-based interconnects have the potential to promote 3D integration, heterogeneous integration, and other directions to achieve high-performance, low-cost, high-energy-efficiency architectures.

Author Contributions: Conceptualization, J.L. and B.X.; writing—original draft preparation, B.X. and J.L.; writing—review and editing, J.L., B.X., J.Z. and R.C. All authors have read and agreed to the published version of the manuscript.

Funding: This research was funded by the National Natural Science Foundation of China (grant number 62104138).

Institutional Review Board Statement: Not applicable.

Informed Consent Statement: Not applicable.

Conflicts of Interest: The authors declare no conflict of interest.

References

1. Buchanan, K. The evolution of interconnect technology for silicon integrated circuitry. In Proceedings of the GaAs MANTECH Conference, San Diego, CA, USA, 8–11 April 2002.
2. Auth, C.; Allen, C.; Blattner, A.; Bergstrom, D.; Brazier, M.; Bost, M.; Buehler, M.; Chikarmane, V.; Ghani, T.; Glassman, T.; et al. A 22 nm high performance and low-power CMOS technology featuring fully-depleted tri-gate transistors, self-aligned contacts and high density MIM capacitors. In Proceedings of the 2012 Symposium on VLSI Technology (VLSIT), Honolulu, HI, USA, 12–14 June 2012; pp. 131–132.
3. Gall, D. The search for the most conductive metal for narrow interconnect lines. *J. Appl. Phys.* **2020**, *127*, 050901. [CrossRef]
4. Rossnagel, S.; Wisnieff, R.; Edelstein, D.; Kuan, T. Interconnect issues post 45nm. In Proceedings of the IEEE International Electron Devices Meeting, 2005. IEDM Technical Digest, Washington, DC, USA, 5 December 2005; pp. 89–91.
5. Croes, K.; Adelmann, C.; Wilson, C.; Zahedmanesh, H.; Pedreira, O.V.; Wu, C.; Leśniewska, A.; Oprins, H.; Beyne, S.; Ciofi, I. Interconnect metals beyond copper: Reliability challenges and opportunities. In Proceedings of the 2018 IEEE International Electron Devices Meeting (IEDM), San Francisco, CA, USA, 1–5 December 2018; pp. 5.3.1–5.3.4.
6. Hu, C.K.; Gignac, L.; Lian, G.; Cabral, C.; Motoyama, K.; Shobha, H.; Demarest, J.; Ostrovski, Y.; Breslin, C.M.; Ali, M.; et al. Mechanisms of Electromigration Damage in Cu Interconnects. In Proceedings of the 2018 IEEE International Electron Devices Meeting (IEDM), San Francisco, CA, USA, 1–5 December 2018; pp. 5.2.1–5.2.4.
7. Gall, D.; Cha, J.J.; Chen, Z.; Han, H.-J.; Hinkle, C.; Robinson, J.A.; Sundararaman, R.; Torsi, R. Materials for interconnects. *MRS Bull.* **2021**, *46*, 959–966. [CrossRef]
8. Adelmann, C.; Wen, L.G.; Peter, A.P.; Siew, Y.K.; Croes, K.; Swerts, J.; Popovici, M.; Sankaran, K.; Pourtois, G.; Van Elshocht, S. Alternative metals for advanced interconnects. In Proceedings of the IEEE International Interconnect Technology Conference, San Jose, CA, USA, 20–23 May 2014; pp. 173–176.
9. Zhou, C.; Vyas, A.A.; Wilhite, P.; Wang, P.; Chan, M.; Yang, C.Y. Resistance Determination for Sub-100-nm Carbon Nanotube Vias. *IEEE Electron Device Lett.* **2015**, *36*, 71–73. [CrossRef]
10. van der Veen, M.H.; Heyler, N.; Pedreira, O.V.; Ciofi, I.; Decoster, S.; Gonzalez, V.V.; Jourdan, N.; Struyf, H.; Croes, K.; Wilson, C. Damascene benchmark of Ru, Co and Cu in scaled dimensions. In Proceedings of the 2018 IEEE International Interconnect Technology Conference (IITC), Santa Clara, CA, USA, 4–7 June 2018; pp. 172–174.
11. Zhang, X.; Huang, H.; Patlolla, R.; Wang, W.; Mont, F.W.; Li, J.; Hu, C.-K.; Liniger, E.G.; McLaughlin, P.S.; Labelle, C. Ruthenium interconnect resistivity and reliability at 48 nm pitch. In Proceedings of the 2016 IEEE International Interconnect Technology Conference/Advanced Metallization Conference (IITC/AMC), San Jose, CA, USA, 23–26 May 2016; pp. 31–33.
12. Adelmann, C.; Sankaran, K.; Dutta, S.; Gupta, A.; Kundu, S.; Jamieson, G.; Moors, K.; Pinna, N.; Ciofi, I.; Van Eishocht, S. Alternative metals: From ab initio screening to calibrated narrow line models. In Proceedings of the 2018 IEEE International Interconnect Technology Conference (IITC), Santa Clara, CA, USA, 4–7 June 2018; pp. 154–156.
13. Wang, N.C.; Sinha, S.; Cline, B.; English, C.D.; Yeric, G.; Pop, E. Replacing copper interconnects with graphene at a 7-nm node. In Proceedings of the 2017 IEEE International Interconnect Technology Conference (IITC), Hsinchu, Taiwan, 16–18 May 2017; pp. 1–3.
14. Yu, T.; Liang, C.-W.; Kim, C.; Song, E.-S.; Yu, B. Three-dimensional stacked multilayer graphene interconnects. *IEEE Electron Device Lett.* **2011**, *32*, 1110–1112. [CrossRef]
15. White, C.T.; Todorov, T.N. Carbon nanotubes as long ballistic conductors. *Nature* **1998**, *393*, 240–242. [CrossRef]
16. Hone, J.; Whitney, M.; Piskoti, C.; Zettl, A. Thermal conductivity of single-walled carbon nanotubes. *Phys. Rev. B* **1999**, *59*, R2514. [CrossRef]
17. Wei, B.; Vajtai, R.; Ajayan, P. Reliability and current carrying capacity of carbon nanotubes. *Appl. Phys. Lett.* **2001**, *79*, 1172–1174. [CrossRef]
18. Yu, M.-F.; Lourie, O.; Dyer, M.J.; Moloni, K.; Kelly, T.F.; Ruoff, R.S. Strength and breaking mechanism of multiwalled carbon nanotubes under tensile load. *Science* **2000**, *287*, 637–640. [CrossRef]
19. Wei, X.; Li, S.; Wang, W.; Zhang, X.; Zhou, W.; Xie, S.; Liu, H. Recent Advances in Structure Separation of Single-Wall Carbon Nanotubes and Their Application in Optics, Electronics, and Optoelectronics. *Adv. Sci.* **2022**, *9*, 2200054. [CrossRef]
20. Bruke, P.J. Luttinger liquid theory as a model of the gigahertz electrical properties of carbon nanotubes. *IEEE Trans. Nanotechnol.* **2002**, *1*, 129–144. [CrossRef]
21. Srivastava, N.; Banerjee, K. Performance analysis of carbon nanotube interconnects for VLSI applications. In Proceedings of the ICCAD-2005. IEEE/ACM International Conference on Computer-Aided Design, San Jose, CA, USA, 6–10 November 2005; pp. 383–390.
22. Li, H.; Yin, W.Y.; Banerjee, K.; Mao, J.F. Circuit Modeling and Performance Analysis of Multi-Walled Carbon Nanotube Interconnects. *IEEE Trans. Electron Devices* **2008**, *55*, 1328–1337. [CrossRef]
23. Srivastava, N.; Joshi, R.V.; Banerjee, K. Carbon nanotube interconnects: Implications for performance, power dissipation and thermal management. In Proceedings of the IEEE InternationalElectron Devices Meeting, 2005. IEDM Technical Digest, Washington, DC, USA, 5 December 2005; pp. 249–252.
24. Srivastava, N.; Li, H.; Kreupl, F.; Banerjee, K. On the applicability of single-walled carbon nanotubes as VLSI interconnects. *IEEE Trans. Nanotechnol.* **2009**, *8*, 542–559. [CrossRef]

25. Kreupl, F.; Graham, A.P.; Duesberg, G.S.; Steinhögl, W.; Liebau, M.; Unger, E.; Hönlein, W. Carbon nanotubes in interconnect applications. *Microelectron. Eng.* **2002**, *64*, 399–408. [CrossRef]
26. Ahmad, M.; Silva, S.R.P. Low temperature growth of carbon nanotubes—A review. *Carbon* **2020**, *158*, 24–44. [CrossRef]
27. Xu, T.; Wang, Z.; Miao, J.; Chen, X.; Tan, C.M. Aligned carbon nanotubes for through-wafer interconnects. *Appl. Phys. Lett.* **2007**, *91*, 042108. [CrossRef]
28. Sun, S.; Mu, W.; Edwards, M.; Mencarelli, D.; Pierantoni, L.; Fu, Y.; Jeppson, K.; Liu, J. Vertically aligned CNT-Cu nano-composite material for stacked through-silicon-via interconnects. *Nanotechnology* **2016**, *27*, 335705. [CrossRef]
29. Subramaniam, C.; Yamada, T.; Kobashi, K.; Sekiguchi, A.; Futaba, D.N.; Yumura, M.; Hata, K. One hundred fold increase in current carrying capacity in a carbon nanotube–copper composite. *Nat. Commun.* **2013**, *4*, 1–7. [CrossRef]
30. Feng, Y.; Burkett, S.L. Modeling a copper/carbon nanotube composite for applications in electronic packaging. *Comput. Mater. Sci.* **2015**, *97*, 1–5. [CrossRef]
31. Zhao, W.-S.; Zheng, J.; Hu, Y.; Sun, S.; Wang, G.; Dong, L.; Yu, L.; Sun, L.; Yin, W.-Y. High-Frequency Analysis of Cu-Carbon Nanotube Composite Through-Silicon Vias. *IEEE Trans. Nanotechnol.* **2016**, *15*, 506–511. [CrossRef]
32. Xu, C.; Li, H.; Suaya, R.; Banerjee, K. Compact AC Modeling and Performance Analysis of Through-Silicon Vias in 3-D ICs. *IEEE Trans. Electron Devices* **2010**, *57*, 3405–3417. [CrossRef]
33. Xie, R.; Zhang, C.; Van der Veen, M.; Arstila, K.; Hantschel, T.; Chen, B.; Zhong, G.; Robertson, J. Carbon nanotube growth for through silicon via application. *Nanotechnology* **2013**, *24*, 125603. [CrossRef] [PubMed]
34. Ghosh, K.; Verma, Y.K.; Tan, C.S. Implementation of carbon nanotube bundles in sub-5 micron diameter through-silicon-via structures for three-dimensionally stacked integrated circuits. *Mater. Today Commun.* **2015**, *2*, e16–e25. [CrossRef]
35. Jiang, D.; Mu, W.; Chen, S.; Fu, Y.; Jeppson, K.; Liu, J. Vertically Stacked Carbon Nanotube-Based Interconnects for Through Silicon Via Application. *IEEE Electron Device Lett.* **2015**, *36*, 499–501. [CrossRef]
36. Uhlig, B.; Liang, J.; Lee, J.; Ramos, R.; Dhavamani, A.; Nagy, N.; Dijon, J.; Okuno, H.; Kalita, D.; Georgiev, V.; et al. Progress on carbon nanotube BEOL interconnects. In Proceedings of the 2018 Design, Automation & Test in Europe Conference & Exhibition (DATE), Dresden, Germany, 19–23 March 2018; pp. 937–942.
37. Zhang, Z.; Ding, Y.; Yang, B.; Ren, A.; Chen, Z. A Low-Cost and Low-Temperature Method to Realize Carbon Nanotube Conductor in Through-Silicon-Via. In Proceedings of the 2021 IEEE International 3D Systems Integration Conference (3DIC), Raleigh, NC, USA, 26–29 October 2021; pp. 1–4.
38. Naeemi, A.; Meindl, J.D. Compact physical models for multiwall carbon-nanotube interconnects. *IEEE Electron Device Lett.* **2006**, *27*, 338–340. [CrossRef]
39. Naeemi, A.; Meindl, J.D. Design and Performance Modeling for Single-Walled Carbon Nanotubes as Local, Semiglobal, and Global Interconnects in Gigascale Integrated Systems. *IEEE Trans. Electron Devices* **2007**, *54*, 26–37. [CrossRef]
40. Pu, S.-N.; Yin, W.-Y.; Mao, J.-F.; Liu, Q.H. Crosstalk prediction of single-and double-walled carbon-nanotube (SWCNT/DWCNT) bundle interconnects. *IEEE Trans. Electron Devices* **2009**, *56*, 560–568. [CrossRef]
41. Liang, F.; Wang, G.; Ding, W. Estimation of time delay and repeater insertion in multiwall carbon nanotube interconnects. *IEEE Trans. Electron Devices* **2011**, *58*, 2712–2720. [CrossRef]
42. Yin, W.-Y.; Zhao, W.-S. Modeling of carbon nanotube (CNT) interconnects. In Proceedings of the 2011 IEEE 15th Workshop on Signal Propagation on Interconnects (SPI), Naples, Italy, 8–11 May 2011; pp. 79–82.
43. Das, D.; Rahaman, H. Analysis of Crosstalk in Single- and Multiwall Carbon Nanotube Interconnects and Its Impact on Gate Oxide Reliability. *IEEE Trans. Nanotechnol.* **2011**, *10*, 1362–1370. [CrossRef]
44. Li, H.; Liu, W.; Cassell, A.M.; Kreupl, F.; Banerjee, K. Low-resistivity long-length horizontal carbon nanotube bundles for interconnect applications—Part I: Process development. *IEEE Trans. Electron Devices* **2013**, *60*, 2862–2869. [CrossRef]
45. Ceyhan, A.; Naeemi, A. Cu Interconnect Limitations and Opportunities for SWNT Interconnects at the End of the Roadmap. *IEEE Trans. Electron Devices* **2013**, *60*, 374–382. [CrossRef]
46. Kumar, V.R.; Majumder, M.K.; Alam, A.; Kukkam, N.R.; Kaushik, B.K. Stability and delay analysis of multi-layered GNR and multi-walled CNT interconnects. *J. Comput. Electron.* **2015**, *14*, 611–618. [CrossRef]
47. Zhao, W.S.; Liu, P.W.; Yu, H.; Hu, Y.; Wang, G.; Swaminathan, M. Repeater Insertion to Reduce Delay and Power in Copper and Carbon Nanotube-Based Nanointerconnects. *IEEE Access* **2019**, *7*, 13622–13633. [CrossRef]
48. Dhillon, G.; Sandha, K.S. Mixed CNT bundles as VLSI interconnects for nanoscale technology nodes. *J. Comput. Electron.* **2021**, *20*, 248–258. [CrossRef]
49. Naeemi, A.; Meindl, J.D. Physical Modeling of Temperature Coefficient of Resistance for Single- and Multi-Wall Carbon Nanotube Interconnects. *IEEE Electron Device Lett.* **2007**, *28*, 135–138. [CrossRef]
50. Close, G.F.; Wong, H.P. Fabrication and Characterization of Carbon Nanotube Interconnects. In Proceedings of the 2007 IEEE International Electron Devices Meeting, Washington, DC, USA, 10–12 December 2007; pp. 203–206.
51. Close, G.F.; Yasuda, S.; Paul, B.; Fujita, S.; Wong, H.S.P. A 1 GHz Integrated Circuit with Carbon Nanotube Interconnects and Silicon Transistors. *Nano Lett.* **2008**, *8*, 706–709. [CrossRef]
52. Pop, E.; Mann, D.; Wang, Q.; Goodson, K.; Dai, H. Thermal conductance of an individual single-wall carbon nanotube above room temperature. *Nano Lett.* **2006**, *6*, 96–100. [CrossRef]

53. Chen, R.; Liang, J.; Lee, J.; Georgiev, V.P.; Ramos, R.; Okuno, H.; Kalita, D.; Cheng, Y.; Zhang, L.; Pandey, R.R. Variability study of MWCNT local interconnects considering defects and contact resistances—Part I: Pristine MWCNT. *IEEE Trans. Electron Devices* **2018**, *65*, 4955–4962. [CrossRef]
54. Chen, R.; Liang, J.; Lee, J.; Georgiev, V.P.; Ramos, R.; Okuno, H.; Kalita, D.; Cheng, Y.; Zhang, L.; Pandey, R.R. Variability study of mwcnt local interconnects considering defects and contact resistances—Part II: Impact of charge transfer doping. *IEEE Trans. Electron Devices* **2018**, *65*, 4963–4970. [CrossRef]
55. Liang, J.; Lee, J.; Berrada, S.; Georgiev, V.P.; Pandey, R.; Chen, R.; Asenov, A.; Todri-Sanial, A. Atomistic- to Circuit-Level Modeling of Doped SWCNT for On-Chip Interconnects. *IEEE Trans. Nanotechnol.* **2018**, *17*, 1084–1088. [CrossRef]
56. Chen, R.; Chen, L.; Liang, J.; Cheng, Y.; Elloumi, S.; Lee, J.; Xu, K.; Georgiev, V.P.; Ni, K.; Debacker, P. Carbon Nanotube SRAM in 5-nm Technology Node Design, Optimization, and Performance Evaluation—Part II: CNT Interconnect Optimization. *IEEE Trans. Very Large Scale Integr. (VLSI) Syst.* **2022**, *30*, 432–439. [CrossRef]
57. Todri-Sanial, A.; Dijon, J.; Maffucci, A. *Carbon Nanotubes for Interconnects*; Springer: Berlin/Heidelberg, Germany, 2017.
58. International Roadmap for Devices and Systems. Available online: https://irds.ieee.org/editions/2021 (accessed on 11 July 2022).
59. Magen, N.; Kolodny, A.; Weiser, U.; Shamir, N. Interconnect-power dissipation in a microprocessor. In Proceedings of the 2004 International Workshop on System Level Interconnect Prediction, Paris, France, 14–15 February 2004; pp. 7–13.
60. Hu, C.K.; Liniger, E.G.; Gignac, L.M.; Bonilla, G.; Edelstein, D. Materials and scaling effects on on-chip interconnect reliability. *MRS Proc.* **2013**, *1559*, mrss13-1559-aa07-01. [CrossRef]
61. Chern, J.H.; Huang, J.; Arledge, L.; Li, P.C.; Yang, P. Multilevel metal capacitance models for CAD design synthesis systems. *Electron Device Lett. IEEE* **1992**, *13*, 32–34. [CrossRef]
62. Sathyakam, P.U.; Mallick, P.S. *Design and Crosstalk Analysis in Carbon Nanotube Interconnects*; Springer: Berlin/Heidelberg, Germany, 2017.
63. Anglada, R.; Rubio, A. An approach to crosstalk effect analysis and avoidance techniques in digital CMOS VLSI circuits. *Int. J. Electron.* **1988**, *65*, 9–17. [CrossRef]
64. Zhao, W.-S.; Fu, K.; Wang, D.-W.; Li, M.; Wang, G.; Yin, W.-Y. Mini-Review: Modeling and Performance Analysis of Nanocarbon Interconnects. *Appl. Sci.* **2019**, *9*, 2174. [CrossRef]
65. Liang, J.; Todri-Sanial, A. Importance of Interconnects: A Technology-System-Level Design Perspective. In Proceedings of the 2019 IEEE International Electron Devices Meeting (IEDM), San Francisco, CA, USA, 7–11 December 2019; pp. 23.21.21–23.21.24.
66. Chen, R.; Chen, L.; Liang, J.; Cheng, Y.; Elloumi, S.; Lee, J.; Xu, K.; Georgiev, V.P.; Ni, K.; Debacker, P.; et al. Carbon Nanotube SRAM in 5-nm Technology Node Design, Optimization, and Performance Evaluation—Part I: CNFET Transistor Optimization. *IEEE Trans. Very Large Scale Integr. (VLSI) Syst.* **2022**, *30*, 432–439. [CrossRef]
67. Li, H.; Liu, W.; Cassell, A.M.; Kreupl, F.; Banerjee, K. Low-resistivity long-length horizontal carbon nanotube bundles for interconnect applications—Part II: Characterization. *IEEE Trans. Electron Devices* **2013**, *60*, 2870–2876. [CrossRef]
68. Liang, J.; Chen, R.; Ramos, R.; Lee, J.; Okuno, H.; Kalita, D.; Georgiev, V.; Berrada, S.; Sadi, T.; Uhlig, B.; et al. Investigation of Pt-Salt-Doped-Standalone- Multiwall Carbon Nanotubes for On-Chip Interconnect Applications. *IEEE Trans. Electron Devices* **2019**, *66*, 2346–2352. [CrossRef]
69. Ismail, Y.I.; Friedman, E.G. Effects of inductance on the propagation delay and repeater insertion in VLSI circuits. *IEEE Trans. Very Large Scale Integr. (VLSI) Syst.* **2000**, *8*, 195–206. [CrossRef]
70. Banerjee, K.; Mehrotra, A. A power-optimal repeater insertion methodology for global interconnects in nanometer designs. *IEEE Trans. Electron Devices* **2002**, *49*, 2001–2007. [CrossRef]
71. Kong, J.; Yenilmez, E.; Tombler, T.W.; Kim, W.; Dai, H.; Laughlin, R.B.; Liu, L.; Jayanthi, C.; Wu, S. Quantum interference and ballistic transmission in nanotube electron waveguides. *Phys. Rev. Lett.* **2001**, *87*, 106801. [CrossRef]
72. Schoenenberger, C.; Bachtold, A.; Strunk, C.; Salvetat, J.-P.; Forro, L. Interference and Interaction in multi-wall carbon nanotubes. *Appl. Phys. A* **1999**, *69*, 283–295. [CrossRef]
73. Koo, K.-H.; Kapur, P.; Saraswat, K.C. Compact performance models and comparisons for gigascale on-chip global interconnect technologies. *IEEE Trans. Electron Devices* **2009**, *56*, 1787–1798. [CrossRef]
74. Ward, J.W.; Nichols, J.; Stachowiak, T.B.; Ngo, Q.; Egerton, E.J. Reduction of CNT interconnect resistance for the replacement of Cu for future technology nodes. *IEEE Trans. Nanotechnol.* **2011**, *11*, 56–62. [CrossRef]
75. Li, H.; Xu, C.; Srivastava, N.; Banerjee, K. Carbon Nanomaterials for Next-Generation Interconnects and Passives: Physics, Status, and Prospects. *IEEE Trans. Electron Devices* **2009**, *56*, 1799–1821. [CrossRef]
76. Nihei, M.; Kondo, D.; Kawabata, A.; Sato, S.; Shioya, H.; Sakaue, M.; Iwai, T.; Ohfuti, M.; Awano, Y. Low-resistance multi-walled carbon nanotube vias with parallel channel conduction of inner shells [IC interconnect applications]. In Proceedings of the IEEE 2005 International Interconnect Technology Conference, Burlingame, CA, USA, 6–8 June 2005; pp. 234–236.
77. Pasricha, S.; Dutt, N.; Kurdahi, F.J. Exploring carbon nanotube bundle global interconnects for chip multiprocessor applications. In Proceedings of the 2009 22nd International Conference on VLSI Design, New Delhi, India, 5–9 January 2009; pp. 499–504.
78. Berber, S.; Kwon, Y.-K.; Tománek, D. Unusually High Thermal Conductivity of Carbon Nanotubes. *Phys. Rev. Lett.* **2000**, *84*, 4613. [CrossRef]
79. Yu, W.; Liu, C.; Fan, S. Advances of CNT-based systems in thermal management. *Nano Res.* **2021**, *14*, 2471–2490. [CrossRef]
80. Prasher, R.S.; Hu, X.; Chalopin, Y.; Mingo, N.; Lofgreen, K.; Volz, S.; Cleri, F.; Keblinski, P. Turning carbon nanotubes from exceptional heat conductors into insulators. *Phys. Rev. Lett.* **2009**, *102*, 105901. [CrossRef]

81. Kim, P.; Shi, L.; Majumdar, A.; McEuen, P.L. Thermal transport measurements of individual multiwalled nanotubes. *Phys. Rev. Lett.* **2001**, *87*, 215502. [CrossRef]
82. Liew, K.; Wong, C.; He, X.; Tan, M. Thermal stability of single and multi-walled carbon nanotubes. *Phys. Rev. B* **2005**, *71*, 075424. [CrossRef]
83. Li, S.; Yu, Z.; Yen, S.-F.; Tang, W.C.; Burke, P.J. Carbon Nanotube Transistor Operation at 2.6 GHz. *Nano Lett.* **2004**, *4*, 753–756. [CrossRef]
84. Todri-Sanial, A.; Ramos, R.; Okuno, H.; Dijon, J.; Dhavamani, A.; Widlicenus, M.; Lilienthal, K.; Uhlig, B.; Sadi, T.; Georgiev, V. A survey of carbon nanotube interconnects for energy efficient integrated circuits. *IEEE Circuits Syst. Mag.* **2017**, *17*, 47–62. [CrossRef]
85. Kim, S.; Kulkarni, D.D.; Rykaczewski, K.; Henry, M.; Tsukruk, V.V.; Fedorov, A.G. Fabrication of an UltraLow-Resistance Ohmic Contact to MWCNT–Metal Interconnect Using Graphitic Carbon by Electron Beam-Induced Deposition (EBID). *IEEE Trans. Nanotechnol.* **2012**, *11*, 1223–1230. [CrossRef]
86. Chai, G.; Chen, Q. Characterization study of the thermal conductivity of carbon nanotube copper nanocomposites. *J. Compos. Mater.* **2010**, *44*, 2863–2873. [CrossRef]
87. Purewal, M.S.; Hong, B.H.; Ravi, A.; Chandra, B.; Hone, J.; Kim, P. Scaling of Resistance and Electron Mean Free Path of Single-Walled Carbon Nanotubes. *Phys. Rev. Lett.* **2007**, *98*, 186808. [CrossRef] [PubMed]
88. Berger, C.; Yi, Y.; Wang, Z.L.; de Heer, W.A. Multiwalled carbon nanotubes are ballistic conductors at room temperature. *Appl. Phys. A* **2002**, *74*, 363–365. [CrossRef]
89. Shefali, M.; Fatima, K.; Sathyakam, P.U. Performance Analysis of CNT Bundle Interconnects in Various Low-k Dielectric Media. *ECS J. Solid State Sci. Technol.* **2022**, *11*, 061003. [CrossRef]
90. Hosseini, A.; Shabro, V. Thermally-aware modeling and performance evaluation for single-walled carbon nanotube-based interconnects for future high performance integrated circuits. *Microelectron. Eng.* **2010**, *87*, 1955–1962. [CrossRef]
91. Mohsin, K.M.; Srivastava, A.; Sharma, A.K.; Mayberry, C. A Thermal Model for Carbon Nanotube Interconnects. *Nanomaterials* **2013**, *3*, 229–241. [CrossRef]
92. Rai, M.K.; Kaushik, B.K.; Sarkar, S. Thermally aware performance analysis of single-walled carbon nanotube bundle as VLSI interconnects. *J. Comput. Electron.* **2016**, *15*, 407–419. [CrossRef]
93. Hau-Riege, C.S. An introduction to Cu electromigration. *Microelectron. Reliab.* **2004**, *44*, 195–205. [CrossRef]
94. Li, B.; Christiansen, C.; Badami, D.; Yang, C.-C. Electromigration challenges for advanced on-chip Cu interconnects. *Microelectron. Reliab.* **2014**, *54*, 712–724. [CrossRef]
95. Radosavljević, M.; Lefebvre, J.; Johnson, A. High-field electrical transport and breakdown in bundles of single-wall carbon nanotubes. *Phys. Rev. B* **2001**, *64*, 241307. [CrossRef]
96. Noguchi, J.; Ohashi, N.; Yasuda, J.; Jimbo, T.; Yamaguchi, H.; Owada, N.; Takeda, K.; Hinode, K. TDDB improvement in Cu metallization under bias stress. In Proceedings of the 2000 IEEE International Reliability Physics Symposium Proceedings. 38th Annual (Cat. No.00CH37059), San Jose, CA, USA, 10–13 April 2000; pp. 339–343.
97. Liao, Z.; Gall, M.; Yeap, K.B.; Sander, C.; Mühle, U.; Gluch, J.; Standke, Y.; Aubel, O.; Vogel, N.; Hauschildt, M.; et al. In-situ study of the TDDB-induced damage mechanism in Cu/ultra-low-k interconnect structures. *Microelectron. Eng.* **2015**, *137*, 47–53. [CrossRef]
98. Zou, Y.S.; Gan, C.L.; Chung, M.-H.; Takiar, H. A review of interconnect materials used in emerging memory device packaging: First- and second-level interconnect materials. *J. Mater. Sci. Mater. Electron.* **2021**, *32*, 27133–27147. [CrossRef]
99. Dijon, J.; Chiodarelli, N.; Fournier, A.; Okuno, H.; Ramos, R. Horizontal carbon nanotube interconnects for advanced integrated circuits. *MRS Proc.* **2013**, *1559*, mrss13-1559-aa02-01. [CrossRef]
100. Li, H.; Srivastava, N.; Mao, J.-F.; Yin, W.-Y.; Banerjee, K. Carbon nanotube vias: Does ballistic electron–phonon transport imply improved performance and reliability? *IEEE Trans. Electron Devices* **2011**, *58*, 2689–2701. [CrossRef]
101. Sable, K.; Sahoo, M. Electrical and thermal analysis of cu-CNT composite TSV and GNR interconnects. In Proceedings of the 2020 International Symposium on Devices, Circuits and Systems (ISDCS), Howrah, India, 4–6 March 2020; pp. 1–6.
102. Singh, A.; Prabhu, T.R.; Sanjay, A.; Koti, V. An overview of processing and properties of Cu/CNT nano composites. *Mater. Today Proc.* **2017**, *4*, 3872–3881. [CrossRef]
103. Lee, J.; Berrada, S.; Adamu-Lema, F.; Nagy, N.; Georgiev, V.P.; Sadi, T.; Liang, J.; Ramos, R.; Carrillo-Nunez, H.; Kalita, D. Understanding electromigration in Cu-CNT composite interconnects: A multiscale electrothermal simulation study. *IEEE Trans. Electron Devices* **2018**, *65*, 3884–3892. [CrossRef]
104. Zhao, W.S.; Hu, Q.H. Carbon Nanotube Through-Silicon Via: Modeling, Design and Applications. In Proceedings of the 2020 International Conference on Microwave and Millimeter Wave Technology (ICMMT), Shanghai, China, 20–23 September 2020; pp. 1–3.
105. Milowska, K.Z.; Ghorbani-Asl, M.; Burda, M.; Wolanicka, L.; Ćatić, N.; Bristowe, P.D.; Koziol, K.K. Breaking the electrical barrier between copper and carbon nanotubes. *Nanoscale* **2017**, *9*, 8458–8469. [CrossRef]
106. Milowska, K.Z.; Burda, M.; Wolanicka, L.; Bristowe, P.D.; Koziol, K.K. Carbon nanotube functionalization as a route to enhancing the electrical and mechanical properties of Cu–CNT composites. *Nanoscale* **2019**, *11*, 145–157. [CrossRef]
107. Kumari, B.; Pandranki, S.; Sharma, R.; Sahoo, M. Thermal-Aware Modeling and Analysis of Cu-Mixed CNT Nanocomposite Interconnects. *IEEE Trans. Nanotechnol.* **2022**, *21*, 163–171. [CrossRef]

108. Fu, K.; Zheng, J.; Zhao, W.S.; Hu, Y.; Wang, G. Analysis of Transmission Characteristics of Copper/Carbon Nanotube Composite Through-Silicon Via Interconnects. *Chin. J. Electron.* **2019**, *28*, 920–924. [CrossRef]
109. Sundaram, R.M.; Sekiguchi, A.; Sekiya, M.; Yamada, T.; Hata, K. Copper/carbon nanotube composites: Research trends and outlook. *R. Soc. Open Sci.* **2018**, *5*, 180814. [CrossRef]
110. Chen, G.; Sundaram, R.; Sekiguchi, A.; Hata, K.; Futaba, D.N. Through-Silicon-Via Interposers with Cu-Level Electrical Conductivity and Si-Level Thermal Expansion Based on Carbon Nanotube-Cu Composites for Microelectronic Packaging Applications. *ACS Appl. Nano Mater.* **2020**, *4*, 869–876. [CrossRef]
111. Wan, D.; Paolillo, S.; Rassoul, N.; Kotowska, B.K.; Blanco, V.; Adelmann, C.; Lazzarino, F.; Ercken, M.; Murdoch, G.; Bömmels, J.; et al. Subtractive Etch of Ruthenium for Sub-5nm Interconnect. In Proceedings of the 2018 IEEE International Interconnect Technology Conference (IITC), Santa Clara, CA, USA, 4–7 June 2018; pp. 10–12.
112. Xu, K.; Cheng, Y. Fault Testing and Diagnosis Techniques for Carbon Nanotube-Based FPGAs. In Proceedings of the 2022 27th Asia and South Pacific Design Automation Conference (ASP-DAC), Taipei, Taiwan, 17–20 January 2022; pp. 479–484.
113. Das, D.; Rahaman, H. Modeling of IR-Drop induced delay fault in CNT and GNR power distribution networks. In Proceedings of the 2012 5th International Conference on Computers and Devices for Communication (CODEC), Kolkata, India, 17–19 December 2012; pp. 1–4.
114. Patil, N.; Deng, J.; Lin, A.; Wong, H.P.; Mitra, S. Design Methods for Misaligned and Mispositioned Carbon-Nanotube Immune Circuits. *IEEE Trans. Comput. -Aided Des. Integr. Circuits Syst.* **2008**, *27*, 1725–1736. [CrossRef]
115. Radway, R.M.; Sethi, K.; Chen, W.C.; Kwon, J.; Liu, S.; Wu, T.F.; Beigne, E.; Shulaker, M.M.; Wong, H.S.P.; Mitra, S. The Future of Hardware Technologies for Computing: N3XT 3D MOSAIC, Illusion Scaleup, Co-Design. In Proceedings of the 2021 IEEE International Electron Devices Meeting (IEDM), San Francisco, CA, USA, 11–16 December 2021; pp. 25.24.21–25.24.24.
116. Aly, M.M.S.; Gao, M.; Hills, G.; Lee, C.S.; Pitner, G.; Shulaker, M.M.; Wu, T.F.; Asheghi, M.; Bokor, J.; Franchetti, F.; et al. Energy-Efficient Abundant-Data Computing: The N3XT 1000x. *Computer* **2015**, *48*, 24–33. [CrossRef]
117. Wang, T.; Chen, S.; Jiang, D.; Fu, Y.; Jeppson, K.; Ye, L.; Liu, J. Through-silicon vias filled with densified and transferred carbon nanotube forests. *IEEE Electron Device Lett.* **2012**, *33*, 420–422. [CrossRef]
118. Mu, W.; Hansson, J.; Sun, S.; Edwards, M.; Fu, Y.; Jeppson, K.; Liu, J. Double-Densified Vertically Aligned Carbon Nanotube Bundles for Application in 3D Integration High Aspect Ratio TSV Interconnects. In Proceedings of the 2016 IEEE 66th Electronic Components and Technology Conference (ECTC), Las Vegas, NV, USA, 31 May–3 June 2016; pp. 211–216.

MDPI
St. Alban-Anlage 66
4052 Basel
Switzerland
Tel. +41 61 683 77 34
Fax +41 61 302 89 18
www.mdpi.com

Micromachines Editorial Office
E-mail: micromachines@mdpi.com
www.mdpi.com/journal/micromachines

www.ingramcontent.com/pod-product-compliance
Lightning Source LLC
Chambersburg PA
CBHW061841210926
51850CB00006B/938
9783036567334